CAMBRIDGE TROPICAL BIOLOGY SERIES

EDITORS:

Peter S. Ashton *Arnold Arboretum, Harvard University*

Stephen P. Hubbell *University of Iowa*

Daniel H. Janzen *University of Pennsylvania*

Peter H. Raven *Missouri Botanical Garden*

P. B. Tomlinson *Harvard Forest, Harvard University*

The botany of mangroves

Hinchinbrook Island, Queensland, at the turn of the tide. Mangrove detritus carried by the ebbing tide is visible as scum on the surface of the sea.

The botany of mangroves

P. B. TOMLINSON

Harvard University, Harvard Forest
Petersham, Massachusetts

The right of the University of Cambridge to print and sell all manner of books was granted by Henry VIII in 1534. The University has printed and published continuously since 1584.

CAMBRIDGE UNIVERSITY PRESS

Cambridge
London   New York   New Rochelle
Melbourne   Sydney

Published by the Press Syndicate of the University of Cambridge
The Pitt Building, Trumpington Street, Cambridge CB2 1RP
32 East 57th Street, New York, NY 10022, USA
10 Stamford Road, Oakleigh, Melbourne 3166, Australia

First published 1986

Printed in the United States of America

Library of Congress Cataloging in Publication Data
Tomlinson, P. B. (Philip Barry), 1932–
The botany of mangroves.
(Cambridge tropical biology series)
Bibliography: p.
Includes index.
1. Mangrove plants.   2. Mangrove swamp ecology.
I. Title.   II. Series
QK938.M27T66   1986   583'.42   84–23274
ISBN 0 521 25567 8

British Library Cataloging-in-Publication applied for

Contents

Section B. Detailed description by family

Preface

Mangrove plants are encountered almost inadvertently by visitors to the tropics because they occupy all but the most exposed or rockiest shorelines. In sheltered estuaries and lagoons they are usually extensive and may even form a community (mangal) up to several kilometers wide with a gradual transition to terrestrial vegetation. Early travelers certainly were familiar with them, but avoided them because they had an unwholesome reputation. They were tropical swamps, therefore mosquito- and fever-ridden, that harbored unpleasant animals (notably crocodiles) and were difficult to traverse on foot. People entered to gather wood or to fish, but lived only at the edge of the swamp. Most of the dangers of the mangal are imaginary, however, and the modern tropical ecologist finds it intriguing because it represents an interphase between two contrasting types of community: terrestrial, as represented by lowland forests of varying kinds; and marine, as represented by distinctive littoral ecosystems, notably seagrass meadows and coral reefs. There is an abrupt transition from mangal to marine communities, but transitions to terrestrial communities, such as fresh-water swamps, are gradual.

Mangroves are trees, but their form is very versatile: In marginal habitats they are low, scrubby plants, whereas under favorable conditions they can form majestic forests with the canopy 30 to 40 m tall. The adaptations necessary for trees to become established in and occupy the tidal habitat are varied and not always obvious. Even though these adaptations have been developed, seemingly independently, by only a few species, the biological peculiarities of mangroves have intrigued botanists for many generations and much has been written about them. There is a strong commercial significance to the study of mangroves because of their economic value, although this is often assessable only indirectly. For example, commercial fisheries are ultimately dependent on mangroves, which initiate the food chain that ends in a marketable product. The community interphase of the mangal is matched by an industrial interphase, since departments of forestry, agriculture, and fisheries can all claim jurisdiction over the management and use of mangrove swamps.

This combination of lengthy commercial and scientific interest has generated an almost unmanageable literature on the mangal. A recent bibliography (Rollet 1981) cites more than 6,000 references to research reports and journal articles for the period 1600–1975. Because this bibliography is indexed, it is a primary source of information. Where evaluations of the mangrove literature have been attempted, they usually analyze the community collectively and seek interactions between plants and animals in generalized ecological terms – describing zonation and attempting to interpret it in terms of succession, competition, and differential adaptation to the subtle mosaic of geomorphological features that exist in tidal estuaries in the tropics.

In this account a descriptive overview of mangrove plants is presented, with emphasis on the biology of individual species. Section A defines mangroves and refers to associated communities; briefly summarizes geographical, floristic, architectural, morphological, reproductive, and anatomical features of mangroves; and considers some of the physiological specializations that allow mangroves to grow in sea water subject to tidal influences and usually in unstable and often anaerobic substrates. Only a very limited discussion of the ecology of mangroves is given, since this is outside the author's area of expertise and is well treated in several recent reviews. The only discussion of the mangrove fauna focuses on those animals that interact with mangrove plants as predators or pollinators. The commercial value of mangroves is assessed, emphasizing the biological features of mangrove plants and the peculiarities of mangals that relate to economic exploitation.

Section B, the larger part of this book, is a systematic survey of mangrove plants and mangrove associates by family. It will serve as a universally applicable florula for students and research workers and covers the more recent relevant botanical literature. The treatment of the strict mangrove departs from the usual systematic description found in a standard flora in its more extended consideration of the dynamics and biological aspects of plant behavior: growth, reproduction, and morphological adaptation. Detailed illustrations are provided, mainly based on fluid-preserved specimens I have collected.

In preparing this work, the biggest problem has been to decide what to leave out. In association with workers active in mangrove research, I have long been aware of the need for a concise treatment of the botany of mangroves, but to provide the complete compendium that might be needed would have made an interminable task. By using a rather subjective point of view and adhering to topics with which I have had some firsthand experience, a consistent but limited format has been adopted. The beginning student and nonspecialist need a starting point in tropical botany, and the mangroves, in my opinion, provide a useful introduction: They are not too diverse floristically; they have morphological and structural features whose adaptive function can often be demonstrated by simple experiment; they are relatively accessible and cosmopolitan; and their ecological zonation is obvious but not easily explained.

I hope the book succeeds in its simple but multiple objectives. If it pleases nobody, at least I know where I can go to hide!!

P. B. TOMLINSON

Acknowledgments

An introduction to mangrove ecosystems followed almost naturally from my interest in field botany and a position at Fairchild Tropical Garden (1960–71), which allowed reasonable freedom of research objectives with one of the largest areas of mangroves at hand in South Florida. Dr. A. M. Gill on the staff of Fairchild Garden (1967–70) did much to stimulate my interest through his own ecologically orientated research. Financial support from 1965 to the present has come from the Maria Moors Cabot Foundation for Botanical Research and subsequently from the Atkins Garden Funds, both of Harvard University. In addition I have received travel support from the National Geographic Society (Tomlinson 1982a) and the National Science Foundation, Office of International Programs (United States–Australia Scientific Collaborative Program). These awards facilitated the limited knowledge I had gained in the Caribbean area to be broadened by visits to the richer mangal of the Indo-Pacific region. Several institutions and many individuals have contributed facilities and field assistance. They include the Australian Institute of Marine Science, Townsville, Queensland (AIMS), and the Division of Botany, Department of Forests, Lae, Papua New Guinea (PNG). Individuals who have made fieldwork possible include J. S. Bunt, B. Clough, N. C. Duke (AIMS), J. S. Womersley, E. E. Henty, M. Galore, R. J. Johns (PNG), and Dr. J. Davey (University of Brisbane). Professor Engkik Soepadmo of the University of Malaysia, Kuala Lumpur, Dr. Paul Chai and the staff of the Forestry Department, Sarawak, and Mr. J. M. Maxwell of the Singapore Botanic Garden all were helpful on a visit to Malaysian mangroves. Jorge Jiménez was my host when I visited Costa Rican mangroves. Richard Primack of Boston University was a stimulating field companion and a guide to the theoretical and applied aspects of floral biology. The libraries and collections of the Harvard University Herbaria served as an indispensable source of systematic, geographic, and historical information. Peter Stevens, Curator of the Harvard University Herbaria, was particularly helpful in a critical review of the systematic descriptions. Adrian Juncosa made helpful comments on early drafts and supplied some material for illustration. John Sperry provided discussions of mangrove physiology. The morphological illustrations in Section B are the work of Priscilla Fawcett, Botanical Illustrator at Fairchild Tropical Garden, supplemented in Massachusetts by the drafting skills of Susan White and Elizabeth Bullock. Ms. D. R. Smith worked without complaint to maintain some coherence throughout the many draft versions of the book.

In addition to the figures prepared especially for this book, a number of sources have been used for illustration. Permission to reproduce these figures has been given by the following authors and/or journal editors: *American Fern Journal* (Figs. B.60, B.61); *American Journal of Botany* and A. M. Juncosa (Fig. 8.5); *Annals of Botany, London* (Figs. 4.4, B.51, B.52); *Biotropica* (Figs. 5.10, 5.11, B.66, B.67, B.71, B.72, B.74, B.76); *Biological Journal of the Linnean Society* and A. G. Marshall (Fig. 7.2); *Botanical Journal of the Linnean Society* (Fig. 4.8); *Bulletin of the Fairchild Tropical Garden* and I. Olmsted (Figs. 1.10, 1.11, 1.12); *Contributions Herbarium Australiense* (Figs. B.66, B.70); *Journal of the Arnold Arboretum* (Fig. B.21); *Vegetatio* and V. Semeniuk (Fig. 1.7); J. Cramer and Helen Correll (Figs. B.6, B.15, B.29, for figures that first appeared in D. S. Correll and H. B. Correll, *Flora of the Bahama Archipelago*); and Figures B.8, B.19, B.24, B.25, B.26, B.28, B.37, B.38 first appeared in P. B. Tomlinson *Biology of Trees Native to Tropical Florida*, copyright by the author. In addition, previously unpublished diagrams and photographs were supplied by J. S. Bunt and N. C. Duke (Figs. 1.8, 3.5); A. M. Gill (Figs. 5.4, B.41, B.62, B.75); and A. M. Juncosa (Figs. 1.3, 1.4, 4.10, 5.5, 8.6, B.22, B.23, B.55, B.57, B.58, B.80). All other figures are either original or redrawn.

Illustrations and tables

Figures in Section A (General Account) are numbered according to chapter (e.g., Fig. 4.6 is the sixth illustration in Chapter 4); figures in Section B (Detailed Description by Family) are numbered consecutively with the prefix B (i.e., B.1– B.84). Tables in Section A are also numbered according to chapter (e.g., Table 2.2 is the second table in Chapter 2).

Section A

General account

1 Ecology

Mangrove and mangal

The word "mangrove" has been used to refer either to the constituent plants of tropical intertidal forest communities or to the community itself. Usually there is no contextual confusion; otherwise, some qualification, such as "mangrove plants" or "mangrove community," is needed. MacNae (1968) proposed "mangal" as a term for the community, leaving "mangrove" for the constituent plant species, and this usage is increasingly adopted. "Mangal associate," "mangrove associate," and "back mangal" may also be used as a logical extension of this distinction between species and community. Mangroves in the more limited sense may thus be defined as tropical trees restricted to intertidal and adjacent communities.

Another definition of mangal is a community that contains mangrove plants. This is less nonsensical in its circularity than it at first appears, because strict mangroves are characterized by their high fidelity to the ecotone influenced by tides (Figs. 1.1, 1.2). Tidal influence can be interpreted narrowly, simply to mean the shoreline inundated by the extremes of tides, or it can more widely refer to river-bank communities where tides cause some fluctuation but no salinity. Mangroves can penetrate inland extensively along river banks.

This book is about mangroves and not mangal. This restriction is appropriate because there is need for a uniform systematic, morphological, geographic, and biological treatment of mangroves distinct from those numerous general treatments of mangal that emphasize community aspects, though frequently with attention to details of the biology of individual species.

Mangal is a discrete community but with a close physiographic relationship to other strand communities. Schimper (1891), in his description of coastal communities in the Indo-Malayan region, distinguished "mangrove" from "*Nypa* formation," "*Barringtonia* formation," and "*Pes-caprae* formation" (from *Ipomoea pes-caprae*). The transition between these is usually abrupt, but elements of all communities are sometimes intermixed and some species from the nonmangal are treated here as "mangal associates." A more gradual transition between mangal and fresh-water swamp forest may again intermingle species from contrasted habitats. Corner (1978) describes this interrelationship well for South Johore and Singapore. Schimper's emphasis on mangal as only one of a number of discrete coastal

3

Figure 1.1. The mangrove habitat. *Sonneratia* and *Avicennia* at low tide. Semetan, Sarawak.

Figure 1.2. Same mangrove habitat shown in Figure 1.1, but at high tide.

communities is appropriate, since he considered that many biological processes that are particularly well expressed in mangroves, such as dispersal, salt tolerance, xerophytism, and gas exchange, also occur more generally.

From the mangrove forest one can move easily to contrasted marine and shoreline communities. Toward the sea on shallow sloping shores there may be a submerged seagrass meadow with scattered mangrove seedlings in the shallowest parts. Along the coast one may move to beach communities with *Ipomoea*, *Suriana*, or *Tournefortia* in front, but trees like *Barringtonia*, *Hibiscus tiliaceus*, and *Thespesia* behind. *Casuarina* forest may grow behind mangal but only when the soil is sandy and well aerated. Rocky coasts or coasts with exposed coral are unsuited to mangroves except where plants can root in silt-filled depressions.

On drier coasts (with low levels of rainfall) salt accumulation may produce a sterile salt flat or "salt desert," with a fringe of strict halophytes (Fig. 1.3). The back mangal is very complex, since the possible adjacent terrestrial communities are numerous and diverse and the transition may be abrupt or gradual. Consequently the number of mangal-associate species is large. I have referred to or illustrated those that are most commonly mentioned (see Rollet 1981).

Mangal subtypes

Within the strict mangal a complex subdivision is possible, even with the relatively few species available. The *Nypa* formation of Schimper (1891) is most distinctive and easily recognized. Competitive exclusion of other mangroves seems related to the rhizomatous habit of *Nypa*. Much of the description of other mangrove communities is concerned with the relative abundance or dominance of given species or even contrasted vegetation types distinguished by differing habits but made up of one species.

The ecological literature seems incapable of being reduced to a simple set of rules to account for the diversity of vegetation types within the broad generic concept of mangal. Lack of uniformity is, in fact, a measure of the plasticity of mangroves and their ability to colonize such an enormous range of habitats. Instability and change are the most consistent characteristics of mangal.

Mangal concentration

Although mangroves may grow throughout the tropics in suitable areas, with the exception of the Central Pacific, particular regions are noted for the broad extent of mangal. Typically these are the estuaries of large rivers that run over a shallow continental shelf. Examples are the mouths of the Ganges and Brahmaputra rivers (the Sundarbans) mainly in Bangladesh, the Fly and Purari rivers in Papua

Figure 1.3. Salt desert in the back mangal. Low *Ceriops tagal* in the background; halophytes (e.g., *Batis* and *Sesuvium*) in the middle ground fairly abruptly transitional to area from which plants are totally excluded by persistent high salinity. This corresponds to the salt flat in Figure 1.7. (From a color transparency by A. M. Juncosa)

New Guinea, and the Mekong Delta in Vietnam. The Florida Everglades is a drainage basin that gradually changes from fresh water to an extensive mangal at its seaward margin. The two largest tropical rivers, the Amazon and the Congo, do not develop extensive estuarine mangal for physiographic reasons.

Mangal characteristics

A distinctive character of mangal is its diversity: The most consistent feature is the vegetation itself, easily recognized because there are few species. Nevertheless, in its most commonly expressed features, the community has a distinct aspect and atmosphere. The substrate is usually a firm to soft mud into which the traveler may sink, so that walking is difficult if not impossible. Progress is chiefly hindered, however, by the looping aerial roots of *Rhizophora*, which is usually the most abundant component (Fig. 1.4). The traveler has the choice of walking with care on the slippery crowns of these arching stepping-stones or working almost hand over hand around them. Travel into the mangrove is best done using shallow-

Figure 1.4. *Rhizophora* forest, Biscayne Bay, Miami, Florida. "Where this sort of tree grows, it is impossible to march by reason of these stakes, which grow so mixed one among another that I have, when forced to go through them, gone half a mile and never set my foot on the ground, stepping from root to root" (Dampier, *Voyage au nouvelle monde*, 1723). The world record for the 100-m dash through a mangrove swamp (22 min. 30 sec) is held, I believe, by J. S. Bunt, Australian Institute of Marine Science. (From a color transparency by A. M. Juncosa)

draft boats in drainage creeks (Fig. 1.5). The substrate can also be richly organic with peat made up largely of accumulated underground portions of mangrove root systems. In muddy substrates a disturbance produces a strong smell of hydrogen sulfide, indicating the completely anaerobic property of water-logged soil. The oxidation–reduction potential from aerobic (+ 700 mV) to extreme anaerobic (− 300 mV) provides a range of values that influences the soil chemistry (Clough et al. 1983). As the value drops, oxygen is first reduced to water; then at the lowest levels, carbon dioxide is reduced to methane. The ease with which ions function as electron acceptors determines the intermediate progression: nitrate (NO_3-) – nitrogen (N_2), manganous (Mn^{4+}) – manganic (Mn^{2+}), ferrous (Fe^{3+}) – ferric (Fe^{2+}), sulfate (SO_4^{2-}) – sulfide (S^{2-}).

Canopy height depends on climate, topography, and the extent of human disturbance. Mature undisturbed forest develops a high, dense canopy; the trees have tall boles. Aerial roots may be limited in their development, and passage may be relatively easy. There is little stratification and an understorey of at most suppressed

Figure 1.5. Mangal is best explored in a small, shallow-draft boat with an outboard motor.

Figure 1.6. Mangrove at Semetan, Sarawak, with *Rhizophora, Bruguiera,* and *Kandelia.* Uniformity of leaf form is pronounced.

seedlings. The canopy is monotonous because of the uniform leaf shape, size, and texture (Fig. 1.6). In disturbed or impoverished sites, plants are stunted and scrubby and form an open or closed community. To force a way through continuous mangrove scrub is particularly difficult; fortunately, there are few spiny plants, although dead snags of stems and roots can be quite sharp. Crabs and mudskippers may be abundant and conspicuous, even on the trees. There may be nests of the tailor ant (*Oecophylla*) to avoid as well as nests of bees and wasps. Where the community is dense but with a tall canopy, the interior of the mangal can be dark, but there is often a welcome coolness relative to the bright heat of open water. The staff of the Inshore Productivity Group at the Australian Institute of Marine Science wears white dungarees for field clothes because this increases their mutual visibility (Fig. B.82).

Probably the best account of mangal is by MacNae (1968), even though it deals solely with Indo-Pacific mangal, since it is written by somebody with extensive firsthand familiarity with mangal and its plants and animals. It not only is a valuable scientific treatment but also communicates the author's enthusiasm for the subject that has obviously sustained him in the field. It is clearly written by a person who saw in detail the vegetation he describes. Other treatments are cosmopolitan or regional in their coverage. Examples of cosmopolitan treatments are Schimper (1891), Haberlandt (1910), Chapman (1976), and Walsh (1974). Regional treatments, often extensively illustrated, are exemplified by Davis (1940) for South Florida, Semeniuk et al. (1978) for western Australia, Karsten (1890, 1891) and Schimper (1891) for Indo-Malaya, Walter and Steiner (1936) for East Africa, Percival and Womersley (1975) for Papua New Guinea, and Bunt and his associates (e.g., Bunt et al. 1982) for Queensland. Chapters with varying degrees of completeness by individual contributors in Chapman (1977a) deal respectively with the mangal of the eastern United States (Reimold 1977), middle and South America (West 1977), Africa, the Indian subcontinent (Blasco 1977, Chapman 1977b, Zahran 1977), Indo-Malesia (Chapman 1977c), the Pacific Islands (Hosokawa et al. 1977), and Australasia (Saenger et al. 1977). A general account of the ecology of mangroves written by van Steenis is included in Ding Hou's account of the Rhizophoraceae for *Flora Malesiana* (1958); Lugo and Snedaker (1974) provide a more modern ecological summary. A number of international symposia have been held, some of them leading to extensive published proceedings (e.g., Walsh et al. 1975).

Mangrove associates

The scope of individual treatments of mangal varies enormously. Some accounts deal with cryptogams (algae, bryophytes, lichens). Fungi are rarely discussed, even though there may be some significant pathogens of mangroves in the fungal flora, for example, a gall disease *Cylindrocarpon didyum* (Hartog) Wallenw. (Olexa and Freeman 1978); in the Gambia a related species is lethal and estimated to have lost firewood valued at $40 million (Teas 1982). The study of saprophytic fungi is important because these organisms (together with bacteria) convert lignocellulose into energy sources for other organisms in the food web (Ulken 1983). Epiphytic vegetation, usually only the vascular epiphytes, is occasionally described.

Fauna

Treatments of the animals found in mangal are usually limited to a list of species without much discussion of the interaction between animals and mangroves. MacNae is exceptional in this respect. He discusses the environment of the mangal in terms of animal microhabitats. He distinguishes the tree canopy, rot holes in trunks and branches, the surface of the soil, the soil subsurface, and permanent and semiper-manent pools, to which should be added, of course, water courses themselves. This gives some idea of habitat diversity. Faunistic lists may be long because both terrestrial and marine faunas should be included. Saenger et al. (1977), for example, list the following numbers of animals for the Australasian mangal and tidal salt marshes: mollusca, 95 species; crustacea, 65 determined species; worms, 97 species; and birds, 242 species. Little attempt has been made to list insects in mangroves, but the numbers would be proportionately large. The artificiality of such lists is well illustrated by the number of bird species. Almost all the species in the census are listed as ''visitors'' (seen in mangal or salt marshes) or ''associated'' (utilizing the community but not restricted to it). One species of noddy (*Anous*) is recorded as nesting in mangroves. The significant bird fauna is the one that is ''exclusive'' to the plant community; only 14 species of birds are listed in this category. Direct plant–animal interaction can best be perceived in nectarivorous birds. Other birds may feed in the mangrove habitat, mainly on insects and other invertebrate faunas. An example of direct dependence pointed out by these authors is the mistletoe bird (*Dicaeum hirundinaceum*), which feeds on the fruits of mistletoe (*Amyema* spp., Loranthaceae). The mistletoe bird is common in mangal in northern and northeastern Australia only where *Amyema* species are parasitic in mangroves; it occurs elsewhere in terrestrial communities. The bird presumably is a major cause of infection of mangroves by vascular parasites. Large animals that occupy the canopy are some leaf-eating monkeys and flying foxes, which may roost.

In contrast, marine animals are associated with mangal largely because they share the same substrate requirement. Some larger animals may be characteristic: mud-skippers, crabs, oysters, and snails. MacNae (1968) discusses these animals in terms of their adaptation to the mangal environment. Mudskippers, which are gobioid fish, are characteristic of mangal and related areas. Several genera and species are involved; some can even climb trees. Crocodiles are probably the most dangerous large animals, even though now they are rare and often protected as endangered because their populations have been reduced enormously by hunters. There are a few poisonous snakes. There are few large terrestrial animals, but tigers are said to characterize the mangal of the Ganges Delta, which for them is almost a refugium. Other mammals are deer, raccoons, and bats, but these should all be described as visitors. Bats play an important role in the reproduction of *Sonneratia* in some parts of its range.

Indirectly a number of marine animals may influence mangal vegetation. Crabs destroy seedlings and prevent regeneration. *Acrostichum* and other plants are characteristic associates on the mounds made by the lobster *Thalassina anomala*, which is common throughout Southeast Asia.

There is some controversy about the significance of faunal predators on the aerial roots of *Rhizophora* (Ribi 1982). These roots branch opportunistically; that is, they remain unbranched until they take root distally or until some environmental stress induces regenerative or proliferative branching by a process that is not understood developmentally (see Fig. 5.11). The stress may be climatic (a simple drying out of the apical meristem) but seems more often to be the result of attacks of predatory animals. Common examples in the Florida–Caribbean region, where the phenomenon has been studied, are the isopod *Sphaeroma terebrans* and the beetle *Poecilips rhizophorae*. Where these attacks are extensive (e.g., *Sphaeroma*) the mangroves have been considered to be threatened (Rehm and Humm 1973). Other authors (e.g., Simberloff et al. 1978) have suggested that the attacks are beneficial, since the roots branch if attacked and an extensive branched root system is necessary to the survival of the mangrove. The ultimate question is: When does enough presumed beneficial effect become too much and a detrimental effect? The example specifies the general question of the effect of the total epifauna on the aerial roots of mangroves.

In general there is very little precise information about plant–animal interactions in mangroves; the most significant exception relating to pollination is discussed in Chapter 7. Even here the lack of close correlation is indicated by the observation that pollinators are generalized for mangroves. The honey eaters that visit the flowers of *Bruguiera gymnorrhiza* in Queensland (Tomlinson et al. 1979) are replaced by sunbirds in East Africa (Davey 1975). One peculiar association suggested by Primack and Tomlinson (1980) is that honey eaters may be attracted to the shoots of *Rhizophora stylosa* by the sugary secretion that originates in the stipules. In addition the birds may eat insects on the leaves and branches; this is presumably beneficial if the insects are phytophagous. One further reason for limited plant–animal interaction in mangal is that mangroves are almost exclusively dispersed by water; fruits and seeds are not attractive to animals.

The insect fauna of the mangal has not been studied extensively, although insects are probably the main predators of mangrove leaves. Mosquitoes and midges are the most notable and offensive insects of the mangal and may make the community totally inhospitable to humans, as in South Florida in the summer months. Where the mosquitoes are hosts to diseases like malaria and dengue fever, the mangal is truly fever-ridden. Again MacNae is exceptional in his discussion of the insect fauna. He describes the activities of tailor ants (*Oecophylla* species), which may be beneficial to the tree because they eat (but also tend) scale insects. He discusses the behavior of fireflies in some detail, a characteristic feature of Asian mangroves. Johnstone (1981), however, reports that *Oecophylla* does not affect the amount of leaf tissue eaten by herbivores.

The herbivory of mangroves by insects varies considerably. The results obtained by Johnstone (1981) from a study in the Port Moresby region of Papua New Guinea, which showed that as much as 20 percent of leaf tissue could be consumed by herbivores, also showed no correlation between the amount eaten and such factors as specific diversity and density, seasonality, nitrogen pollution, and the calorific value of leaf tissue. His observation that the amount was also independent of the chloride concentration of leaves suggests that the leaf salinity of mangroves does not discourage predators. The results are of interest because it is usually assumed that mangrove biomass finds its way into food chains largely via detritus feeders; alternatives to that particular food chain must exist. There seem to be no studies of the leaf consumption of larger herbivores, even though MacNae (1968) mentions that leaf-eating monkeys occupy mangal. One specialized example is *Aratus pisonii*, a Brazilian tree-climbing crab that has been observed to consume mangrove wood pulp. This species makes a virtue of necessity, since it seems to be confined by its terrestrial predators to the mangal canopy during its adult life (Lacerda 1981).

The overwhelming conclusions are that mangal does not provide good examples of close interdependence between plant and animal and that coevolutionary processes do not occur. There are very few natural history studies, however, and this conclusion may have to be changed when life history studies of specific animals are initiated. The fact remains that although much has been written about the botany of mangal, there is no definitive zoology of mangal. The account by MacNae is a model and a guide should anybody make the attempt.

Analytical methods

Quantified analyses of mangal are numerous. They rely largely on profile diagrams based on transects usually at right angles to the shore, so that the most extreme transition is portrayed. In profile diagrams the species are usually so few and distinct in architecture and morphology (e.g., aerial root systems) that an accurate visual impression of zonation can be conveyed. Often the transects carry further infor-

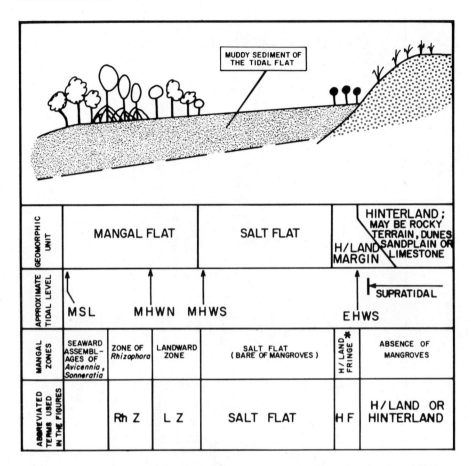

Figure 1.7. Schematic profile of tidal flat in northwestern Australia, illustrating some geomorphological and mangrove terms. Asterisk indicates that the zone may be absent from drier regions. MSL, mean sea level; MHWN, mean high water at neap tide; MHWS, mean high water at spring tide; EHWS, extreme high water at spring tide. (From Semeniuk 1983)

mation that is important in ecological interpretation. Figure 1.7 is a stylized example. Tidal ranges are almost always indicated. Particularly helpful is the substrate composition. Gradients in salinity and pH are less helpful unless there is some indication of ranges and frequencies, that is, they are based on repeated measurements. Sometimes the distribution of marine animals (shellfish and crabs) is included, or the distribution of mosquito eggs and their larva can be shown. The range of plant species within the zonations can also be added but not represented in the profile diagram. Figure 1.8 is a schematic profile from the wet coast of Queensland, transitional inland to wet forest.

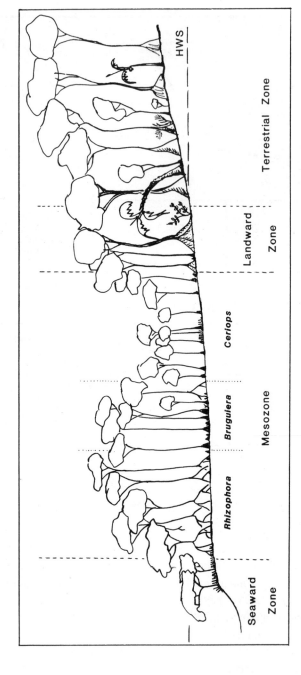

Figure 1.8. Schematic and generalized profile of a tidal flat in northeastern Australia (Queensland) to contrast zonation in a region of high rainfall with the region shown in Figure 1.7. The seaward zone is usually *Avicennia*. The salt flat in Figure 1.7 corresponds to the *Ceriops* zone in Figure 1.8. It is assumed that rainfall is the controlling factor, but site (like distance up river), latitude, tidal range, and substrate affect the zonation. Profiles similar to Figure 1.7 therefore can occur in eastern Australia in drier areas. (Courtesy of N. C. Duke)

Figure 1.9. *Avicennia* and *Sonneratia* saplings established on a mud flat in open water. High tide.

Profile diagrams are particularly helpful for comparative purposes. They have demonstrated that there is no constant sequence for mangroves of contrasted regions, even though species tend to prefer a more seaward or a more landward site. However, the example that shows *Rhizophora* as the most seaward species in one locality may be contrasted with adjacent localities where *Avicennia* or *Sonneratia* occupies the seaward limit of the mangal (Fig. 1.9). Profile diagrams are sometimes used as evidence for succession, since the zonation in space is often accepted as a zonation in time. Perhaps there is no more confusing or controversial topic in mangrove ecology than succession (Lugo 1980). It would be inappropriate to attempt to discuss the issue, least of all in this book, but the essay by Snedaker (1982) is relevant.

The transect method is sometimes complemented by a plan view. The one used by Watson (1928), which shows zones following shoreline contours, has been reproduced several times, but the fact that it is purely imaginary and used to illustrate the concept of inundation classes is not usually stated. Watson (1928) mentions the suitability of mangroves for aerial survey because of their level topography and the clarity of discrete types, whereas ground surveys are difficult. The whole west coast of peninsular Malaysia was surveyed in 1927–8 from a height of 2000 ft.; if extant, these photographs would be of enormous ecological value. The studies by Semeniuk

(1980, 1983) of mangroves in northwest Australia used a diversity of approaches, ranging from ground to aerial surveys. These studies show the intricate relation of topography, substrate, and tidal influence.

Dynamic approaches to the study of mangal underlay most analyses. They attempt to locate the complex of characters that influence distribution and estimate the relative successional status of different species. The frequent clear-cut zonation of mangroves (Fig. 1.8) fosters this approach and provides the stimulus for such studies. Interpretative analyses often lead to the production of flow charts that connect zones within the mangal, with mangal-associated vegetation, and often with certain aspects of the ecotone, such as salinity, substrate composition, and topography. These kinds of diagram have been proliferated by Chapman (1976).

Zonation and succession

The existence of zones, often monospecific, is evident in mangal even to the most superficial observer. Profile diagrams produce the impression that zonation is a regular series of vegetational bands parallel to the coastline. However, any regular zonation is modified by local topography, which determines tidal and fresh-water runoff, and by sediment composition and stability (Semeniuk 1980, 1983). Lugo and Snedaker (1974) have illustrated how community structure may be elaborated in the absence of much floristic diversity. They recognized six different mangal types in South Florida determined by topography, even though only four mangrove species are involved (Figs. 1.10, 1.11, 1.12). In view of the meandering and often intersecting runoff creeks that cut irregularly across the zonation, it is not surprising that the vegetation becomes more readily interpretable as a mosaic. Where mangroves run along river banks or front steep shores, there is insufficient space for zonation to develop. Where conditions are such that one species dominates a large area, as in a *Nypa* swamp, there may be only a truncated zonation at the margin of the swamp. Generalized descriptions of the ecological distribution of mangroves indicate that most common species can form pure stands in at least some locations (e.g., van Steenis 1958). Species of the genus *Rhizophora* that grow together show pronounced zonation in West Africa (Savory 1953), Queensland, and New Guinea, but the zonation is easily disrupted.

Further instability in the vegetation pattern is introduced by variation in the substrate, which ranges from coarse coral fragments to peat. Along high-energy shores or in estuaries subject to irregular runoff sediment redistribution is highly variable and the environment very unstable.

Because of the diversity of mangrove environments it is difficult to present a consistent view of the relative position of each species in terms of ecological requirements. A precise description of mangal in one region can easily be contra-

Figure 1.10. Islands of *Rhizophora mangle* (Rhizophoraceae) from the air with transition beyond to back mangal with *Avicennia, Laguncularia*, and *Conocarpus*. Lane River, Everglades National Park, Florida. (From a color transparency by Ingrid Olmsted)

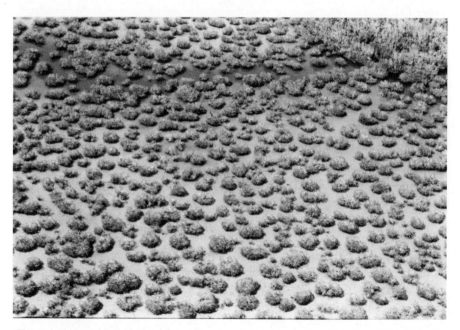

Figure 1.11. Dwarf individuals of *Rhizophora mangle* from the air. Joey Bay, Everglades National Park, Florida, in region of transition to fresh-water marsh. (From a color transparency by Ingrid Olmsted)

Figure 1.12. Fringing mangroves lining creeks through fresh- and salt-water marsh. West side of Shark Slough, Everglades National Park, Florida. (From a color transparency by Ingrid Olmsted)

dicted by an equally precise analysis in an adjacent region. A common assumption is that the zonation of species in space represents their succession in time. Chapman (1976) has been the chief advocate of the zonation-equals-succession school of thought, but he includes numerous cautionary disqualifiers. The idea of mangroves representing a successional stage in the development of some climax terrestrial community is strongly embedded in the ecological literature and comes from the analogy of a fresh-water "hydrosere" leading from open lake to terrestrial system (e.g., Johnstone 1983). This concept is an ecological straw man put there for historical as much as hypothetical reasons, since it derives from the Clementsian school of plant ecology. Johnstone, for example, uses it to make the point that the mangal is a community with its own climax and denies alternative "steady-state" notions of mangal, apparently because for him a climax must be found. Curiously, in the same volume in which his paper appears there is an analysis by Woodroffe (1983), which documents a steady-state *Rhizophora*-dominated community on Grand Cayman established on a Pleistocene substrate and persisting and expanding by the formation of mangrove peat, which can be roughly dated by radiocarbon analysis. This community has persisted while sea levels have risen at least 2 m and for about

2000 years, building up deposits of peat exceeding 4 m in some places. In Biscayne Bay, Florida, mangrove peat is found beyond the seaward limit of the existing *Rhizophora* forest, showing that the zone has either shifted landward or shrunk. Davis (1940) has suggested that the mangroves themselves determine geomorphological processes because of an assumed ability to trap and retain sediments. Other authorities (e.g., Thom 1967, 1975) maintain that geomorphological processes are the primary cause of any vegetational migration.

The chief deficiency seems to be lack of a direct demonstration of real successional processes in time (rather than imaginary ones in space). One reason for this is the absence of datable growth rings in mangrove woods so that a static analysis cannot be transcribed into a dynamic one by comparing the relative ages of trees in the community. This deficiency is stultifying to the whole of tropical ecology (see Bormann and Berlyn 1981).

Multifactorial analysis

The complexity of the mangal ecosystem is dramatically portrayed by Lugo (1980), who summarizes interrelationships by means of a diagram that incorporates marine and terrestrial ecosystems; substrate, climatic, and topographic variables; recurrent cycles; and infrequent (cataclysmic) events in an intersecting reticulum. The diagram is not reproduced here, but its complexity can be appreciated by the fact that there are 35 interconnected factors and the pathways between them seem almost limitless. Bunt and Williams (1981) define 29 ''association groups'' in their complex analysis of 35 species that occur in 1391 sites in northeastern Australia. They emphasize that any unidimensional approach to species associations is misleading.

The study by Semeniuk (1983) of the ecological distribution of mangroves in northwestern Australia is particularly instructive. Although it represents a rather simple (i.e., floristically depauperate) system (Fig. 1.7), it raises most of the questions concerning zonation. The habitat is somewhat marginal for mangroves because of relatively low levels of rainfall and runoff. In general there is a seaward assemblage of *Avicennia* and *Sonneratia* followed by a *Rhizophora* zone transitional to the zone of greatest diversity, which itself grades into an extensive salt flat, always devoid of mangroves. At the landward fringe of the salt flat there may or may not be a second reduced mangrove zone finally transitional to terrestrial communities beyond the influence of tides. Semeniuk's analysis of physiography and hydrology provides evidence that ground-water salinities are an important influence on mangrove distribution. The simplest evidence is that where ground-water seepage is abundant, there was a landward fringe, otherwise it was absent. The explanation of distribution by correlation is strong here, but what, we might ask, is the limiting factor? Is it rainfall, which is the source of ground water? Is it topography, which directs ground-water flow, or edaphic conditions, which modify it? Is it the dispersibility of mangrove propagules, which determines initially whether or not a species can find a given site? Is it tidal influence, which determines the initial supply of sodium chloride? Is it the physiological adaptation of the species to salt concentrations over a given range of values? Is it coastal geomorphology, which provides contrasting sediment types and redistributes them during storms?

It is, of course, no one of these factors but their effect in concert that establishes the vegetation pattern. The ecologist needs a sympathetic audience when it is appreciated that he or she is unable to specify the precise effects of 35 variables in determining vegetation zonation.

Brief mention may be made of some factors that can influence mangrove zonation; these might be contrasted on the basis of whether the influence is abiotic or biotic. Most of the factors can work in concert, since they are not mutually exclusive.

Abiotic factors influencing zonation

Geomorphology

Detailed analyses of coastal geomorphology by Thom and co-workers in Mexico and Australia (Thom 1967) established the direct dependence of mangrove distribution and coastal diversity from a dynamic point of view. Semeniuk (1980, 1983) documents the variation of this dependence in considerable detail in north-western Australia. These results may be interpreted to show that mangal distribution follows and does not determine the dynamics of topography so that mangroves in the environments studied do not over-ride abiotic land-building processes. They still leave unresolved any explanation of zonation in terms of differential biological adaptation to contrasted physiographic factors.

Inundation classes

Watson (1928), in his classic studies of mangal in the Malay Peninsula, defined zones as regional sets influenced by a combination of numbers and kinds of tidal inundation (inundation classes), according to the following simple scheme. Watson emphasizes that these classes refer to the conditions in the Klang area, but they have been applied and elaborated elsewhere (e.g., Chai, 1982) because over the long term they can be precisely measured. It becomes possible to ascribe given species to a narrow range of inundation classes and this implies that a species may be more or less restricted to a given class. The restriction can be modified by other

Table 1.1. *Watson's inundation classes (Port Swettenham = Port Klang)*

Inundation class	Flooded by	Height above datum line (feet)		Times flooded per month	
		From	To	From	To
1	All high tides	0	8	56	62
2	Medium high tides	8	11	45	59
3	Normal high tides	11	13	20	45
4	Spring high tides	13	15	2	20
5	Abnormal (equinoctial tides)	15	—	—	2

factors, notably type of substrate. However, the analysis in no way accounts for this restriction in a functional sense.

Physiological responses to gradients

Numerous authors have concentrated on physiological adaptations that would restrict the range of conditions under which a species might grow. The usual approach is to measure an element of the environment that can be presumed to have a direct physiological effect and establish correlations with the distribution of different species. The most usual parameter considered has been salinity, since it can be measured easily and often correlates closely with species distribution. Salinity refers to the numbers of grams of dissolved salts in 1000 g of sea water, with chloride concentration used as an index. Values are usually expressed in parts per thousand (0/00) and range from 33 to 38 0/00 in the open ocean. Another commonly used but not very precise parameter is the dilution (or concentration) relative to sea water. Mangroves can adjust to about 90 0/00 (Cintron et al. 1978) in the more salt-tolerant species like *Avicennia* (i.e., about 2.5 times the concentration of sea water), but higher tolerances have been claimed.

For many ecologically constraining factors it is the extremes that are limiting, rather than the average; this is particularly clear with the salt concentration in the soil. This may determine zonation at the margin of a salt flat, for instance, but the range of salinities within the extended seaward zones subjected to frequent inundation (Classes 1 to 3 in Watson's scheme) may be so small as to be physiologically insignificant. Other limiting factors must then be sought, such as propagule size and competition, which may be defined as biotic factors.

Biotic factors influencing zonation

Propagule sorting

Perhaps the simplest attempt to explain mangal zonation is by Rabinowitz (1978b), who suggests that it results by propagules sorting according to size, with the heavier propagules occupying the seaward habitats because they are less readily dispersed landward and also become established more easily in deeper water and more frequently inundated sites. Support for this hypothesis comes from reciprocal transplant experiments, which suggest that mangroves are interchangeable between zones, since they grow as seedlings equally well in the "wrong" zone; therefore, they must be constrained by some biotic factor to their normally restricted ecological range. The hypothesis is based primarily on the assumption that since there is no precise explanation of physiological adaptation to the physical conditions of a given mangal habitat, a biological explanation must be provided. It also assumes that the establishment of a plant as a seedling is sufficient to account for its persistence as a mature tree.

Contrary evidence to the propagule-sorting hypothesis is easily provided when species with small propagules can be shown to occupy the seaward margin of the mangal. This occurs in many parts of Australasia where *Avicennia* and *Sonneratia* (see Figs. 1.8, 1.9), which have relatively small propagules (see Chapter 8), are the seaward dominants (Chai 1982). Furthermore some species range widely throughout the zones, such as *Avicennia* in northwestern Australia (Semeniuk 1983).

Competition

If two species have exactly the same responses to physical circumstances, can one exclude the other by some biological mechanism? This possibility is implied in the propagule-sorting hypothesis of Rabinowitz because seedlings planted in the "wrong zone" do not survive. Under these circumstances, discussion of whether mangal structure is the result of either a steady-state or a successional process (Lugo 1980) becomes less philosophical and can be discussed in terms of interaction between species. In this sense it seems appropriate to look at the ecological status of mangroves in relation to modern concepts of "gap phase" dynamics and succession in tropical communities.

Ecological status of mangroves

Numerous authors have compared mangroves with plants of other communities, for example, salt marsh, salt desert, swamp forest, and xerophytic vegetation. Their

Table 1.2. *Comparison of pioneer and mature phase species and communities with mangroves and with mangal*

Species	Pioneer	Mature	Mangrove
Seed size	Small	Large	Usually large
Seed number	Numerous	Few	Often numerous
Dispersal agent	Often abiotic (e.g., wind)	Usually biotic	Always abiotic (water)
Dispersibility	Wide	Limited	Wide
Geographic range	Broad	Narrow	Broad
Seed production	Continuous	Discontinuous	Sometimes continuous
Seed dormancy and viability	Long	Short	Short (?)
Seedlings	Light-demanding, not dependent on seed reserves	Not light-demanding, dependent on seed reserves	Light-demanding(?), dependent on reserves
Reproductive maturity	Early	Late	Early
Life span	Short	Long	Probably long
Leaf size	Often large	Medium or small	Medium
Leaf palatability	High	Low	Low
Wood	Soft, light	Hard, heavy	Hard, heavy
Architecture	Model-conforming	Not model-conforming	Model-conforming
Crown shape	Uniform	Varied	Uniform
Competitiveness	For light	For many resources	For light and other resources
Pollinators	Not specific	Highly specific	Not specific
Flowering period	Prolonged or continuous	Short	Prolonged or continuous
Breeding mechanism	Inbreeding favored	Outbreeding favored	Inbreeding favored

Community	Pioneer	Mature	Mangal
Floristic composition	Poor	Rich	Poor
Stratification	Absent	Well developed	Absent
Age composition	Even-aged	Uneven-aged	Even-aged(?)
Large stems	Absent	Present	Usually absent
Undergrowth	Dense	Sparse	Absent
Climbers	Few	Many	Few
Epiphytes	Few	Many	Few

Source: Budowski (1965), Gomez-Pompa and Vazquez-Yanes (1974), UNESCO (1978); see also Ewel (1980), Primack and Tomlinson (1980), Whitmore (1983).

designation as plants subject to "physiological drought" by Schimper (1891) is based on an incomplete knowledge of their water relations (see Chapter 6). It is better to view mangroves as forming unique communities so that mangal is not readily compared with other vegetation in any constructive way. Insufficient attention has been paid to the biology of individual species. A more synthetic approach to their study can be adopted, in which the status of individuals (mangroves) is emphasized at the expense of a consideration of the status of the community (mangal).

I have found it useful to compare mangroves and mangal with pioneer and mature-phase plants in tropical forests (Table 1.2). From this it is clear that mangroves share an interesting mixture of the attributes of the species and community in these contrasted phases. They have clearly pronounced characteristics of *pioneer species* in their *reproductive biology* but of *mature-phase species* in some aspects of their *community structure* and *vegetative growth*. An alternative designation would be to say that they have the properties of *r*-selected species in finding their habitat, but of *K*-selected species in maintaining it, where the concept of *r*- and *K*-selection introduced by MacArthur and Wilson (1967) refers to species that either maximize their intrinsic rate of population increase (r) or maintain populations at the maximum carrying capacity (K) of the environment. Mangroves have their cake and eat it! Some evidence for these designations is anecdotal or even conjectural, but it does suggest areas where further investigation is desirable. Later descriptions of individual taxa may be construed in light of some of these features.

The mixture of characteristics suggest that mangroves initially have problems similar to those of pioneer species in locating a habitat that is patchy, varied, and available for only limited periods. Once established – and establishment has led to the development of unique biological features, notably vivipary – the mangrove community that develops has unusual properties; it has little structure because there is no further successional development. No understorey develops, there is no stratification, and much competition is intraspecific, while species distribution is strongly influenced by edaphic factors such as the degree of salinity and frequency of inundation.

On this basis we might categorize mangroves as pioneer species, primarily because of their reproductive capabilities, but they form a community without succession, as shown by their performance at maturity in the vegetative state. This, at least, is a hypothesis that should be capable of development.

2   Floristics

Categorization

Mangal is a community that is defined as much by its constituent species as by its environment, hence, the original use of the word "mangrove" to define both the plants and the vegetation. This leads to a rather circular definition of mangroves as "trees characteristically found in tidal swamps," since tidal swamps are best defined by the species that grow in them. Excluding herbs is not artificial; the only likely candidate for a herbaceous mangrove is *Crenea patentinervis* (Lythraceae). Just as mangal is difficult to delimit in its transition to terrestrial and other seashore communities, it becomes rather arbitrary to determine the floristic limits of the group of plants that should be included in an account of mangroves. If one establishes the limits of the community by the extent of tidal influence, salinity, or type of substrate, one inevitably includes elements that are more characteristic of adjacent swamp forest, strand, and beach communities, estuarine riverbank forest, salt flats, and so on. In this account I arbitrarily set limits between three groups: major elements of mangal, minor elements of mangal, and mangal associates, using fairly rigid criteria to distinguish them. Of course, the groups are not sharply circumscribed and the assessment is somewhat subjective, since there is a continuum of possibilities.

Major elements of mangal (*"strict or true mangroves"*)

Major elements are recognized because they possess all or most of the following features:

1. Complete *fidelity* to the mangrove environment; that is, they occur only in mangal and do not extend into terrestrial communities.
2. A *major role* in the structure of the community and the ability to form pure stands.
3. *Morphological specialization* that adapts them to their environment; the most obvious are *aerial roots*, associated with gas exchange, and *vivipary* of the embryo, whose functional significance is not clear.
4. Some physiological mechanism for *salt exclusion* so that they can grow in sea water; they frequently visibly excrete salt.
5. *Taxonomic isolation* from terrestrial relatives. Strict mangroves are separated from their relatives at least at the generic level and often at the subfamily or family level. For minor mangroves, the isolation is mostly at the generic level.

25

Although criterion 1 is the most significant, the importance of the group of traits is stressed because many occur singly in other groups of plants. For example, many swamp forest plants develop aerial roots associated with gas exchange, and all halophytes have some degree of salt exclusion though not necessarily the same as that in mangroves. Vivipary among seed plants, on the other hand, is well developed only in mangroves. It occurs occasionally in other flowering plants like the seagrass *Amphibolis*. Criterion 5 might seem artificial, irrelevant, and derived only *a posteriori*, but comparative taxonomic study supports it as a correlative feature so that the more faithful a taxon is to the mangrove community, the more isolated it seems to be from its relatives. This may reflect the age of the taxon as a plant specialized to mangal and its consequent lengthy period for ecological isolation, which leads to genetic isolation and thence evolutionary divergence from terrestrial relatives.

Minor elements of mangal

These minor species are distinguished by their inability to form a conspicuous element of the vegetation. They may occupy peripheral habitats and only rarely form pure communities.

Table 2.1 lists the first two categories of mangroves with some of their attributes relating to the criteria I have erected. It is clear that these plants do not have equivalent ecological roles. *Nypa*, for example, forms rather distinctive pure communities in quiet estuaries. At the same time it is the only palm that has a viviparous fruit as an indication of its degree of specialization. *Kandelia* is an uncommon element in mangal but occasionally does form pure stands.

Mangrove associates

The list of plants in Table 2.2 is arbitrary and could certainly be extended, especially if nonwoody plants were included. It has been drawn from a number of sources with extensive lists of mangrove associates (e.g., Chai 1982, Watson 1928) and from my own experience (e.g., Tomlinson 1980). If herbaceous or subwoody terrestrial plants were included, most members of the tropical beach community would be added, many of which are pantropical. One could then include such plants as *Achyranthes*, *Entada*, *Ipomoea*, *Pluchea*, *Remirea*, *Sesuvium*, and *Suriana*, species of which are dispersed by sea currents and are characteristic of open coastal communities in the tropics. Such a larger list could be drawn from sources that discuss seed dispersal by ocean currents (e.g., Guppy 1917, Ridley 1930). These elements, however, are never inhabitants of strict mangrove communities and may occur only in transitional vegetation. Grasses, rushes, and sedges occur only when they penetrate the more open parts of mangroves from adjacent fresh-water or saline

Table 2.1. *Genera of mangroves on a cosmopolitan basis*

Family	Genus	No. of mangrove species	Aerial roots	Vivipary	Level of taxonomic isolation and status
Major components					
Avicenniaceae	*Avicennia*	8	+ +	+	Monogeneric family related to Verbenaceae
Combretaceae	*Laguncularia*	1	+	−	Tribe Lagunculariae
	Lumnitzera	2	+	−	(+ *Macro-pteranthes* nonmangrove)
Palmae	*Nypa*	1	−	+	Isolated group (subfamily?) within the family
Rhizophoraceae	*Bruguiera*	6	+ +	+ +	Collectively form a
	Ceriops	2	+ +	+ +	natural tribe, Rhizo-
	Kandelia	1	−	+ +	phoreae, within the
	Rhizophora	8	+ +	+ +	family
Sonneratiaceae	*Sonneratia*	5	+ +	−	One of two genera in the family (*Duabanga* nonmangrove)
Total		9	34		

In summary, closest relatives of mangrove taxa are families, subfamilies, tribes, or genera

Minor components					
Bombacaceae	*Camptostemon*	2	+	−	Genus isolated within family
Euphorbiaceae	*Excoecaria*	1 (−2)	−	−	Genus includes about 35 nonman-grove taxa
Lythraceae	*Pemphis*	1	−	−	Genus distinct within family, one species nonmangrove
Meliaceae	*Xylocarpus*	2	+ +	−	One species non-mangrove; forms with *Carapa* tribe Xylocarpeae
Myrsinaceae	*Aegiceras*	2	−	+	Genus isolated within family
Myrtaceae	*Osbornia*	1	−	−	Genus isolated in the family

Table 2.1. *(cont.)*

Family	Genus	No. of mangrove species	Aerial roots	Vivipary	Level of taxonomic isolation and status
Pellicieraceae	*Pelliciera*	1	−	+	Monotypic family related to and some-times included within Theaceae
Plumbaginaceae	*Aegialitis*	2	−	+	Isolated genus, sometimes segre-gated as family Aegialitidaceae
Pteridaceae	*Acrostichum*	3	−	−	Genus isolated in the family
Rubiaceae	*Scyphiphora*	1	−	−	Genus isolated in the family
Sterculiaceae	*Heritiera*	3	−	−	Three mangrove species only; rest (about 26) are terrestrial
Total		11	20		

In summary, closest relatives of mangrove taxa are families, genera, or species

Note. + + means present or well developed; + means present; − means absent.

marshes. Examples are *Paspalum distichum* L.; *Phragmites karka* (Roxb.) Trin. ex Steud. (Gramineae); *Juncus roemerianus* L. (Juncaceae); and *Cyperus javanicus* Houtt. (Cyperaceae). Mangroves otherwise form forests with no understorey except their own seedlings; there is no understory of herbs. The fern *Acrostichum* may become dominant in disturbed sites; it exists in the undisturbed mangrove by virtue of its ability to colonize elevated sites that are not inundated at high tide. In Malaysia for example it occurs on the tall mounds made by the burrowing lobster *Thalassina*. Ferns do not otherwise occur as terrestrial plants in mangroves, but may exist as epiphytes in adjacent communities.

A feature of mangrove associates is the much greater diversity of leaf form, size, and texture, partly shown in Figure 2.1 (cf. Fig. 4.11).

Systematic position of mangroves and their associates

Because mangal associates and specialized groups play so inconspicuous a role in the basic structure of mangrove forests, it may be argued that any discussion or

Table 2.2.*Mangal associates on a cosmopolitan basis*[a]

Family	Genus	Mangal or coastal species[b]	Terrestrial species (approx.)	Characteristic habitat of associate
Acanthaceae	*Acanthus*	3 (OW)	30	Back mangal
Anacardiaceae	*Gluta*	1 (OW)	20	Back mangal
Apocynaceae	*Cerbera*	3 (OW)	3	Back mangal
	Rhabdadenia	1 (NW)	3	Back mangal
Batidaceae	*Batis*	1 (−2) (P)	—	Salt marsh
Bignoniaceae	*Amphitecna*	1 (NW)	1	Back mangal, river banks
	Anemopaegma	1 (NW)	30	Back mangal, river banks
	Dolichandrone	1 (OW)	9	Back mangal, river banks
Celastraceae	*Cassine*	1 (OW)	80	Wet coastal communities
Combretaceae	*Conocarpus*	1 (NW)	1	Back mangal, wet communities inland
	Terminalia	1 (OW)	200	Beach and coastal communities
Compositae	*Tuberostylis*	2 (NW)	—	Back mangal
Ebenaceae	*Diospyros*	1 (OW)	400	Back mangal to rain forest habitat
Euphorbiaceae	*Glochidion*	1 (OW)	300	Back mangal
	Hippomane	1 (NW)	—	Beach communities
Flacourtiaceae	*Scolopia*	1 (OW)	37	Back mangal, river sides, and inland
Goodeniaceae	*Scaevola*	2 (OW, NW)	90	Beach communities
Guttiferae	*Calophyllum*	1 (OW)	250	Beach communities
Lecythidaceae	*Barringtonia*	2 (OW)	40	Back mangal, river banks, and inland
Leguminosae				
(Caesalpinoideae)	*Cynometra*	2 (OW)	70	Back mangal and inland
	Caesalpinia	2 (P)	40	Back mangal
(Papilionoideae)	*Aganope*	1 (OW)	6	Back mangal
	Dalbergia	2 (NW)	300	Back mangal
	Derris	1 (OW)	50	Back mangal
	Inocarpus	1 (OW)	3	Lowland swamps
	Intsia	1 (OW)	8	Back mangal
	Mora	1 (NW)	10	Back mangal

Table 2.2. *(cont.)*

Family	Genus	Mangal or coastal species[b]	Terrestrial species (approx.)	Characteristic habitat of associate
	Pongamia	1 (OW)	2	Coastal environments and back mangal
Malvaceae	*Hibiscus*	1 (P)	200	Beach and coastal communities
	Pavonia	1 (NW)	200	Back mangal and coastal communities
	Thespesia	2 (P)	15	Beach and coastal communities
Melastomataceae	*Octhocharis*	1 (OW)	5	Back mangal
Meliaceae	*Amoora*	1 (OW)	20	Back mangal
Myristicaceae	*Myristica*	1 (OW)	120	Wet forest, back mangal
Myrsinaceae	*Ardisia*	1 (OW)	250	Back mangal
	Myrsine	1 (OW)	10	Back mangal
Palmae	*Calamus*	1 (OW)	400	Back mangal
	Oncosperma	1 (OW)	4	Back mangal
	Phoenix	1 (OW)	12	Coastal swamps
	Raphia	1 (NW, OW)	10	Coastal swamps
Pandanaceae	*Pandanus*	2 (OW)	300	Back mangal, coastal swamps
Rubiaceae	*Rustia*	1 (NW)	12	Coastal communities
Rutaceae	*Merope*	1 (OW)	-	Back mangal
Sapindaceae	*Allophyllus*	1 (OW)	190	Diverse
Sapotaceae	*Pouteria*	1 (OW)	50	Back mangal
Tiliaceae	*Brownlowia*	2 (OW)	30	Swamp forest
Total		46	60	

[a] Lacking aerial roots and vivipary.
[b] OW = Old World; NW = New World; P = pantropical.

description of them is irrelevant. They do interact with mangroves at two important levels, they may share (or compete for) the same pollinators, and they may share the same predators and parasites so that as alternative hosts their influence may not be entirely negligible. In any evolutionary sense also some associates may be capable of further specialization as major mangrove components; they may give clues to the evolutionary pathway by which the highly specialized adaptive syndrome of

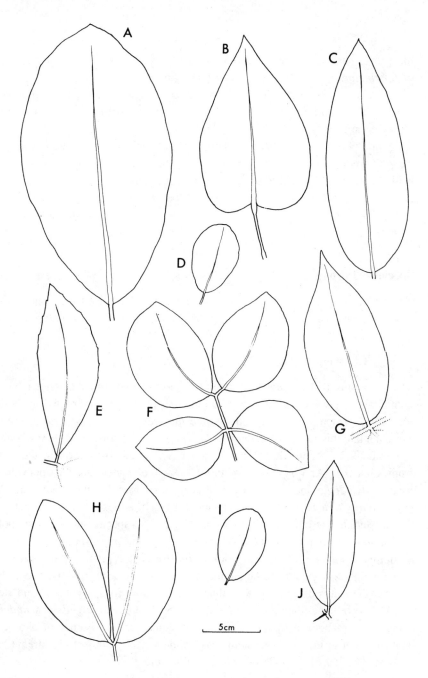

Figure 2.1. Leaf outlines of some species from the back mangal. Range of size and shape is greater than in the front mangal. (A) *Heritiera littoralis*; (B) *Brownlowia argentata*; (C) *Inocarpus fagifer*; (D) *Cassine viburnifolia*; (E) *Allophyllus cobbe* (one leaflet); (F) *Intsia bijuga*; (G) *Mora oleifera* (one leaflet); (H) *Cynometra ramiflora*; (I) *Glochidion littorale*; (J) *Merope angulata*, with axillary spine.

mangroves has been achieved. In this respect the distinctive method of germination of *Barringtonia* suggests a preadaptation to the mangrove habitat.

An additional species excluded from Tables 2.1 and 2.2 because it is not woody is *Crenea patentinervis* (Lythraceae). I have suggested that this is probably the closest to a mangrove herb of any species.

Gymnosperms play no role in these communities, although the cycad *Cycas rumphii* frequently occurs in communities where the topography rises abruptly behind mangroves; it will tolerate some salt spray but is typically a denizen of dry hillsides. Watson (1928) records occasional specimens of *Podocarpus polystachyus* and *Agathis alba* in sandy coastal sites. Conifers in general have little or no tolerance of salt.

Specialized groups

Three additional synusiae or "guilds" characteristic of tropical forests but sparingly represented in mangal deserve mention: climbers, epiphytes, and parasites.

Climbers

Plants that are not self-supporting do not occur in the strict mangrove community. However, a number of scrambling or viny plants rooted in the back mangal can extend their aerial shoots into the seaward zone. Several leguminous vines fall into this category: *Aganope*, *Caesalpinia*, *Dalbergia*, and *Derris*. *Smythea lanceata* (Tul.) Summerh. (Rhamnaceae) has a wide distribution in the eastern tropics but is never very common. In the New World there are the apocynaceous vine *Rhabdadenia* and the woody bignoniaceous lianes *Anaemopegma* and *Phryganocydia*. *Calamus erinaceus* (Palmae) in Malaya is described as a "mangrove rattan" and may be extensive in the back mangal. This species and the cosmopolitan *Caesalpinia bonduc* are grapnel climbers and are the most spiny element of the mangrove community. Spines or prickles are (fortunately) rare, but do occur in the nonclimbing *Merope* and some palms and pandans. *Acanthus* is often scrambling and has the status of a mangrove "thistle" because of its spiny leaves and stems.

Other recorded climbing plants are the orchid *Vanda* and *Lygodium scandens*, a climbing fern, both in Malaysia. We can account for the absence of climbers in mangal because their slender stems have wide vessels. Subject to extreme water tension (Chapter 6) the xylem is highly vulnerable to cavitation.

Epiphytes

Most vascular epiphytic plants are intolerant of salt, so one encounters only a limited range of species, mainly in the back mangal and then relatively high in the canopy and in areas transitional to adjacent terrestrial communities where the epiphytes are more characteristic. Some epiphytic orchids grow within a few feet of the high-water mark, however. The list that could be drawn up is potentially long; the following are simply representative of mangal in Sarawak (from Chai 1982).

Dicotyledons
  Asclepiadaceae
    Dischidia spp.
  Ericaceae
    Vaccinium piperifolium Sleum.
  Melastomataceae
    Medinilla crassifolia Reinw. ex B1.
  Rubiaceae
    Hydnophytum formicarum Jack
    Myrmecodia tuberosa Jack
  Urticaceae
    Poikilospermum suaveolens (B1.) Merr.
Monocotyledons
  Orchidaceae
    Species of *Agrostyphyllum, Dendrobium, Dipodium, Eria, Luisia, Podochilus, Taeniophyllum, Trichoglottis*
Ferns and fern allies
  Aspleniaceae
    Asplenium macrophyllum Sw.
    Asplenium nidus L.
  Davalliaceae
    Humata cf. *repens* Diels
  Lycopodiaceae
    Lycopodium carinatum Desv.
    Lycopodium phlegmaria L.
  Oleandraceae
    Nephrolepis acutifolia (Desv.) Chr.
  Polypodiaceae
    Crypsinus spp.
    Drymoglossum piloselloides (L.) Presl.
    Platycerium coronarium (Koenig) Desv.

In the New World bromeliads (species of *Aechmea*, *Tillandsia*, and *Vriesia*) are occasional epiphytes that transgress from nearby terrestrial communities, indicating that they can withstand a limited amount of salt. A list of mangrove epiphytic ferns in the New World has not been drawn up. Less attention has been given to the epiphytic nonvascular cryptogams, but Nakanischi (1964) lists examples that grow on *Kandelia* in southern Japan.

Parasites

These are not a large component, but mention may be made of species in the Loranthaceae of *Amyema* [e.g., *A. mackayense* (Blakely) Dans.] and *Lysianthus* [e.g., *L. subfalcata* (Hook.) Barlow subsp. *maritima* Barlow] in eastern Australia and adjacent Papua New Guinea (Barlow 1966). These may be described as mangrove "mistletoes" and seem indiscriminate in their choice of host.

Nomenclature

The names of plants provide the key to knowledge about them in the scientific literature; in turn, correct identification is necessary before correct names can be applied. Consequently, precise nomenclature, identification, and systematics are all basic to precise research endeavor. One justification for writing this book is to provide a base line of reliable data for the use of other biologists, and no little time has been spent trying to make this as accurate as possible.

Although the number of plant species growing in mangal is small, the nomenclature is still confused for some of them, either for historical reasons or because of incomplete systematic knowledge. The paper by Booberg (1933), an update of the species listed by Schimper (1891), shows the continual need for systematic revision. Most early systematic work was done entirely on the basis of herbarium specimens, so numerous descriptions of the same species under different names came into being. Fortunately, through the activities of monographic specialists, considerable stability has been achieved. Nevertheless, there can be confusion where older, invalid names have been retained. For example, the nomenclature in the classic, much-cited study of Malaysian mangroves by Watson (1928) is about 50 percent incorrect by modern standards. A reader unaware of this would find a comparison of Watson's results with those in modern works very difficult. In some instances, Watson mentions some taxa he did not recognize as undescribed (e.g., *Acrostichum speciosum* and *Bruguiera hainesii*). The papers by Wyatt-Smith (1953a,b; 1954) were, in part, intended to correct some of Watson's errors and omissions and to update his nomenclature.

History

Mangroves were certainly known to the ancients (MacNae 1968), but significant study began with European colonization in the sixteenth and seventeenth centuries. Since mangroves are common coastal plants that are readily observed and collected, many of them, especially in the East Indies, were familiar to the earliest European naturalists. The earliest published account of them is found in the *Hortus indicus malabaricus* of H. van Rheede tot Drakenstein (Rheede 1678–1703) but especially the *Herbarium amboinense* of Georg Everhard Rumpf (Rumphius 1741–55). The latter did not appear until 39 years after its author's death because of a series of misfortunes to both its author and the book itself in its manuscript format. The Rumphian accounts are mainly of the plants that grew on the Dutch-occupied island of Amboina, which would have included a rich mangrove flora. Species identification on the basis of the combined Dutch–Latin text and illustrations by native artists is possible in most instances (Merrill 1917). The names used by Rheede and Rumphius are of no scientific significance because they predate the publication of Linnaeus's *Species plantarum* (1753), which is the starting point for names of higher plants according to the International Rules of Botanical Nomenclature. The descriptions and identifications are important, however, because they served as a basis for many validly published names by Linnaeus and subsequent authors. In all Rumphius included descriptions of 14 currently accepted mangrove and mangrove-associated genera. Rumphius had a broad, essentially ecological concept of the systematics of mangrove plants, including in his comprehensive genus *Mangium* elements that are now listed in genera as disparate as *Aegiceras*, *Avicennia*, *Bruguiera*, *Pemphis*, *Rhizophora*, and *Sonneratia*. One species, *Mangium montanum*, is referred by Merrill to the modern terrestrial species *Acacia mangium* (an important fuel wood) – ironically the only place in modern nomenclature where the Rumphian mangrove name is still used. Other current names that perhaps can be thought of as persistent Rumphian nomenclature are *Catappa* (in *Terminalia catappa*), *Arbor excoecans* (in *Excoecaria agallocha*), *Granatum* (in *Xylocarpus granatum*), and *Mangium caseolare* (in part, in *Sonneratia caseolaris*).

In using Rumphian descriptions, Linnaeus himself initially retained a similar broad ecological view so that his *Rhizophora* included species that are now included in at least five modern genera. The only Linnaean name for a mangrove that is still valid is *Rhizophora mangle*, based by Linnaeus on the description by Patrick Browne in the *Civil and natural history of Jamaica* (1765). Linnaeus retained a very broad view of plant distribution in his later refinements of some mangrove plant names; his assumption that one species of *Avicennia* was pantropical led to nomenclatural confusion between New and Old World species.

Synonymy

In the preparation of species descriptions for this book an extended synonymy was initially developed, but it has been largely omitted because most of it was of historical interest only. Because the synonymy was largely derivative, there was no point in simply recycling secondhand information and, without checking original sources, introducing error. A recent reliable source of nomenclatural information has been cited where available, however, so that the reader seriously interested in nomenclature may make progress. To assist the nonspecialist, some elementary synonymy has been retained where misapplied names have been extensively used in the literature, largely as indicated in the bibliography by Rollet (1981).

Systematics

Stability in systematic botany is achieved only gradually and is constantly subject to new discovery. Some of this discovery involves the recognition of new species. *Bruguiera hainesii*, a widely distributed and distinct taxon, was not formally recognized until 1950; it was overlooked by Watson (see Wyatt-Smith 1953a). *Rhizophora* × *lamarckii*, described by Montreuzier in 1860, was thought to be endemic to New Caledonia until it was recognized as an apparent hybrid (*R. stylosa* × *apiculata*) and shown to have a wide distribution in the Western Pacific (Tomlinson and Womersley 1976). The recent detailed exploration of Queensland has indicated the likely existence of previously undescribed species of *Sonneratia* (Duke 1984). Some hybrids whose existence has long been suspected have only recently received formal names (e.g., *Rhizophora* × *selala* in Fiji, Tomlinson 1979; *Sonneratia gulngai* in Queensland, Duke 1984). Taxonomic uncertainties still exist despite our apparent familiarity with these common tropical plants, largely because they have wide distributions and few fieldworkers have had the opportunity to explore species throughout their range. Often the characters that separate species are not very obvious; the physiognomic uniformity of mangrove plants is confusing to the beginner. Mistaken identifications in herbaria are common and are made even by specialists. Consequently, much of our systematic knowledge of mangrove plants is surprisingly modern, even though they have been studied for over 300 years. It must be impressed on modern workers that problems still exist: The limits of several *Avicennia* taxa are still uncertain, even though this is the most constant genus in mangal, and the names for its species are not always reliable.

Details of geographic distribution are given in Section B, but brief mention should be made of the relation between geography and systematics. Knowledge of the geographic range of a species is often incomplete, so checklists for given areas are constantly in need of revision (see also the legend to Fig. 3.4). In Queensland, for example, one can contrast the list made by Jones (1971) with the results of a thorough survey by Bunt et al. (1982) to appreciate that the list of mangrove species in this area has grown almost 40 percent in the intervening period. Latitudinal and longitudinal limits of species are often unclear, even though the information is of considerable phytogeographic interest. The range of many species of mangrove eastward into the Pacific from the Indo-Malayan region is not well established; the available information is ably summarized by Fosberg (1980). The discontinuities in the ranges of some species are suspect (e.g., *Bruguiera hainesii*), and the disjunct ranges of different species within one genus [e.g., *Aegialitis* (see Fig. 3.4) and *Camptostemon*] need precise verification. A survey of the distribution of widespread species from an examination of herbarium specimens may be all that is possible initially and provides good preliminary documentation [e.g., *Rhizophora* (see Figs. B.64, B.65)], but it requires detailed confirmation by ground study. For example, we will not know the status of infraspecific taxa in the *Avicennia marina* complex until a thorough ground survey throughout its total range has been carried out – from East Africa to New Zealand, from Indochina to New South Wales. A field-worker could go through quite a few pairs of boots establishing this simple but important systematic base line. Hybridization fortunately is not usually a factor complicating mangrove systematics, but, as we have seen, exceptions to this generalization are known in *Lumnitzera*, *Rhizophora*, and *Sonneratia*.

On this basis, more than modesty recognizes the current account as nothing more than a progress report. Field verification of current systematic and phytogeographic information is a continuing need.

Identification

Since mangal has few species and many of them are illustrated in this account, the beginner may not need much help in learning the common mangroves. The following keys are intended as a preliminary aid to the identification of genera largely using readily visible vegetative characters. After a genus is located, there are detailed keys to species in the individual descriptions in Section B.

Field identification of mangroves is of considerable practical importance to foresters and fieldworkers, and there are numerous regional field keys (e.g., Allen 1956; Jones 1971; Jonker 1959; Percival and Womersley 1975; Semeniuk et al. 1978; Stearn 1958; Watson 1928; Wyatt-Smith 1953a,b, 1954, 1960).

> *Key to dicotyledonous genera of strict mangroves (major and minor) of*
> *the Atlantic-Caribbean-American Pacific region (New World)*
>
> 1A. Leaves opposite; flowers small (1–1.5 cm); aggregated aerial roots usually present.. 2
> 1B. Leaves alternate, flowers showy, large (10–12 cm); terminal; aerial roots absent, trunk base expanded and fluted ....*Pelliciera*
> 2A. Stipules present, overwrapping, leaving a conspicuous annular scar above petiole insertion; flowers small (1 cm), seedlings conspicuous; viviparous; aerial roots forming arching loops from trunk and branches ..*Rhizophora**
> 2B. Stipules absent; seedlings not conspicuously viviparous; aerial roots from underground cable roots 3

* Pantropical genera.

3A. Leaf undersurface with a fine glaucous gray indumentum; leaf base grooved; petiole without glands; aerial roots as pointed pneumatophores. . . . . *Avicennia**

3B. Leaf undersurface glabrous; petiole with a pair of glands just below insertion of blade; pneumatophores, if present, blunt, not pointed. . . . . *Laguncularia*

Since the limits of the back mangal are imprecise, it is not possible to include a comprehensive key to all plants that might be encountered. The mangrove associates mentioned in this book may be distinguished as follows: (1) *vines*, with milky sap (*Rhabdadenia*), with numerous prickles and bipinnate leaves (*Caesalpinia*), with short woody tendrils and pealike flowers (*Dalbergia*); (2) *shrubby* or *creeping plants*, with succulent leaves (*Batis*), with opposite leaves and daisylike flowers (*Tuberostylis*); (3) *trees*, with pinnate leaves and four leaflets (*Mora*); with milky sap (*Hippomane*); with conelike fruits, alternate leaves, and petiolar glands (*Conocarpus*); with cordate leaves and capsular (*Hibiscus*) or globose (*Thespesia*) fruits.

Key to dicotyledonous genera of strict mangroves (major and minor) of the Indo-Pacific region (Old World)

1A. Leaves compound, paripinnate, with two or three pairs of leaflets . . . . *Xylocarpus*

1B. Leaves simple 2

2A. Leaves opposite 3

2B. Leaves alternate 7

3A. Stipules present. 4

3B. Stipules absent 5

4A. Stipules overwrapping, standing above the leaf insertion, falling to leave an annual nodal scar . . . . . mangrove Rhizophoraceae (*Bruguiera, Ceriops, Kandelia, Rhizophora*)

4B. Stipules interpetiolar, not overwrapping, persistent . . . . *Scyphiphora*

5A. Leaves with a basal groove; stem lacking nodal glands 6

5B. Leaves without a basal groove; stem usually with a pair of nodal glands . . . . *Sonneratia*

6A. Leaves aromatic, with translucent punctate dots . . . . *Osbornia* (uncommon)

6B. Leaves not aromatic, without translucent dots, with a fine glaucous gray or olive green indumentum on the lower surface. . . . . . *Avicennia* (abundant)

7A. Leaves with a broad encircling attachment, leaving an annular scar; plants shrubby with basally swollen trunks. . . . . *Aegialitis*

7B. Leaves with a narrow, nonencircling attachment, plants usually trees, without a swollen base but often with buttresses. 8

8A. Stipules present, leaves 15–30 cm long, with a silvery white undersurface . . . . *Heritiera*

8B. Stipules absent, leaves less than 15 cm, not silvery white underneath . . . 9

9A. Milky latex exuding from cut surface, leaf margin slightly notched . . . . *Excoecaria*

9B. Milky latex absent, leaf margin entire 10

10A. Lower leaf surface with an indumentum of minute scales, leaf blade abruptly inserted on petiole .... *Camptostemon*

10B. Lower leaf surface without a scaly indumentum, leaf blade gradually inserted on petiole 11

11A. Leaves with conspicuous hairs on both surfaces, usually small (less than 3 cm); usually shrubby plants at seaward margin of mangal .... *Pemphis*

11B. Leaves glabrous, usually larger than 3 cm; shrubs to tall trees not exclusive to seaward margin of mangal 12

12A. Leaves ovate not fleshy, with extended glandular translucent dots; fruits elongated, one-seeded, cryptoviviparous .... *Aegiceras*

12B. Leaves obovate, fleshy, without translucent dots; fruits not elongated, drupelike, not cryptoviviparous .... *Lumnitzera*

Many species of dicotyledons are likely to be encountered in the back mangal of the Indo-Malayan mangroves, and any synoptic key would be incomplete and probably misleading. The following are the common distinctive genera and their most obvious diagnostic features:

Spiny or *prickly* plants include *Acanthus* (leaf and stem prickles); *Merope* (axillary thorns); *Caesalpinia*, which includes prickly vines with bipinnate leaves; and *Calamus*, which is a prickly climbing palm. Other *climbing plants* are *Derris* and *Aganope*.

Shrubby plants include *Batis* (with succulent leaves); *Ochthocharis* with prominently veined leaves; *Allophyllus*, which has trifoliate leaves with the leaflet margins dentate; *Cassine* with small white flowers; and *Scaevola* with silvery hairy leaves.

Leguminous trees include *Intsia, Pongamia* (with pinnate leaves), *Inocarpus* with simple leaves, and *Cynometra* with bijugate leaves.

Other *trees* with *compound leaves* are *Dolichandrone* and *Amoora*.

Trees with white *latex* include *Cerbera* (with frangipani-like leaves) and *Pouteria* (with round fruits); trees with black-spotted leaves and a clear sap turning black are likely to be *Gluta* and should be avoided.

Trees with a scaly or stellate indumentum on the lower surface include *Brownlowia, Hibiscus*, and *Thespesia*.

3   Biogeography

The main facts of mangrove biogeography are well known and several summaries are available, the most recent and comprehensive by Barth (1982). Details are lacking, however, so that only generalizations are possible. The distribution of *Rhizophora* species is shown fairly precisely in Figures B.64 and B.65. Mangroves are essentially tropical, occupy two separate hemispheric regions, and are more abundant in the Old World than in the New World tropics. Regional summaries are provided in varying detail in Chapman (1977a). Barth points out some discrepancies between different authors who describe the same region; most of these relate to the taxonomic identity of species. Barth takes a broad view and includes in his lists many species here referred to as "mangal associates." Figure 3.1 is a diagrammatic version of general information on the major and minor mangrove species listed in Tables 3.1 and 3.2. I have made additions and corrections to Barth's analysis on the basis of my own field experience and a study of herbarium specimens. Field study complemented by existing collections is still the most consistent source of information, in preference to the published literature, which is not always systematically reliable.

Figure 3.1 shows that there are two main centers of mangrove diversity. They have very dissimilar floristic inventories in both size and composition:

1. *Eastern group* (essentially the Eastern Hemisphere) including East Africa, India, Southeast Asia, Australia, and the Western Pacific. The total number of true mangrove species in this area is 40.
2. *Western group* (essentially the Western Hemisphere) including West Africa, Atlantic South America, the Caribbean, Florida, Central America, and Pacific North and South America. The total number of true mangrove species in this area is only eight, although there is a local concentration of species that are incipient mangroves in western Colombia.

The numbers of species in the two regions can, of course, be expanded if the concept of "mangrove" is made broader, but since the floristic inventory favors the eastern group, the overall conclusion remains the same: There is a five-to-one disparity between the number of species in the two groups.

Longitudinal distribution

The lower part of Figure 3.1 plots the number of species of mangroves in each longitudinal band of 15°, considering only those species that belong to genera in

40

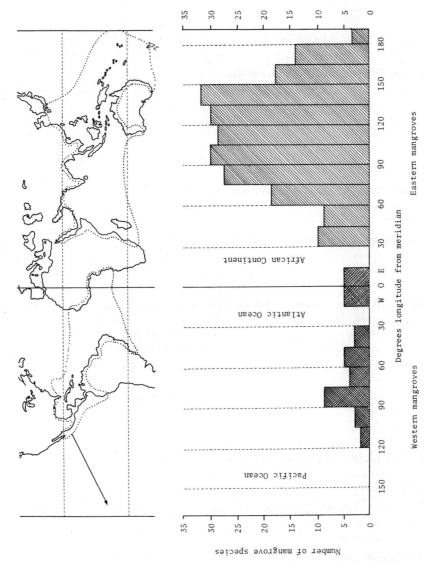

Figure 3.1. Generalized distribution of mangroves. Above: approximate limits for all species; eastern and western groups do not overlap except for possible extension (arrow) in the Western Pacific (*Rhizophora samoensis*, see Fig. B.65d). Below: histogram showing approximate number of species of mangroves per 15° of longitude. This quantifies the floristic richness of the eastern group.

Table 3.1. *Distribution of mangroves at intervals of 15° longitude in western area: West Africa, Atlantic South and North America, Caribbean, Central America, and Pacific South and North America*

Longitude 15°W–15°E (West Africa): spp. = 5
Avicennia germinans   *Rhizophora mangle*
Laguncularia racemosa   *Rhizophora racemosa*
Rhizophora harrisonii

Longitude 15°W–30°W (Atlantic Ocean)

Longitude 30°W–45°W (eastern Brazil): spp. = 3
Avicennia schaueriana   *Rhizophora mangle*
Laguncularia racemosa

Longitude 45°W–60°W (northeastern Brazil, Guyana): spp. = 5
Avicennia germinans   *Rhizophora mangle*
Avicennia schaueriana   *Rhizophora racemosa*
Laguncularia racemosa

Longitude 60°W–75°W (eastern Caribbean, northern South America): spp. = 6
Avicennia germinans   *Rhizophora × harrisonii*
Avicennia schaueriana   *Rhizophora mangle*
Laguncularia racemosa   *Rhizophora racemosa*

Longitude 75°W–90°W (Florida, western Caribbean, Central America, Pacific South America): spp. = 7
Avicennia bicolor   *Rhizophora × harrisonii*
Avicennia germinans   *Rhizophora mangle*
Laguncularia racemosa   *Rhizophora racemosa*
Pelliciera rhizophorae

Longitude 90°W–105°W (Mexico): spp. = 3
Avicennia germinans   *Rhizophora mangle*
Laguncularia racemosa

Longitude 105°W–120°W (Baja California): spp. = 2
Avicennia germinans
Laguncularia racemosa

Longitude 120°W–180°E (Western Pacific)

Table 2.1 (i.e., major and minor mangroves). The lists generated in Tables 3.1 and 3.2 are certainly preliminary and not quite consistent with Figure 3.1. I have made some minor corrections to the listings by Barth (1982). For example, I consider *Xylocarpus* to include only two mangrove species: *Avicennia lanata*, a distinctive species, is omitted from his lists; and *Camptostemon philippinensis* is not counted in his reckoning. The main emphasis is to show the enormous floristic disparity between mangal of the eastern and western groups.

Table 3.2. *Distribution of mangroves at intervals of 15° longitude in eastern area: East Africa, India, Indochina, the Malay Peninsula and Archipelago, Japan, Philippines, Australia, and the Pacific islands*

Longitude 15°E–30°E (African continent)

Longitude 30°E–45°E (East Africa, excluding Madagascar and the Red Sea): spp. = 8

Avicennia marina	Pemphis acidula
Bruguiera gymnorrhiza	Rhizophora mucronata
Ceriops tagal	Sonneratia alba
Heritiera littoralis	Xylocarpus granatum

Longitude 45°E–60°E (Madagascar, Red Sea): spp. = 8

Avicennia marina	Pemphis acidula
Bruguiera gymnorrhiza	Rhizophora mucronata
Ceriops tagal	Sonneratia alba
Lumnitzera racemosa	Xylocarpus granatum

Longitude 60°E–75°E (West India, excluding Andamans and Nicobars): spp. = 19

Aegiceras corniculata	Ceriops tagal
Avicennia alba	Excoecaria agallocha
Avicennia marina	Kandelia candel
Avicennia officinalis	Lumnitzera racemosa
Bruguiera cylindrica	Rhizophora apiculata
Bruguiera gymnorrhiza	Rhizophora mucronata
Bruguiera parviflora	Sonneratia alba
Bruguiera sexangula	Sonneratia apetala
Ceriops decandra	Sonneratia caseolaris
	Xylocarpus granatum

Longitude 75°E–90°E (Ceylon, East India, including Andamans and Nicobars): spp. = 28

Aegialitis rotundifolia	Kandelia candel
Aegiceras corniculata	Lumnitzera littorea
Avicennia alba	Lumnitzera racemosa
Avicennia marina	Nypa fruticans
Avicennia officinalis	Rhizophora apiculata
Bruguiera cylindrica	Rhizophora mucronata
Bruguiera gymnorrhiza	Rhizophora stylosa
Bruguiera parviflora	Scyphiphora hydrophyllacea
Bruguiera sexangula	Sonneratia alba
Ceriops decandra	Sonneratia apetala
Ceriops tagal	Sonneratia caseolaris
Excoecaria agallocha	Sonneratia griffithii
Heritiera fomes	Xylocarpus granatum
Heritiera littoralis	Xylocarpus mekongensis

Table 3.2. *(cont.)*

Longitude 90°E–105°E (Burma, Thailand, Cambodia, Malay Peninsula, Sumatra): spp. = 31

Aegialitis rotundifolia	*Kandelia candel*
Aegiceras corniculatum	*Lumnitzera littorea*
Avicennia alba	*Lumnitzera racemosa*
Avicennia lanata	*Nypa fruticans*
Avicennia marina	*Osbornia octodonta*
Avicennia officinalis	*Rhizophora apiculata*
Bruguiera cylindrica	*Rhizophora mucronata*
Bruguiera gymnorrhiza	*Rhizophora stylosa*
Bruguiera hainesii	*Scyphiphora hydrophyllacea*
Bruguiera parviflora	*Sonneratia alba*
Bruguiera sexangula	*Sonneratia apetala*
Ceriops decandra	*Sonneratia caseolaris*
Ceriops tagal	*Sonneratia griffithii*
Excoecaria agallocha	*Sonneratia ovata*
Heritiera littoralis	*Xylocarpus granatum*
	Xylocarpus mekongensis

Longitude 105°E–120°E (South China, Hainan, Vietnam, Borneo, Java, Lesser Sunda Islands, Western Australia): spp. = 29

Aegiceras corniculatum	*Heritiera littoralis*
Aegiceras floridum	*Kandelia candel*
Avicennia alba	*Lumnitzera littorea*
Avicennia marina	*Lumnitzera racemosa*
Avicennia officinalis	*Nypa fruticans*
Bruguiera cylindrica	*Osbornia octodonta*
Bruguiera gymnorrhiza	*Rhizophora apiculata*
Bruguiera parviflora	*Rhizophora mucronata*
Bruguiera sexangula	*Rhizophora stylosa*
Camptostemon philippensis	*Scyphiphora hydrophyllacea*
Ceriops decandra	*Sonneratia alba*
Ceriops tagal	*Sonneratia caseolaris*
Excoecaria agallocha	*Sonneratia ovata*
Heritiera globosa	*Xylocarpus granatum*
	Xylocarpus mekongensis

Longitude 120°E–135°E (Japan, Ryu-Kyu Islands, Philippines, Celebes, Northern Territories of Australia): spp. = 32

Aegialitis annulata	*Ceriops tagal*
Aegiceras corniculata	*Excoecaria agallocha*
Aegiceras floridum	*Heritiera littoralis*
Avicennia alba	*Lumnitzera littorea*
Avicennia eucalyptifolia	*Lumnitzera racemosa*
Avicennia marina	*Nypa fruticans*
Avicennia officinalis	*Osbornia octodonta*
Bruguiera cylindrica	*Rhizophora apiculata*
Bruguiera exaristata	*Rhizophora mucronata*
Bruguiera gymnorrhiza	*Rhizophora stylosa*
Bruguiera hainesii	*Scyphiphora hydrophyllacea*

Table 3.2. *(cont.)*

Bruguiera parviflora	*Sonneratia alba*
Bruguiera sexangula	*Sonneratia caseolaris*
Camptostemon philippinensis	*Sonneratia ovata*
Camptostemon schultzii	*Xylocarpus granatum*
Ceriops decandra	*Xylocarpus mekongensis*

Longitude 135°E–150°E (Gulf of Carpentaria, north Queensland, New Guinea, Micronesia): spp. = 31

Aegiceras corniculata	*Excoecaria agallocha*
Aegialitis annulata	*Lumnitzera littorea*
Avicennia alba	*Lumnitzera racemosa*
Avicennia eucalyptifolia	*Nypa fruticans*
Avicennia marina	*Osbornia octodonta*
Avicennia officinalis	*Pemphis acidula*
Bruguiera cylindrica	*Rhizophora apiculata*
Bruguiera exaristata	*Rhizophora × lamarckii*
Bruguiera gymnorrhiza	*Rhizophora mucronata*
Bruguiera hainesii	*Rhizophora stylosa*
Bruguiera parviflora	*Scyphiphora hydrophyllacea*
Bruguiera sexangula	*Sonneratia alba*
Camptostemon schultzii	*Sonneratia caseolaris*
Ceriops decandra	*Sonneratia ovata*
Ceriops tagal	*Xylocarpus granatum*
	Xylocarpus mekongensis

Longitude 150°E–165°E (southeastern Australia, Bismarck Archipelago, Pacific Atolls): spp. = 17

Avicennia marina	*Nypa fruticans*
Bruguiera gymnorrhiza	*Pemphis acidula*
Bruguiera parviflora	*Rhizophora apiculata*
Bruguiera sexangula	*Rhizophora × lamarckii*
Ceriops tagal	*Rhizophora mucronata*
Excoecaria agallocha	*Rhizophora stylosa*
Heritiera littoralis	*Sonneratia alba*
Lumnitzera littorea	*Sonneratia caseolaris*
	Xylocarpus granatum

Longitude 165°E–180°E (New Zealand, New Caledonia, Fiji, islands of the West Central Pacific): spp. = 14

Avicennia marina	*Rhizophora apiculata*
Bruguiera gymnorrhiza	*Rhizophora × lamarckii*
Bruguiera parviflora	*Rhizophora samoensis*
Ceriops tagal	*Rhizophora × selala*
Excoecaria agallocha	*Rhizophora stylosa*
Heritiera littoralis	*Xylocarpus granatum*
Lumnitzera littorea	*Xylocarpus mekongensis*

Longitude 180°E–165°W (Samoa): spp. = 4

Bruguiera gymnorrhiza	*Heritiera littoralis*
Excoecaria agallocha	*Rhizophora samoensis*

Rhizophora mangle has a very wide distribution from West Africa to the Pacific coast of South America (Fig. B.65).

The mangrove flora of Samoa is the relict of the great eastern mangrove flora which dies out progressively in the western Pacific (Fosberg, 1980). *Bruguiera gymnorrhiza* has the broadest distribution of any mangrove, from East Africa to Samoa.

The considerable disparity in numbers of species between eastern and western groups is emphasized since the tabulation is based on the premise that no species is common to the two groups. The only possible exception would be *Rhizophora mangle* which, treated in the broad sense includes populations in New Caledonia, Fiji, Tonga and Samoa in addition to its widespread "western" occurrence. In a narrow sense, however, these populations are treated as a separate species, *R. samoensis*. Certainly the two species are very similar and they represent the closest phytogeographical link between the eastern and western mangroves. *Avicennia* and *Rhizophora* are both pan-tropical genera, but the species in the two groups are different, as listed in Table 3.3. Further discussion is provided in the detailed description of each genus.

Wide range

From Tables 3.1 and 3.2 it is clear that most mangroves have a wide range. This is also true of many mangal associates, especially beach plants or plants of coastal scrub, which are sometimes pantropical (e.g., *Caesalpinia bonduc*). No mangrove is pantropical in its distribution, however. In the eastern group five species encompass almost the total longitudinal range of the flora (i.e., *Avicennia marina, Bruguiera gymnorrhiza, Excoecaria agallocha,* and *Heritiera littoralis, Sonneratia alba*). *Rhizophora mucronata* is widespread but does not extend as far into the Pacific as any of these species, and in the eastern end of its range it is not abundant. *Avicennia marina* has a wide latitudinal as well as longitudinal range. This species is probably capable of taxonomic subdivision, but the exact method of subdivision and the geographical range of the subunits still remain uncertain. Wide-ranging species in the western group include *Avicennia germinans, Laguncularia racemosa,* and *Rhizophora mangle*. Species with a narrow range include *Avicennia lanata* in the eastern group and *A. bicolor* and *Pelliciera rhizophoreae* in the western group.

Seagrasses and corals

Coral-reef biotas are very rich; the biogeography of corals themselves is comparable to that of mangroves. There are an Atlantic and an Indo-Pacific province, entirely separate in diversity, with the latter being richer by about 50 percent (Stehli and

Table 3.3. *Distribution of Rhizophora and Avicennia species*

Genus	Eastern group	Western group
Rhizophora	*Rh. apiculata*	*Rh. × harrisonii*
	Rh. × lamarckii	*Rh. mangle*
	Rh. mucronata	*Rh. racemosa*
	Rh. samoensis[a]	
	Rh. × selala	
	Rh. stylosa	
Avicennia	*A. alba*	*A. bicolor*
	A. eucalyptifolia	*A. germinans*
	A. lanata	*A. schaueriana*
	A. marina	
	A. officinalis	

[a] If distinct from *R. mangle*.

Wells 1971). Generic diversity in the two provinces is concentrated around Jamaica on the one hand, and Indonesia – Philippines on the other. The paleontological record for corals is extensive and provides direct information for understanding evolutionary processes because the time of appearance of genera in the record is known. Consequently we know that the centers of diversity represent concentrations of geologically "young" genera, with diversity gradients of progressively "older" genera leading away from the epicenter. This is consistent with these areas of concentration representing centers of evolutionary diversity from which genera have migrated with diminished rates of evolutionary change. Alternatively, however, the centers could be refugia for temperature-sensitive groups that once occupied a larger area, although this is again only possible with differential rates of evolution between peripheral and central regions. By analogy, this latter explanation is consistent with our knowledge of the restriction of range in certain mangroves (e.g., *Pelliciera* and *Nypa*) but without concomitant evolutionary change. Certainly there are enough parallels between the distributions of shallow-water marine organisms within the tropics (corals, seagrasses, and mangroves) to suggest a unified explanation for their generally similar geographic patterns (McCoy and Heck 1976). Unfortunately we have virtually no information about the relative geological ages of the plants as compared with corals.

The distribution of mangroves, is also paralleled by that of the seagrasses, in which there is a similar disjunction between floras of the two regions. More seagrass genera are shared in common, but again there seems to be no species common to the two regions (den Hartog. 1970). This similarity cannot be coincidence and requires an evolutionary explanation, which is in the next section.

Table 3.4. *Distribution of tropical seagrass taxa*

Family[a] and genus	Eastern species	Western species
Cymodoceaceae		
Amphibolis	2	0
Cymodocea	6	0
Halodule	c. 8	c. 3
Syringodium	1	1
Thalassodendron	2	0
Hydrocharitaceae		
Enhalus	1	0
Halophila	c. 8	c. 3
Thalassia	1	1

[a] As in Tomlinson (1982b).

In the tropical seagrasses there are eight genera (Table 3.4), four genera are pantropical, four are unique to the eastern group, but none is unique to the western group. The geographical limits of these two groups coincides with the limits established for mangroves. The four pan-tropical genera have no species common to the two regions. Seagrasses also include species (representing the two families Zosteraceae and Posidoniaceae) that differ in their essentially temperate and bipolar distributions. Tropical and temperate seagrasses overlap only slightly in their distribution. *Amphibolis* of the tropical group is best described as "warm temperate."

Distribution of mangroves in geological time

The distinctive bimodal distribution of modern mangroves is explained in terms of their origin somewhere in the cradle of early angiosperm diversification and their subsequent redistribution via migration and range expansion governed by continental realignment (plate tectonics). Range contraction and diversification have been important (see Fig. 3.2). The principles underlying the hypotheses are those of geological and ecological uniformitarianism; for example, we must accept that mangroves have always occupied tidal habitats in the tropics and have retained constant biological adaptations and that geomorphological processes are consistent in time. A corollary of this is that there has been extinction. Accepting these principles, paleobotanists have argued that the mangroves' habitat is an ancient one and representatives of groups of seed plants have been interpreted as showing the habit of mangroves, for example, stilt-rooted representatives of the Cordaitales (Cridland 1964). No one structural feature characterizes mangroves, however; stilt roots occur in modern fresh-water swamp species. Better evidence comes from the association

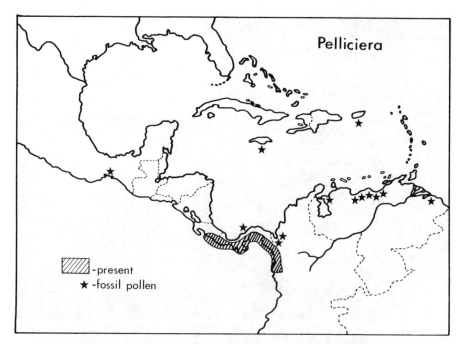

Figure 3.2. *Pelliciera rhizophoreae*. Past and present distribution (Graham 1977). More recently the present distribution has been extended to the Atlantic coast of Colombia (Winograd 1983).

of Upper Carboniferous Cordaitalean remains with marine sediments identified faunistically (Raymond and Phillips 1983). These authors emphasize that mangroves are unlikely to have existed before the evolution of the seed habit because establishment by free-living independent gametophytes may not be possible in salt water. *Acrostichum*, the only modern mangrove with this primitive life cycle, is not exceptional in this respect because, although its gametophyte will tolerate some salt, the plant never becomes established in open tidal water. However, Retallack and Dilcher (1981) consider the fossil leptosporangiate fern *Weichsella* to have formed a pantropical mangal in the early Cretaceous; if so, we would have to think of it as a back-mangal species in modern terms. Retallack and Dilcher further discuss examples of fossil mangroves and even suggest the extreme view that angiosperms may have radiated from coastal environments. They do not imply, of course, that modern mangroves are ancestors of other flowering plants. Before elaborating biogeographic theories of mangrove redistribution, it is necessary to review briefly the relevant geographical and ecological facts.

1. Mangroves are tropical and require climatic conditions appropriate for an ever-growing state. The distribution of mangroves in time will therefore depend on the distribution of hospitable tropical shores in time and space.

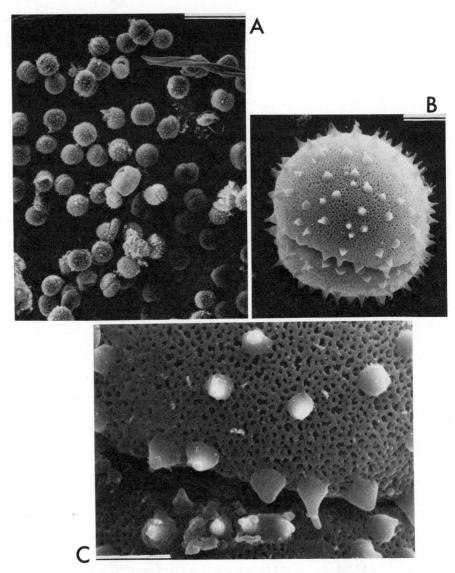

Figure 3.3. *Nypa fruticans* (Palmae) pollen. (A) Grains at low power ($\times$ 200, scale = 10 μm); (B) single grain ($\times$ 1500, scale = 100 μm), oblique view "resembling a rather forbidding hamburger"; (C) detail of circumpolar sulcus and sculpturing ($\times$ 5000, scale = 250 μm). Pollen is sticky and can be transported only by an animal vector.

2. Mangroves are restricted to intertidal regions. The minor incursion of mangroves into river mouths does not obscure overall constraints. Mangroves cannot cross land barriers but must migrate along coasts or archipelagoes with continual climatic suitability.

3. Mangroves are dispersed exclusively by sea currents, seemingly widely and efficiently. There are limits to their establishment, however, independent of dispersibility. Mangrove propagules are frequently dispersed into regions where they cannot grow or persist for any length of time (allochthonous occurrence).

4. As a corollary of fact 3, the distribution of fossil mangroves does not necessarily represent their range of occurrence as growing plants (autochthonous occurrence). Propagules, pollen, and plant fragments are readily carried beyond the natural range (e.g., the London Clay Flora, Wilkinson 1981).

5. Distributions have changed with time. The fossil occurrence of *Nypa* is much more extensive than its present distribution (Tralau 1964); the evidence comes from the wide occurrence of pollen (Fig. 3.3) and seeds. *Pelliciera* is a species whose original wider range in the Caribbean has been reduced as a consequence of Pleistocene glaciation (see Fig. 3.2). Consequently, its present range reflects part of a larger refugium for species in tropical America (Gentry 1982). Even though this interpretation is based on fossil pollen, it is unlikely that the fossil occurrence is allochthonous because *Pelliciera* pollen is not capable of wide dispersal. Whether or not a species can migrate completely from one area to another without morphological change remains a matter for speculation.

6. Modern mangrove taxa can be of appreciable geological age (Muller 1981). The oldest is *Nypa* (end of the Cretaceous, 69×10^6 years B.P.); records for *Pelliciera* and *Rhizophora* go back to the Eocene (30×10^6 years B.P.); other genera appear at progressively later periods. On this evidence, the *possible* previous range of mangroves older than 30 million years is dependent on the *probable* distribution of mangal climates as determined by plate tectonics and independent factors such as atmospheric carbon dioxide concentrations and solar input variations. So we have to sum several uncertainties in estimating past mangrove distribution. There has been a tendency, however, to interpret the present distribution of all mangroves in terms of tectonic events too ancient to influence relatively modern phytogeography.

From these generally accessible facts and principles, several often sharply contrasting hypotheses have been constructed to account for the present distribution of mangroves, for example, Aubréville (1964), Chapman (1976), van Steenis (1962), and Mepham (1983). In the early Cretaceous there was an extensive tropical sea: the Tethys separating the northern land mass of Laurasia from the southern Gondwanaland. The western margin of the Tethys became the Mediterranean by the juxtaposition of Africa and Arabia, no longer hospitable to mangroves. The present Indo-Malayan concentration of mangroves is usually regarded as a relict of their origin somewhere on the southern margin of the continent Laurasia, on the eastern shores of the Tethys Sea. The persistence to the present of a rich mangrove flora in this area is consistent with an equable climate. From this eastern center migration had earlier proceeded via the western end of the ancient sea into the Atlantic, the future Caribbean, and eventually the Pacific side of South America. The total past and present distribution of *Nypa* could exemplify the results of this migration. The isthmus of Panama remained open until only 3 million years ago, the period of former continuity between the Atlantic and Pacific oceans, sufficiently long to establish the western (but depauperate) mangrove flora. Subsequently there has been no evident divergence between the floras of the Atlantic and Pacific coasts of Central America. *Pelliciera* seems to be an example of a genus "trapped" by this process (see Fig. 3.2). Further westward migration from the western group reintroduced *Rhizophora mangle* to the Western Pacific in the form we now call *R. samoensis*. Apart from *Rhizophora*, the other migrants from the original Tethyan stock were *Avicennia* (to diversify as its western group of species), *Laguncularia* (from some protocombretaceous ancestor), and possibly *Carapa* (a fresh-water swamp species), a close a relative of *Xylocarpus*. This series of events is much as proposed by Chapman (1976) and Aubréville (1964). The idea that the western group is an immigrant and therefore recent flora could account for its relative poverty, but the impoverishment could be geomorphological because of the limited availability of mangal sites. Morphological information is not helpful in this analysis, however.

In contrast, van Steenis (1962) is of the opinion that the migration from the Tethyan stock was primarily eastward across the Pacific; he claims that the shores of the Tethys were too dry to support mangroves. In his view fossil mangroves are evidence of the dispersal ability of mangroves or their propensity to be carried as fragments in drift rather than an indication of the existence of a mangrove community; this would be his explanation of the London Clay mangroves. The London clay flora is very rich in tropical species, however (Chandler 1957); it is difficult to accept this explanation for all species. According to van Steenis, the Atlantic was stocked via the open isthmus of Panama from an easterly direction, achieving some secondary speciation in the Caribbean region; the only present evidence for this trans-Pacific migration is the relict *R. samoensis*. As Mepham (1983) points out, however, faunistic fossil evidence renders van Steenis's objections untenable and most authors would agree that van Steenis's arguments are not very parsimonious. McCoy and Herk (1976) do not accept that the center of diversity is equivalent to the center of origin but consider that plate tectonics alone account for the present biogeographic patterns. They reach this conclusion by considering the distribution of *genera*; a consideration of *species* distribution raises unexplained problems.

Both hypotheses depend to a greater or lesser extent on trans-Pacific migration, which the present distribution of mangroves (and the current distribution of currents!) suggests is impossible in either direction. The long distances to be traversed seem beyond the range of dispersal of mangrove propagules and there are east–west countercurrents. Eastern and western mangal have therefore diversified in isolation. A common feature of all theories is that southerly extensions beyond the Cape of Good Hope and Cape Horn have always been impossible for mangroves. The present barriers of these two poleward extensions of Africa and South America presumably have always existed.

In reviewing much of the biogeographical evidence on this subject, Mepham

(1983) shows how difficult it is to arrive at universally acceptable ideas, since so much depends on individual speculative opinion. He argues, however, that the Australian mangrove flora was stocked at a time when the continent had a position more southerly than its present one and that the present concentration of species in Queensland represents a refugium for the older flora as the Australian continent became drier, with the addition of new recruits via the Torres Strait by the existing Indo-Malayan mangrove flora. This view is scarcely supported by a study of the distribution of the modern mangrove flora alone, with the present southerly and westerly decline in species numbers (Figs. 3.5, 3.6). Fossil evidence for Mepham's idea is very limited.

A consistently applicable hypothesis to account for the distribution of modern mangrove floras still has to be developed. Existing hypotheses certainly suggest reasonable approaches. An understanding based on more extensive records of both fossil and modern mangroves, integrated with general biogeographic knowledge and paleoclimatic analysis, has still to be made. This undoubtedly will provide a stimulus for biogeographic exploration well into the future. In addition, we have almost no knowledge of interspecific variation in mangroves based on nonmorphological features. Development of this information is highly desirable, because clines in cytological or biochemical features, for example, could reflect directions of migrations and radiation.

Distribution of species

Discontinuities

Most mangroves have a continuous distribution, but there are a few exceptions. *Bruguiera hainesii* seems discontinuous between western Malaysia and New Guinea, although this may simply reflect the lack of discrimination by collectors. *Pemphis acidula* is recorded in East Africa but is absent from a broad intervening area from South India to Sumatra, reappearing at about 137°E in eastern Malaysia.

Vicariants

Different species of one genus with nonoverlapping ranges are found in *Aegialitis* and *Camptostemon*. *Aegialitis annulata* in Australia and eastern Malaysia, contrasted with *A. rotundifolia* in Burma, Bengal, and the Andamans, provides a unique example where the total range of the genus is comparable to that of several other mangroves but the ranges of the two species are different (Fig. 3.4). This suggests that the present range of the species represents a diversification of an

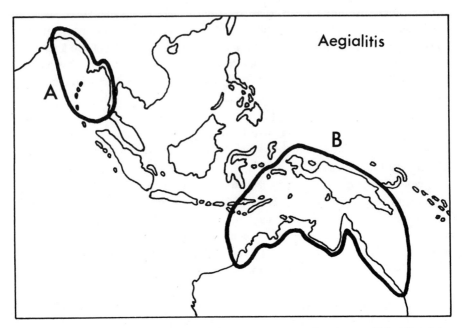

Figure 3.4. *Aegialitis*. Disjunct distribution of its two constituent species: *A. rotundifolia* (A) and *A. annulata* (B). (After van Steenis 1949). *A. annulata* is now known to range much farther south in western Australia (see the detailed distribution in Fig. 3.6).

originally continuous population after geographical isolation. *Camptostemon* may reflect the same process, with *C. philippinensis* in Borneo and the Philippines and *C. schultzii* in northern Australia and New Guinea. In both examples it is not known whether the species of the same genus are interfertile.

Hybridization

Where, as is usual, different species of the same genus partly occupy the same range it is of interest to know what are interspecific barriers and whether these are always maintained. Hybrids are known in *Sonneratia*, *Rhizophora*, and *Lumnitzera*. These examples are discussed in detail when reproductive biology is considered.

Floristic decline

The floristic richness of the central portion of the eastern group (longitude 135°E – 165°E) is well documented in Figure 3.1. Of interest is the progressive decline in the number of species toward the Pacific. This is discussed by Fosberg

Table 3.5. *Distribution of mangrove species in the isthmian region of tropical America*

Occuring on both Atlantic and Pacific coasts	
Avicennia bicolor	Rhizophora mangle
Avicennia germinans	Rhizophora × harrisonii
Laguncularia racemosa	Rhizophora racemosa
Occuring on only the Pacific coast	
Avicennia tonduzii (if distinct from A. bicolor)	Pelliciera rhizophoreae
Crenea patentinervis[a]	Phryganocydia phellosperma[a]
Mora megistosperma (if distinct from M. oleifera)[a]	Tabebuia palustris[a]

[a] Mangrove associates with limited distribution. *Source*: after Gentry (1982).

(1975), who finds the absence of mangroves from eastern Micronesia difficult to account for, since there are suitable substrates and dispersal to them should be possible. Fossil mangrove vegetation is recorded in the northern Marshall Islands, and it has been suggested that sea-level changes could have eliminated previously existing mangal. In the genus *Rhizophora*, first *R. apiculata*, then *R. mucronata*, and finally *R. stylosa* disappear in an easterly direction (see Fig. B.64). Why one species (*R. stylosa*) should be the most persistent in the Pacific while *R. mucronata* is the most persistent in the Indian Ocean is not at all apparent. Details of the distribution of *Rhizophora* are discussed in Section B.

Floristics of New World mangroves

The well-established disparity between the rich flora of the eastern mangal and the poor flora of the western mangal is somewhat tempered if one accepts more than the traditional number of mangrove-associate taxa in the western community. The increase in the number of New World mangrove flora has been suggested by Gentry (1982). This is set against a background of knowledge that suggests that much of the present distribution of mangroves in the New World is relict. Gentry argues that mangroves provide some evidence for the Chocó (the Pacific province of Colombia) as a refuge for tropical species during earlier glacial periods, since the mangal of the Pacific coast of Central and South America is floristically much richer than that of the Atlantic coast of northern South America (Table 3.5). However, his list of "true mangroves" includes species (as in the footnote) that, in this account, are considered to be mangrove associates because they lack complete fidelity to mangal.

Table 3.5 is based on what the author refers to as "rather casual personal observation and is perhaps not complete." In addition, Gentry lists *Conocarpus erectus*, *Amphitecna latifolia*, and *Muellera moniliformis* as "mangrove fringe species," whereas *Cassipourea elliptica*, *Pterocarpus officinalis*, and *Acrostichum aureum* are mentioned as mangrove components that are less specialized ecologically and also occur in other habitats. Gentry's argument is based on the concept that the present distribution of these species must have a phytogeographic, historical explanation rather than an ecological explanation. He also emphasizes the extreme wetness of the Chocó region, however, and on the basis of the contrast made later between the east (wet) and west (dry) coasts of Australia, it could be argued that the difference between Atlantic and Pacific mangrove floras is ecologically, and ultimately climatically, determined. Lot-Helgueras et al. (1975) describe mangroves near their limit of distribution in the Gulf Coast of Mexico, and they list many characteristic mangrove associates, including among the woody genera species in *Acacia*, *Clusia*, *Dalbergia*, *Ficus*, *Inga*, *Lycium*, *Mouriria*, *Pachyra*, *Pithecellobium*, *Randia*, and *Ternstroemia*, for example. None is restricted to coastal habitats.

Discussion

A number of additional points need to be made before a synthesis can be attempted.

Incomplete knowledge

Although it is possible to discuss the distribution of mangroves in general terms, precise information is often lacking and our present knowledge is often deficient in significant detail. For example, knowledge about the mangal flora of Queensland has been enormously extended by the detailed surveys carried out by John Bunt and Norman Duke of the Australian Institute of Marine Sciences. In little over 10 years they increased the number of known species recorded for eastern Australia from 28 (Jones 1971) to 45 (Bunt et al. 1982), an increase of 38 percent. Some of these seem to represent new taxa (e.g., in *Sonneratia*); others are new records for Australia and considerably broaden the known range of species. The discovery of *Pelliciera rhizophorae* on the Atlantic coast of Colombia (Winograd 1983) illustrates not so much a range extension as a distribution more in keeping with its fossil record (see Fig. 3.2).

Existing ranges of mangroves may be modified by human activities. Blasco (1977) estimates that species of Rhizophoraceae in western India have been virtually exterminated by overexploitation, since they do not coppice and cannot regenerate in the absence of seedling parents. Similarly, *Nypa* in the Ganges Delta has been largely eliminated. In contrast, several species have been established well outside their natural range, such as *Nypa* near Lagos, Nigeria, and *Rhizophora* in Hawaii. The range of *Kandelia* in southern Japan seems to have been artificially extended. Where they are newly introduced into areas that do not naturally support mangroves, the introductions may be regarded as pests, such as *Rhizophora* in the Society Islands.

Latitudinal limits

Mangal is essentially a tropical vegetation, with some outliers in subtropical latitudes, notably South Florida, South Africa, Victoria, and southern Japan (see Fig. 3.1). The outliers are all a consequence of warm ocean currents moving from rich mangrove regions that provide a renewable source of propagules, plus a continuous coastline or island chain. At the limits of distribution the formation is represented by scrubby, usually monotypic *Avicennia*-dominated vegetation, for example, in Westonport Bay and Corner Inlet, Victoria, Australia. The latter locality is the highest latitude (38° 45′S) at which mangroves occur naturally. The mangroves in New Zealand, which extend as far south as 37°, are of the same type; they start as low forest in the northern part of the North Island but become low scrub toward their southern limit. In both instances the species is *Avicennia marina*, although probably referrable to different varieties (var. *australis*, var. *resinifera*, respectively). In Western Australia, *A. marina* extends as far south as Bunbury (33° 19′S). In the Northern Hemisphere, scrubby *Avicennia germinans* in Florida occurs as far north as St. Augustine on the east coast and Cedar Point on the west, although there are records of *A. germinans* and *Rhizophora mangle* for Bermuda. In Japan the most northerly record is at a higher latitude and is not represented by *Avicennia*. *Kandelia candel* occurs to about 31°N in southern Japan (Tagawa in Hosakawa et al. 1977).

Temperature influence

The distribution of mangroves is clearly correlated with the sea-surface temperature, and their limits show a close correlation with the 24°C (75°F) isotherm; beyond that the surface temperature is always colder than 24°C. Exceptions to this occur on the North Atlantic coast of America and Africa where the limit is closer to 27°C, and in southern Japan as just mentioned. This suggests that *Avicennia* species in the Western Hemisphere are less tolerant of cold than the species in the Eastern Hemisphere. Mangroves, of course, endure air temperatures much below these values, but they tolerate very little or no frost and even temperatures around 5°C are inimical to the growth of most of them. Minimum sea temperatures may be lower than these average values for extended periods of the year in limiting regions.

The dependence of the limits of mangrove distribution on the sea temperature is clearly shown by the influence of cold currents with a northerly trend in the Southern Hemisphere. The southern limit of mangal in Pacific South America is 3° 40'S at Punta Malpelo, Tumbes, Peru, a latitude at which in Southeast Asia there are perhaps 30 mangrove species. This limited distribution in tropical America is conditioned by the cold Humboldt current, which runs toward the equator, further accentuated by the arid coast, as discussed later. On the Atlantic side of South America, mangal occurs as far south as Florianopolis, Brazil. Similarly, the West African mangroves have a much lower latitudinal limit (12°S) at Lobito in Angola, under the influence of the South Atlantic cold current, whereas in eastern Africa mangal extends south to the Nahoon River (33°S) in the Cape Province (Ward and Steinke 1982). Some of these South African communities are artificially established, however.

The varying tolerances of different mangroves to presumed temperature ranges are well exemplified by the detailed information for eastern Australia (in Figure 3.5). The temperature effect is modified by aridity, as discussed later. Nevertheless, we should not assume a constant response to temperature by each species; this is hardly likely with the broad range of mangroves and some assumption about genetic diversity. A within-species variation in the chilling response has been demonstrated experimentally (e.g., Markley et al. 1982; McMillan 1975) for *Avicennia* and *Rhizophora*. Samples of each species from poleward localities suffer less leaf damage at cold temperatures than samples from more equatorial latitudes. *Laguncularia*, on the other hand, showed a uniform chilling tolerance. Differential resistance could explain the skewed latitudinal distribution of *Kandelia*, which is asymmetric about the equator. Selection for cold tolerance may have occurred in northern, but not southern, populations. Transplant experiments would be needed to verify this hypothesis. The physiological basis for differential cold tolerance remains unexplained but has been discussed by Markley et al. (1982).

Aridity

In addition to air and sea-surface temperatures determining absolute limits, the floristic abundance of mangroves is strongly conditioned by coastal aridity: The mangrove is much richer in species along coasts that receive high levels of rainfall and heavy runoff and seepage. This is primarily related to the fact that many mangroves are typically estuarine and occur in greatest abundance in areas where there is extensive sedimentation. Abundant sedimentation provides a diversity of substrate types and nutrient levels higher than that of sea water. It is probably the maintenance of a balance between mineral nutrients and substrate salinity that is relevant, rather than absolute nutrient levels.

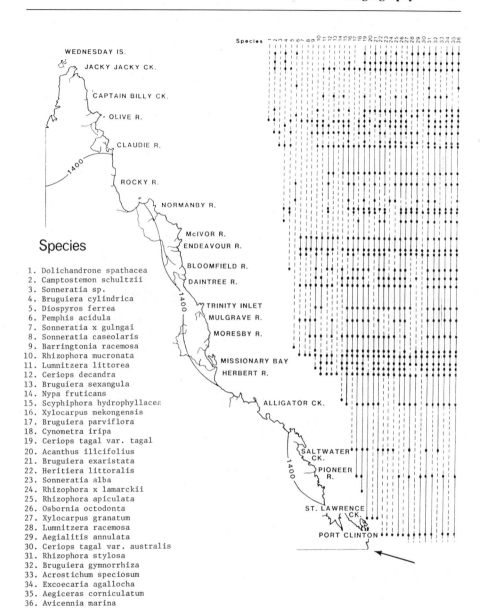

Figure 3.5. Distribution of mangroves in Queensland. Several species continue their distribution south of the Tropic of Capricorn (arrow); *Avicennia marina* extends to Corner Inlet, Victoria. Many areas receive over 1400 mm of rainfall annually. Place names are commonly rivers. (From an unpublished diagram by J. S. Bunt and N. C. Duke)

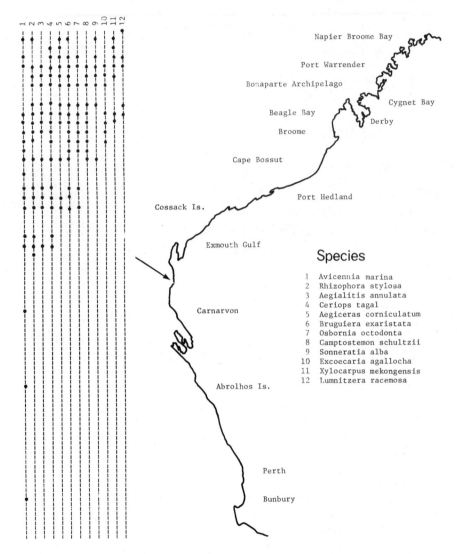

Figure 3.6. Distribution of mangroves in western Australia. All species except one are tropical (arrow, level of Tropic of Capricorn). Rainfall above 1000 mm per year only in the northernmost region (north of Derby). Place names are not distinguished by rivers. (From information in Semeniuk et al. 1978)

A good example on a continental scale of this contrast between wet and dry coasts is found by comparing the mangrove floras of the (wet) eastern and (dry) western coasts of Australia. The contrasted floras are shown diagrammatically in Figures 3.5 and 3.6. At the same latitude (e.g., 20°S) Queensland has 18 species, Western Australia has only 4; and at the Tropic of Capricorn, the comparative figures are 8 and 10. This diagram also shows the progressive decline in the number of species at higher latitudes on both coasts. This decline is not necessarily due entirely to decreasing temperature; the Queensland coast becomes drier in a southerly direction. More complete documentation of mangrove distribution in Australia is provided by Wells (1983), who similarly concludes that a major factor limiting their range is seasonal aridity.

Similar mangrove impoverishment is found in the dry areas of the Red Sea, which has only two or three species, the limit at 27° 14′N (Zahran, 1977). This impoverished mangrove flora is nevertheless in the area where mangrove trees were first described. The name *Avicennia*, after all, commemorates the Arabian philosopher Aba-bin-senna. In other parts of the tropics, arid coasts may have few or no mangroves; examples are the Horn of Africa, northwestern Mexico, the Bahia region of Brazil, and parts of Western Australia.

An interesting microcosmic reflection of this correlation of mangrove distribution with climate is shown by the island of New Caledonia. There are 12 species of mangroves on the northeastern coast ("côte est") (including 5 species of *Rhizophora*!) but only 7 on the southwestern coast ("côte ouest"). The eastern coast receives the heaviest rainfall. Nevertheless, why a species as common as *Rhizophora apiculata* on the east coast should not be able to extend its range a few miles around the northern end of the island is a good measure of the mysteries governing mangrove distribution.

Mangroves and pollution

We have discussed the biogeography of mangroves almost entirely in terms of the influence of natural events. However, the chief factor that currently modifies mangrove distribution is the activity of industrial man. Apart from the direct destruction of mangrove habitats, there is growing concern for the effects of pollutants like oil (Lewis 1983) and toxic metals. The degradation of the mangal may produce an entirely new community, as with the appearance of a *Sesuvium–Iresine* association in place of the *Rhizophora–Avicennia–Laguncularia* association at Guanabara Bay, Rio de Janeiro, Brazil (Lacerda and May 1982).

4   Shoot systems

The structure and especially the microscopic anatomy of mangroves have received considerable attention, which is summarized by Chapman (1976). Information about form is necessary to an understanding of functional processes, particularly those that are specialized and adapt plants to mangal environments. The interpretation of observed structures in terms of function is, however, often based on assumption rather than experimental verification. Physiological investigations of mangroves are few, which is remarkable when one considers that it is the suite of functional characteristics that allows mangroves to survive in tidal environments. In this account, where possible, functional aspects are integrated with structural considerations, but water relations and salt balance are discussed in Chapter 6. Chapter 4 emphasizes crown structure, leaf morphology and anatomy, and wood structure; the root is dealt with in Chapter 5.

Crown form and canopy structure

Vegetative survival and competitiveness of trees are dependent on a number of factors, including an efficient display of foliage and the ability to respond to environmental change and stress (e.g., mechanical damage). The result of the response to these factors can be translated into crown shape. The principles of the responsive processes are well understood, although an appreciation that crown form is a dynamic expression of deterministic and opportunistic processes is slow in developing. A useful descriptive background for tropical trees, where most principles are enunciated, is provided by Hallé et al. (1978). They provide a convenient series of reference points in the set of 23 "architectural models" they define. These models are abstract, but dynamic; they constitute for each species the deterministic component of form; the visible expression of the architectural model of a tree at any time is referred to as its "architecture" (Figs. 4.1, 4.2). Trees may conform more or less precisely to their architectural model (i.e., they may or may not be "model conforming"). They can deviate considerably from the idealized model because the architecture is modified by environmental influences (mechanical damage, predator attack, shading) that divert the normal course of development. Shaded mangrove seedlings remain suppressed, with little or no branching and short internodes. The most usual response in the form of the branched tree to other environmental

62

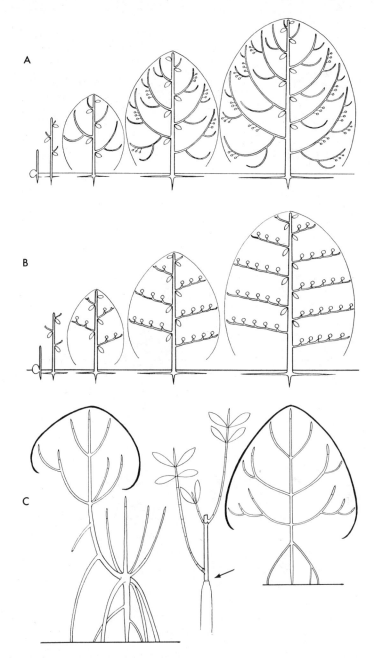

Figure 4.1. Common architectural tree models in mangroves. (A) Attims's model, with continuous growth of the trunk repeated in the branches, which have lateral inflorescences (e.g., *Rhizophora* spp.). (B) Pettit's model, with rhythmic growth of the trunk; the branches have terminal inflorescences and develop by substitution growth (e.g., *Lumnitzera littorea*). (C) Reiteration in *Rhizophora mangle*. Right: the tree conforming to its architectural model (Attims's model, here with pronounced branch tiers); left: reiteration by reorientation of an existing branch to form a reiterated crown; center: reiteration by prolepsis, significant only in seedlings. Reiteration in this species may occur from adventitious buds on the hypocotyl (i.e., below the cotyledonary node – arrow). (After Hallé et al. 1978)

63

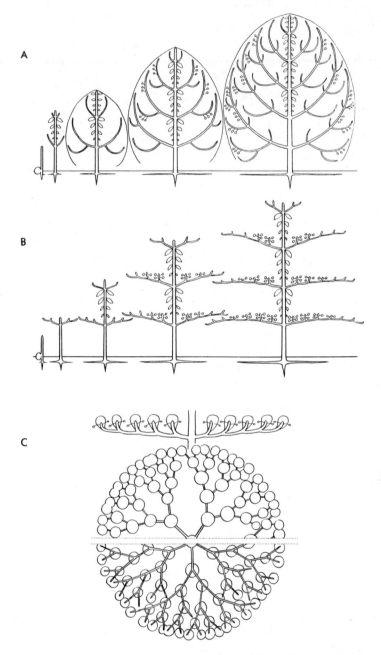

Figure 4.2. Other common architectural models in mangroves. (A) Rauh's model, with rhythmic growth, undifferentiated branches, and lateral inflorescences (e.g., *Xylocarpus* spp). (B) Aubréville's model, with rhythmic growth of the trunk, branches plagiotropic by apposition, and inflorescences lateral (e.g., *Terminalia catappa*). (C) A single branch tier in part B from the side (top), from above (middle), and from below (bottom) to show the distribution of leaf rosettes (circles) that can fill space in an economical and photosynthetically efficient manner. (After Hallé et al. 1978)

Figure 4.3. *Nypa fruticans* (Palmae) at mid-tide. Kuching River, Sarawak.

perturbations is "reiteration," a partial or more usually complete repetition of the architecture of the tree (but not necessarily the root system). Reiteration may involve, for example, either regeneration of new trunk units from previously dormant meristems ("reserve meristems") or reorientation of existing axes (see Fig. 4.1C). These processes are the opportunistic component of form. Regeneration of new units is the usual response to mechanical damage and depends on the availability of reserve meristems, which are developed very unequally in different mangroves. Axes can reorient in the absence of reserve meristems; this may be a response to mechanical damage but it also allows an existing axis to exploit an increased light level where there is a gap in the canopy. *Avicennia*, which coppices readily, and *Rhizophora*, which reorientates branches readily, represent extremes. Both, however, show a remarkable plasticity of form, and the same architectural model can result in tall trees with a narrow crown in a closed canopy as well as the tiny but broad crown of a scrubby mangrove only a meter tall in transitional areas. *Nypa* (Fig. 4.3) is an example of a "tree" that is precisely model conforming in relation to its unique ability to branch terminally (equal dichotomy) with an absence of lateral (axillary) vegetative meristems (Figs. 4.4, 4.5). Here there is no plasticity of form; the *Nypa* swamp has little diversity of aspect (Fig. 4.6).

Crown form, in summary, is the result of deterministic and opportunistic processes. Surprisingly little attention has been devoted to the subject of mangrove

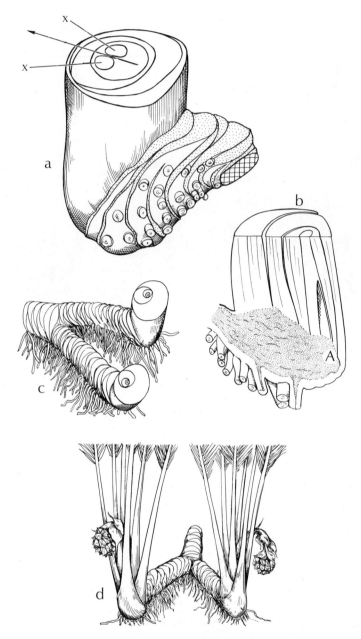

Figure 4.4. *Nypa fruticans* (Palmae) habit (diagrammatic, magnifications are approximate). (a) Young plant (× ¼) with older leaves removed to show late stage of first dichotomy. The growth direction of the daughter axes (X) diverges from the original growth direction (arrow). (b) L.S. (longitudinal section) rhizome (× ⅛), position of shoot apex at A. (c) Older, dichotomized rhizome (× ¹⁄₆₀). (d) Rhizome from front with lateral (axillary) fruiting heads (× ¹⁄₆₀). (After Tomlinson 1971)

Figure 4.5. "Resembling nothing so much as a series of overlapping cowplats": *Nypa fruticans* (Palmae) exposed old stem at low tide showing stem dichotomy. Kuching River, Sarawak.

Figure 4.6. *Nypa fruticans* (Palmae). Aerial view of Labutale Lagoon, Papua New Guinea. The palms form an extensive fringe transitional to the swamp forest behind.

form by ecologists. There is scattered information in the classic texts by German botanists. Schmidt (1903) described bud structure and the development of branch complexes in mangroves in a remarkably complete way, but in a publication (in Danish) that has rarely been cited. Some summary information is provided in the examples of different models in Hallé et al. (1978). Gill and Tomlinson (1969) gave a basic description of *Rhizophora mangle*, which they later presented in a chronological context via phenological study (Gill and Tomlinson 1971b). The later taxonomic descriptions in this book emphasize the dynamic aspects of vegetative morphology wherever possible. It seems to me that botanists neglect this kind of information at their interpretative peril. The tree throughout its vegetative life span continually unfolds and displays its leaves on the developing framework of axes. This vegetative activity is its most continuously expressed attribute. It seems unlikely that the processes involved and the resulting displays have no adaptive significance.

Elements of crown construction

The components of crown form may be treated abstractly and reduced to a series of discrete units of decreasing size and complexity. The crown of a tall tree consists of a series of reiterated complexes reduced progressively in a distal direction. The underlying architecture of each complex is more or less evident, especially by comparison with strong leaders or saplings, which express most clearly the architectural model for the species. Within each complex, differentiated axes can be recognized; the main distinction is between orthotropic (erect) and plagiotropic (horizontal) axes, that is, basically between "trunk axis" and "branch axis." The branches themselves are the constituents of branch complexes, which may have developed sympodially either by substitution (replacement of determinate structures, e.g., inflorescences) or by apposition (displacement of indeterminate, initially terminal, shoots). Where orthotropy is very pronounced, the shoot remains monopodial. The flower or inflorescence position is an important component of crown structure; it may influence vegetative development considerably. The chronology of branch initiation is important. One may contrast syllepsis (the synchronous development of lateral and parent axes) with prolepsis (the delayed, nonsynchronous development of lateral compared with parent axis).

Shoot morphology itself is determined by phyllotaxis and the degree of internodal extension. A contrast between long and short shoots is an essential component of the branch complex in *Bruguiera*, for example (Fig. 4.7). Bud morphology is often complicated. Many species have buds that would be described as "naked" because they lack specialized protective organs. Nevertheless protection is afforded by such features as grooved petioles (*Avicennia*), hairs (*Conocarpus*), and varnish (*Cerbera*). Bud protection is otherwise one function of stipules, which temporarily

Figure 4.7. *Bruguiera gymnorrhiza* (Rhizophoraceae). Branch complex in the lower canopy shows pronounced plagiotropy by apposition. Klang, Malaysia.

enclose younger primordia. Phyllotaxis determines not only leaf arrangement, but also branch arrangement because lateral meristems are almost always axillary. Branch periodicity is largely an expression of growth periodicity. Branch orientation is determined by a combination of primary and secondary effects; secondary orientation or reorientation may be the result of the development of reaction wood (Fisher and Stevenson 1981).

This set of structural features must be perceived in a chronological context; shoot growth and crown development must be studied developmentally. The phenology of vegetative organization in mangroves is being studied increasingly (e.g. Duke et al. 1984, Gill and Tomlinson 1971b, Wium-Anderson 1981, Wium-Anderson and Christensen 1978). Some inferences about developmental processes can be made from structural characteristics, for example, the periodicity of shoot growth from variations in internode length. Proleptic and sylleptic branches, which contrast developmental differences, can be distinguished, after the event, by the presence or absence of basal bud scales (or their scars).

Bud morphology

Bud morphology in the tribe Rhizophoreae is distinctive and constant. Each shoot ends in a bud with a limited number of primordia (usually three to five pairs) and associated stipule pairs. Each stipule pair stands below and encloses the

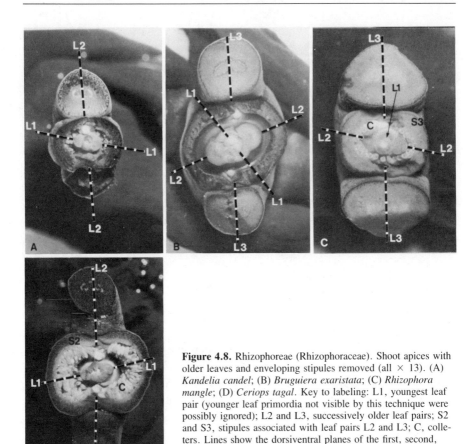

Figure 4.8. Rhizophoreae (Rhizophoraceae). Shoot apices with older leaves and enveloping stipules removed (all × 13). (A) *Kandelia candel*; (B) *Bruguiera exaristata*; (C) *Rhizophora mangle*; (D) *Ceriops tagal*. Key to labeling: L1, youngest leaf pair (younger leaf primordia not visible by this technique were possibly ignored); L2 and L3, successively older leaf pairs; S2 and S3, stipules associated with leaf pairs L2 and L3; C, colleters. Lines show the dorsiventral planes of the first, second, and third pairs of leaves, estimated by eye, the method used for measuring angular divergence between successive leaf pairs. (From Tomlinson and Wheat 1979)

associated leaves but falls as the leaf pair expands. Phyllotaxis is a modified decussate arrangement (bijugate) with each pair at an angle less than 180° to the preceding pair (Fig. 4.8). This arrangement was pointed out by Schmidt in 1903 but not documented in detail until much later (Tomlinson and Wheat 1979). The difference between bijugate and decussate is not trivial because the bijugate arrangement reduces shading and produces branch systems with a greater diversity of orientation compared with a decussate arrangement (Fig. 4.9). A feature of the buds of Rhizophoreae is the secretion of fluid (the viscosity varies with the taxa), apparently emanating from the palisade of glandular colleters within each stipule base (Lersten and Curtis 1974). Bud components are usually loosely packed and the secretion can fill space that otherwise would be unoccupied, notably in *Ceriops*. Similar secretions from colleters associated with either enclosed or exposed pri-

Figure 4.9. *Rhizophora apiculata* (Rhizophoraceae). Seedling from above to show bijugate phyllotaxis, which minimizes self-shading; part of aerial root with characteristic bark left.

mordia occur in *Aegiceras, Cerbera, Osbornia,* and *Scyphiphora. Aegialitis* has mucilaginous buds, with the mucilage accumulating in the cavity of the deep petiolar channel.

Architectural models

The range of architectural models exhibited by mangroves is small; this may simply reflect floristic poverty, but since the observed models are structurally similar, a more likely explanation is that architecture is constrained by the environment. Hallé et al. (1978) comment on the high incidence of Attims's model in mangroves. It is found, for example, in all Rhizophoreae (except some species of *Bruguiera*).

Attims's model (see Fig. 4.1A) is defined by the presence of a trunk with continuous growth, all orthotropic branches with lateral flowers, and continuous or discontinuous branching. This seems an all-purpose, generalized model with considerable plasticity in its expression. The absence of pronounced rhythmic (episodic) growth seems to be correlated with the relatively uniform growing conditions for mangroves (but capable of appreciable seasonal modification in strongly seasonal climates, as shown for *Rhizophora* by Gill and Tomlinson 1971b). Inherent orthotropy of all axes allows for the rapid substitution of damaged leaders by adjacent existing shoots. Branch axes are readily converted into trunk axes, as is notable in *Rhizophora*. Branch expression is very opportunistic; vigorous trees may show continuous branching, but usually branches appear in loosely organized tiers.

Other architectural models found in mangal may be described as variants on the basic *Rhizophora* pattern. Rauh's model (See Fig. 4.2A), which has rhythmic branching, is found in *Xylocarpus* and possibley *Heritiera*. Massart's model differs in that the rhythmically produced branches have a strong endogenous plagiotropy; *Myristica* is probably the best example. It is notable that these three genera are all elements of the back mangal.

Pettit's model (see Fig. 4.1B) is common in species of the front mangal; it differs from Attims's model only in the presence of terminal inflorescences so that the development of branches is sympodial by flowering. *Avicennia* and *Sonneratia* approach this, but much of their branching is independent of flowering and corresponds to Attims's model. *Lumnitzera* shows contrasted architecture in its two species: *L. racemosa* (Attims's model) with lateral inflorescences, and *L. littorea* (Pettit's model) with terminal inflorescences. Here the difference in flower position may relate to contrasted pollinators, small insects that easily penetrate the crown in the former species and birds that do not easily penetrate the crown in the second species.

Aubréville's model (see Fig. 4.2B) is well expressed in the large-flowered species of *Bruguiera* (e.g., *B. gymnorrhiza*). It differs from Attims's model (e.g., *Rhizophora*) in the pronounced plagiotropy by the apposition of its branches, developed *abruptly*, resulting in branching of the familiar *Terminalia* type, best represented by *Terminalia catappa* (Fisher 1978). The developmental difference between Aubréville's and Attims's models is somewhat subtle; apposition growth is achieved *gradually* by branch complexes in *Rhizophora*, which are initially more or less orthotropic. *Terminalia* branching, whether by correlative control of the *Rhizophora* type or by some endogenous mechanism of the *Bruguiera gymnorrhiza* type, has been interpreted as functionally beneficial in minimizing self-shading within a tier (e.g., Fisher and Honda 1979a, b). The hexagonal pattern of structural units that supports the leafy rosettes itself represents a minimum path length arrangement (Fig. 4.2C).

Interpretating crown structure in terms of self-repeating units that maximize photosynthetic efficiency as far as light interception is concerned is but a preliminary step in understanding competitiveness in mangroves. Nevertheless, the characteristic dense canopy of tall mangrove seems understandable in terms of leaf arrangement. In undisturbed mangroves little or no understorey develops except for suppressed, unbranched seedlings.

Vegetative propagation

Where forests are disturbed or clear-felled regeneration may come from seeds or some form of vegetative regrowth, usually in the form of stump sprouts ("coppicing"). An examination of mangrove plants shows that they have little or no capacity for vegetative regeneration or vegetative spread, and no natural capacity for vegetative dispersal. This generalization may be taken as a conservative view and perhaps can be contradicted in some localities and by some examples. Blasco (1977), for example, suggests that *Avicennia* and *Excoecaria* in western India have persisted in the face of over exploitation because they coppice, whereas mangrove Rhizophoraceae have been eliminated through their lack of such ability.

Vegetative dispersal

No mangrove produces detachable vegetative structures that indiscriminately lead to establishment at a distance. Chai (1982) reports that it is possible for uprooted clumps of *Nypa* to float away and become reestablished. Craighead (1971, p. 17) reports that large clumps of *Rhizophora*, detached during a hurricane, can survive in a new location. This kind of vegetative dispersal is accidental and seems of little significance in mangrove distribution. Biological requirements for dispersal and establishment at a distance are met by floating seeds or seedlings, that is, by sexual means. Agamospermy (i.e., the production of seeds by nonsexual processes) is not reported for any mangrove.

Vegetative spread

In a few specialized forms, vegetative spread is possible by either rhizomes or recumbent stems. A limited capacity for localized spread is found in trees whose lower branches root distally. *Nypa* (see Fig. 4.3) spreads by the dichotomy of its rhizome apices, the simplest form of vegetative branching known in flowering plants (see Figs. 4.4, 4.5). This leads to clonal development (see Fig. 4.6). However, it is not know to what extent the individual leafy shoots within a *Nypa* swamp originate sexually (from seed) or asexually (by branching). *Nypa* is nevertheless unique in its ability to colonize its habitat by vegetative means.

Acrostichum is also rhizomatous. It has a potentially unlimited capacity for spread because a branch meristem is developed on the back of each leaf base in adult shoots (Fig. 4.10). However, most of these either abort or remain inhibited, since rhizomes branch little if at all. The vegetative spread of *Acrostichum* within mangal seems restricted by its rather specialized edaphic requirements. In Southeast Asia,

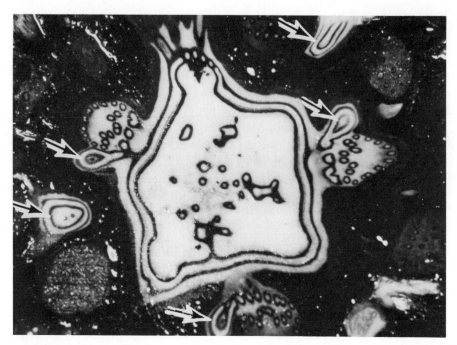

Figure 4.10. *Acrostichum* species (Pteridaceae). Surface view of planed rhizome to show bud or branch traces (arrows) at different levels diverging from the base of each leaf. The section is cut obliquely, somewhat obscuring the phyllotactic pattern. (From a negative by A. M. Juncosa)

for example, it characteristically occurs on the drier pinnacles of mounds thrown up by burrowing lobsters. In disturbed habitats, where these special requirements are more extensively accommodated, it may become quite weedy, in part by rhizomatous growth. In fresh-water swamps *Acrostichum* is long-lived and may form very large clumps, but still with few branches.

Acanthus species are somewhat sprawling plants, since they never develop woody, strongly self-supporting axes. Spread of the recumbent stems is facilitated by the development of aerial adventitious roots. The plants may thus form quite extensive thickets.

Reclining branches

A number of mangrove genera (e.g., *Avicennia*, *Rhizophora*, and *Sonneratia*) have a limited ability to spread because their lower branches may recline under their own weight and root distally. The best documented example of this is provided by Holbrook and Putz (1982) for *Sonneratia alba* in peninsular Malaysia. Individual axes (clones) were traced to a maximum length of 37 m ($\bar{x}$ = 23.6, s.d. = 11.0). The direction of spread was always seaward, and it is suggested that this is the result of the rather exposed rocky coastline at the study site, in which seedlings could not be established and landward spread is prevented by the inshore *Rhizophora* community.

Rhizophora itself has a marked capacity to extend its lower branches (See Fig. B.62) because its continued horizontal growth is supported by aerial roots ("stilt roots"). Varying claims are made about how important these branches are in the vegetative spread of the trees. Davis (1940) emphasized them as playing a major role in the spread of *Rhizophora*, especially in a seaward direction, and consequently as having a major role in land building. Egler (1952) discounted this idea. It is possible that the morphological ability of branches to persist in this way after the parent trunk dies may account for the development of otherwise almost sterile colonies of putative *Rhizophora* hybrids (like *R.* × *lamarckii, R.* × *selala*, and × *harrisonii*) in parts of the tropics. Architecturally it is possible for *Rhizophora* to persist indefinitely in this way if upturned ends of branch axes dedifferentiate to become trunk axes.

It is appropriate to mention here that no mangrove spreads vegetatively by root suckers. Reports of this occurring from the aerial roots of *Rhizophora* result from an observer's inability to distinguish an old root from a branch.

Stump sprouts

Complete regeneration of the crown is possible, at least theoretically, after trunk damage or decapitation, from "resting" or "reserve" buds on the persistent trunk portion (epicormic branches or stump sprouts). This is one expression of the general process of reiteration defined in architectural terms by Hallé et al. (1978). The forester's term "coppicing" is often used, but this relates to regeneration from stumps left after felling, an entirely artificial circumstance. Mangroves have a limited capacity for regeneration from stump sprouts; extremes may be contrasted.

Mangrove Rhizophoraceae are distinctive because they early lose the ability to produce reserve meristems. *Rhizophora* is the best studied example (e.g., Gill and Tomlinson 1969, 1971b) but other genera seem identical. *Rhizophora mangle* seedlings can regenerate after loss of the plumular apex by the development of adventitious buds on the hypocotyl. In contrast, in older branched trees, dormant axillary buds are produced at most nodes but they abort within two or three years, as can be shown by pruning experiments. Consequently, older wood lacks reserve meristems. The inability of damaged *Rhizophora* to develop sprouts from older axes,

which is familiar to mangrove sylviculturists (e.g., Watson 1928), thus has a simple morphological basis.

Most other common mangrove genera (e.g., *Avicennia, Laguncularia, Sonneratia*) retain reserve meristems and develop epicormic sprouts when damaged. Where mangal is clear-felled, however, regeneration of the forest from stump sprouts is not usual. One possible reason is that the delicate root–shoot relationship that supports the tree in its normal architectural configuration is severely disturbed beyond the regenerative ability of the tree.

Natural grafts

There are no reports of natural stem grafts in mature mangroves. Within the intertwining arching aerial roots of a *Rhizophora* forest, however, natural grafts among roots of the same species can develop (e.g., *R. mangle* in South Florida) and even between roots of different species (e.g., *R. apiculata* and *R. stylosa* in Queensland; J. S. Bunt, personal communication). The demographic significance of this is not known.

The grafting ability of *Rhizophora* at the earliest stage of development is pronounced; Larue and Muzik (1954) showed that seedling hypocotyls of *R. mangle* could be joined readily by artificial grafts in a variety of ways. The remarkable regenerative ability or plasticity of this genus at the seedling stage because of hypocotylar adventitious buds is contrasted with its adult inflexibility.

In summary, mangroves have a limited capacity for vegetative spread, *Nypa* being the most conspicuous exception. Regeneration, whether artificial or natural, almost always comes from seeds. Seedlings may have a more pronounced regenerative ability than adults, notably in the Rhizophoraceae. The management of mangroves depends on these biological restrictions on reestablishment; where mangroves are destroyed, seeds or seedlings are the normal mechanism for regeneration of the community. This strong propensity for the establishment of populations solely from seed is part of the ''pioneer'' status of mangroves.

Leaf age and phenology

Leaf age

We can illustrate the interdependence between shoot growth and photosynthetic ability by considering leaf age. How long do leaves persist on mangroves? What determines their longevity? What is the effect of their longevity? These questions cannot be answered in any general way, but Gill and Tomlinson (1971b) in a pilot study provide representative values for *Rhizophora mangle* growing in the seasonal environment of South Florida that illustrate some of the methods needed. This preliminary discussion prepares the way for a presentation of phenological events. Leaf longevity cannot be derived by casual observation, since the trees are evergreen; individual shoots must be marked and monitored at regular intervals. Values by Gill and Tomlinson are for three contrasted populations of shoots (contrasted in their location, vigor, and reproductive status). The average leaf age was 6 to 12 months, with a maximum of 17. A number of factors account for this variation, but notably the time of year when a leaf is initiated. Those that expand in winter have a shorter life span than those that expand in summer. One reason is that growth is intrinsically continuous, but modified by the climate so that leaves are produced faster in summer than in winter. Leaf abscission also is environmentally influenced, again with leaves falling faster in summer than in winter. This synchrony of leaf development and loss results in a fairly constant number of leaves (four or five) on each shoot. Leaves produced in winter are likely to be caught in the peak of leaf loss in summer. It must be emphasized that the leaf turnover rate can be different on different shoots by virtue of the inherent continuous potential for growth in any shoot and the place of that shoot in the branch complex. This seems a general property of mangroves. Leaf longevity then can be seen as a compromise between leaf retention to maximize on the investment of biomass that goes into the construction of the appendage and leaf abscission, which can be beneficial because in mangroves it is a method of salt elimination. Salts are not *concentrated* in older leaves, but seemingly accumulate via the mechanism of leaf succulence (see Chapter 6). We can therefore perceive an interdependence among overall architecture (which determines shoot location), shoot vigor, leaf age, and maintenance of the salt balance.

Evergreens characterize not only mangroves but also mangal. Exceptionally, species of the back mangal may be briefly deciduous; the term "leaf exchanging" of Longman and Jenik (1974) seems appropriate here. *Dolichandrone, Excoecaria, Terminalia*, and *Xylocarpus* are examples. Front-mangal species are necessarily evergreen if leaves are important for maintaining the metabolic and physical processes involved in salt exclusion and stable water potentials. The balance may be delicate; Gill and Tomlinson (1971b) indicate that leaf development and leaf loss are synchronized so that the shoot composition is constant. Defoliation of mangroves by hurricanes may be one factor in their death following such storms (root asphyxiation could be another). No experimental defoliations seem to have been attempted.

Mangroves are somewhat analogous to conifers at high altitudes and latitudes; that is, evergreenness is part of their survival kit in stressed environments, though for different reasons (maximizing assimilation versus maintaining stable salt balances). Additional stress cannot be accommodated by mangroves, so they are excluded from climates with cold winters. Here the mangal is replaced by a treeless salt marsh.

Phenology

The phenological observations of Gill and Tomlinson (1971b) are preliminary and relate to a single species in a region with a markedly seasonal climate. A subsequent series of observations in the monsoon climate of southern Thailand (Christensen 1978, Christensen and Wium-Anderson 1977, Wium-Anderson and Christensen 1978, Wium-Anderson 1981) provides comparative data on more species. These authors again tagged individual shoots and measured the rates of leaf production and leaf loss for at least a year. Most species can be described as growing continuously but not necessarily with a uniform rate of leaf production. For *Rhizophora mucronata* there was no seasonal pattern of leaf production and shedding, so the species could be described as strictly "evergrowing." Shoot construction remained constant, since the "mean leafing rate" (0.33, expressed as a percentage of the leaves produced in relation to the total number of leaves initially present per day) was not significantly different from the "mean shedding rate" (0.30 percent, expressed on the same basis). The average lifetime for the leaves was 330 days. Other species showed either a unimodal peak of leaf production (e.g., *Scyphiphora hydrophyllacea, Lumnitzera littorea*) or a bimodal peak (*Rhizophora apiculata, Ceriops tagal*, and *Avicennia marina*). The unimodal pattern seemed related to the time of heaviest rainfall, whereas it was suggested that the bimodal pattern characterizes species with a more landward distribution that are subjected to least frequent tidal flooding. The average leaf lifetime varied from as short as 9 months (*Lumnitzera littorea*) to as long as 24 months (*Avicennia marina*).

The much longer leaf ages obtained in Thailand compared with South Florida may be explained in part by the deliberate selection of more vigorous shoots in some Florida populations. That the Thailand observations may be more generally applicable, however, is supported by the observations of Duke et al. (1984) in Queensland. These authors, in part, based their findings on the number of appendages (leaves and stipules) that fell into litter traps. This gives values for leaf loss, but since in the Rhizophoraceae there is a stipule for each leaf, stipule fall can be used as a measure of leaf production. This provides a very objective means of assessment. Values were converted to rates of leaf production per 100 shoots per day, a more precise expression than used in the Thailand studies. The Queensland studies show some unimodal species (*Bruguiera gymnorrhiza* and *Ceriops tagal*) and some bimodal species (*Rhizophora stylosa* and *R.* × *lamarckii*) and it was concluded that *R. apiculata* was trimodal. These peaks correlate mainly with rainfall and temperature. Leaf age values fell within the range observed in Thailand.

All these regional studies considered flower and fruit phenology, since in most species flowers are lateral appendages and their development is intimately dependent on leaf production and shoot extension. The most extensive observations are by Duke et al. (1984) in north Queensland, where litterfall studies and generalized observations are combined. Species may be described as having continuous flowering but with peaks of flower production (usually related to peaks of leaf production) or as having definite seasonality of flower production, that is, with flowerless periods that are usually short. Peaks of flower initiation can be followed like populations so that values can be given for a reproductive "cycle." *Rhizophora apiculata* has the longest period, since it takes about 2 ½ years from the appearance of visible flower buds to the maturation and release of the seedling. The shoot morphology of this species accords well with these values, since seedlings always mature much later than the leafy rosette. Other *Rhizophora* species have a shorter cycle (12 to 18 months), whereas *Bruguiera cylindrica* has the shortest (7 to 8 months).

The actual flowering period is difficult to estimate in a population, but the generalized figures for Queensland demonstrate broad flowering periods that differ somewhat in different species. Certainly for animal-pollinated species, the pollinator reward is widely available.

The conclusion from these pilot studies is that although mangroves tend to be ever-growing and ever-flowering, there can be marked seasonality and the values obtained for a limited region should not be too extensively generalized. Clearly the physiological responses of mangroves to climate and substrate conditions need to be assessed. We know that phenological events are correlated with environmental factors in ways that make mangroves highly adaptable to diverse growing conditions. Consequently a given region and climatic regime must at the moment be assessed empirically. The three sets of studies in Florida, Thailand, and Queensland provide excellent guides. All studies demonstrate a need for intimate knowledge of shoot morphology as a preliminary base line.

Leaf morphology and texture

The rather dull, uninspiring, and unchanging aspect of mangal is largely due to the uniform leaf shape and texture of mangroves, coupled with the evergreen habit. Figure 4.11 presents the outlines of leaves of 18 species of mangrove (both New and Old World) to illustrate the uniform size and generally ovate to elliptic outline, with blunt to pointed apex and usually entire margins or with, at most, obscure crenations. Even the leaflets of the compound leaves of *Xylocarpus* match this general pattern well. Nevertheless, there is sufficient variation that shape can be diagnostic; it can be used to distinguish species of *Avicennia* where the range of leaf form in the Indo-Pacific species closely matches that of the Atlantic species

A
A. alba   Av. mar.   Av. lan.   Av. off. (m.)   Av. off. (j.)

4cm

B
Ce. tag.   Br. hai.   Br. cyl.   Br. par.

C
Rh. muc.   Rh. api.   Rh. man.   Ka. can.

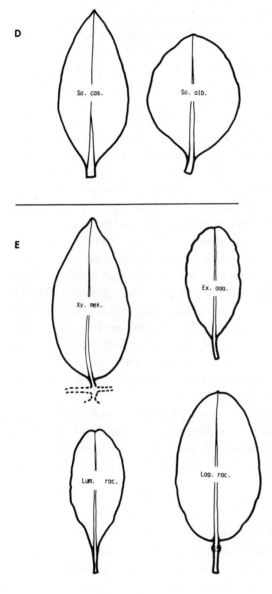

Figure 4.11. Mangrove leaf shape outlines to show uniform size and shape. (A) *Avicennia*: from left to right, *A. alba, A. marina, A. lanata, A. officinalis* (mature), *A. officinalis* (juvenile); (B) Rhizophoreae: from left to right, *Ceriops tagal, Bruguiera hainesii, B. cylindrica, B. parviflora*; (C) Rhizophoreae: from left to right, *Rhizophora mucronata, R. apiculata, R. mangle, Kandelia candel*; (D) *Sonneratia*: left, *S. caseolaris*; right, *S. alba*; (E) above, *Xylocarpus mekongensis* (one leaflet), *Excoecaria agallocha*; below, *Lumnitzera racemosa, Laguncularia racemosa*.

(Figs. B.9–B.11). Leaf morphology is particularly uniform in the mangrove Rhizophoraceae, although there are differences in texture and average size. In *Rhizophora* the mucronate versus nonmucronate shape of the leaf apex segregates species on a broad geographical basis. A hybrid between species of the contrasted groups (*Rhizophora* × *selala*) has an intermediate leaf tip morphology showing that leaf apex morphology is genetically controlled (Tomlinson 1979). This feature seems trivial in functional terms, however.

Leaf texture is fairly consistently firm to almost coriaceous, but never rigid. The veins are obscure and never prominent; this is related to the usual absence of vein sheaths and vein sheath extensions. Succulence is a character that varies according to the degree of salinity and also leaf age. The anatomical basis for this is the differential expansion of mesophyll cells (e.g., Walter and Steiner 1936) and is discussed later. Constancy of leaf appearance is not matched by structural characteristics; leaf dry weight per unit area in different species, for example, varies over the range 6.4 to 21.6 mg/cm^2, according to Johnstone (1981).

A conspicuous indumentum does not usually develop. Conspicuous waxy deposits do not occur. Sometimes young leaves are somewhat hairy in contrast to adult leaves (e.g., *Lumnitzera*). The most notable exception to the glabrous condition is *Avicennia*, with a dense palisade of short uniseriate capitate hairs covering the lower leaf surface and giving it a characteristic grayish or greenish brown color (see Fig. 4.15). *Pemphis* has an indumentum of pointed uniseriate hairs, whereas a scaly indumentum characterizes *Camptostemon* as well as the back mangroves *Brownlowia, Glochidion*, and *Heritiera*. Differences in leaf texture and shape are emphasized in the diagnostic key later in this chapter.

Other surface features (Stace 1965a) are so-called cork warts in *Rhizophora* (Fig. 4.12) and various glandular structures and especially salt glands, which in most instances function to maintain the salt balance. Salt glands are described elsewhere, more appropriately in the discussion of salt balance in Chapter 6.

Secretory structures

Structures regarded as hydathodes have been described in mangrove leaves. These are areas either where the epidermis is interrupted or where there is an enlarged stoma ("water stoma" of Stace 1965b) or where a modified epithelial tissue or epithelium is developed (Fahn 1979). The presumed hydathodes subsequently become occluded, often with the development of a periderm. This may account for the cork warts, which are evident in *Rhizophora* as minute black lesions on the lower leaf surface (see Fig. 4.12). It must be emphasized that there is no record of guttation from mangrove leaves, but this is not surprising because positive pressures in the xylem are hardly possible with plants rooted in sea water. Never-

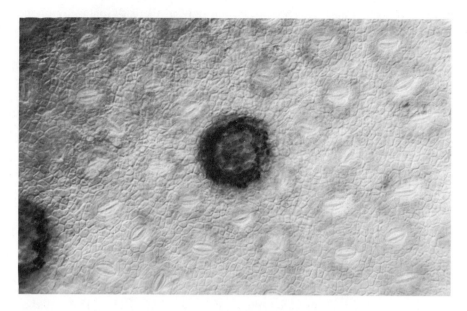

Figure 4.12. *Rhizophora mangle*, surface view (× 10). Detail of a single "cork wart" from lower epidermis.

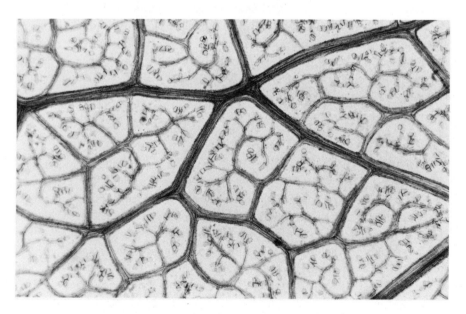

Figure 4.13. *Avicennia germinans*. Portion of cleared leaf (× 65) in surface view to show well-developed system of isodiametric terminal tracheids at free vein endings.

theless, in some plants hydathodes function only in the early ontogeny of the individual leaf or leaflet, as in some ferns (Sperry 1983). Consequently in addressing the problem of the functional morphology of leaves, one should consider ontogenetic events related to the life span of the individual leaf. If one examines mangrove leaves, even quite casually, one sees structures, especially on the surface and margin, that are prominent in the developing primordium and may persist as inconspicuous structures in mature leaves. For example, in leaves with a slightly irregular outline, marginal indentations usually represent the site of structures that are conspicuous only in young leaves (e.g., in Combretaceae and Euphorbiaceae).

Extrafloral nectaries are not recorded in strict mangroves, but the lateral petiolar glands of *Laguncularia* may have this function. Extrafloral nectaries certainly occur on the outside of the spathulate calyx of *Dolichandrone spathacea*.

Leaf anatomy

The uniformity of leaf morphology is matched by a suite of features common to most mangroves, including the usual development of "colorless" or "water-storage" tissue, short tracheids terminating vein endings (Fig. 4.13), and the marked absence of sclerotic vein sheaths (Fig. 4.14A, B–D). Sclereids of various shapes are, on the other hand, quite common (Figs. 4.14B, C; 4.16A). Despite this suite of common characters, leaves of different mangrove genera can be readily distinguished by anatomical features. These are presented in synoptic form in the key ending this chapter. The characteristics are generic; specific differences are usually small so that, for example, there are no obvious structural features by which different species of *Avicennia* or *Rhizophora* can be distinguished. The most variable genus in this respect is probably *Bruguiera*.

Stomata

The stomatal structure in mangroves shows no high degree of specialization. Stomata are scarcely sunken or not sunken at all, but guard cells are somewhat thick walled, often with prominent or even elaborated ledges, suggesting some increased resistance to stomatal transpiration. *Nypa* has probably the most complex stomata, since the guard cells have several parallel ledges on the inner face. Restriction of nonstomatal water loss is imposed by the thickened outer epidermal wall, which is always strongly cutinized (Arzt 1936). The cuticle is usually smooth and, together with the usual absence of hairs, increases reflectivity, since the "finish" of the leaf is glossy rather than matt. *Aegialitis*, however, is an example of a genus with a sculptured cuticle.

Mesophyll

The mesophyll in mangrove leaves is thick, largely because of well-developed nonassimilatory layers, so that the strictly assimilatory tissue makes up a relatively small fraction of the total leaf thickness (Figs. 4.14, 4.16). The nonassimilatory tissue may be described as "colorless," since it has no or few chloroplasts, but this description is not exactly appropriate because tannin is common; the cell contents are then brown. "Water-storage tissue" is a descriptive term commonly used, but this implies a function that is not definitely proved.

Leaves may be described as dorsiventral (Figs. 4.14, 4.15) if there is a marked dissimilarity between the upper and lower halves, with the colorless tissue represented by a thick adaxial hypodermis, or as isolateral (Figs. 4.16, 4.17) when the halves of the leaf are about equal, with the colorless tissue occupying the middle region of the mesophyll. In thick dorsiventrally symmetrical leaves there is often a concentration of palisade layers toward each surface. A common feature of most mangrove leaves is the frequent development of groups of enlarged terminal tracheids at vein endings (see Fig. 4.13); this is found in all common genera, but not in all species of *Bruguiera*. Branched sclereids are not uncommon in the mesophyll and are abundantly developed in *Aegiceras, Rhizophora*, and *Sonneratia* and especially in *Aegialitis*, where they constitute about 15 percent of total leaf tissue. Sclereids in *Pelliciera* are elongated, fiberlike cells. In many genera two or more kinds of sclereid can be distinguished according to size, shape, and location, for example, *Aegiceras* (Rao 1971). Both sclereids and tracheids may be involved in capillary water storage (see Zimmermann 1983), and sclereids may provide mechanical support to leaves with diminished turgor or discourage herbivores. The pliable but firm texture of mangrove leaves is a direct consequence of the limited development of bundle sheath fibers and the absence of bundle sheath extensions. This can be appreciated by contrasting the stiff, almost brittle leaf of *Heritiera littoralis* with that of any true mangrove. *Heritiera* is distinctive in its well-developed bundle sheath extensions (see Fig. 4.14A).

Interesting contrasts and similarities can be found in the anatomy of the leaves of mangroves that belong to the same family. The mangrove Rhizophoraceae, for example, have diverse anatomical features, which is surprising in view of their uniform leaf shape and texture. The Combretaceae are much more uniform in this respect; *Laguncularia* and *Lumnitzera* are distinguished chiefly by their indumentum. We can ascribe their similarity to both a close systematic relationship and an adaptation to similar environments.

Most of the distinctive structural features shared by mangroves probably have functional explanations, although these are not always directly demonstrable. Ep-

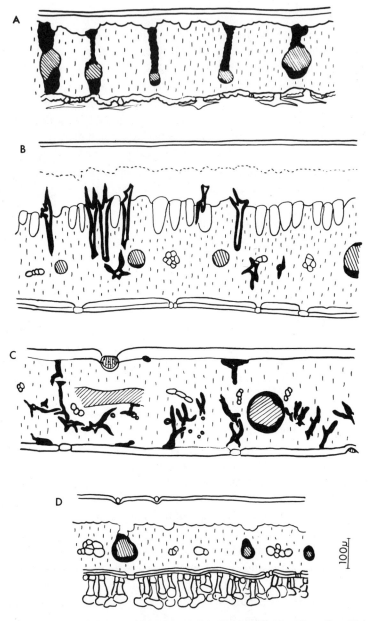

Figure 4.14. Leaf anatomy, diagrammatic, dorsiventral leaves. (A) *Heritiera littoralis*, with well-developed bundle sheath extensions; (B) *Rhizophora stylosa*, adaxial hypodermis with three distinct layers; (C) *Aegialitis annulata*; (D) *Avicennia marina*. Key to symbols: Lined area, assimilating tissue; black, bundle sheath sclerenchyma or sclereids; hatched, veins; open circles, terminal tracheid clusters.

Figure 4.15. *Avicennia eucalyptifolia.* T.S. (transverse section) lamina ($\times$ 65) to show dorsiventral structure. Note dense indumentum of hairs on lower surface and terminal tracheids in middle of mesophyll.

idermal features relate to transpirational control, salt glands and colorless mesophyll layers to the maintenance of a salt balance, and lack of extensive sclerenchymatous structural support to high turgor. My suggestions regarding the capillary storage function of terminal tracheids and sclereids need experimental verification.

Ideoblasts

The complex leaf anatomy is reflected in the various internal structurally (and presumably functionally) specialized cells that often have an ideoblastic distribution, that is, are isolated and not associated to form a distinct tissue. Among these features (in addition to sclereids already mentioned) are oil cells (e.g., in *Osbornia*), mucilage or mucous cells (e.g., subhypodermal in *Rhizophora*, hypodermal in *Sonneratia*, Fig. 4.17), crystalliferous cells (e.g., druse crystals are abundant in the Rhizophoreae), tannin cells (abundant in most genera and often concentrated in certain tissues, e.g., the hypodermal tissue of Rhizophoraceae), and laticifers (e.g., in *Excoecaria*). The precise chemical composition of the contents of these cells is rarely known in any detail; most functional attributes are entirely conjectural. The student of mangrove anatomy is advised to approach this subject empirically, since freehand sections of fresh leaves give a clearer impression of microscopic structural diversity than does a lengthy search of the literature. There is considerable variation within individual species; and most authors have looked at only limited samples.

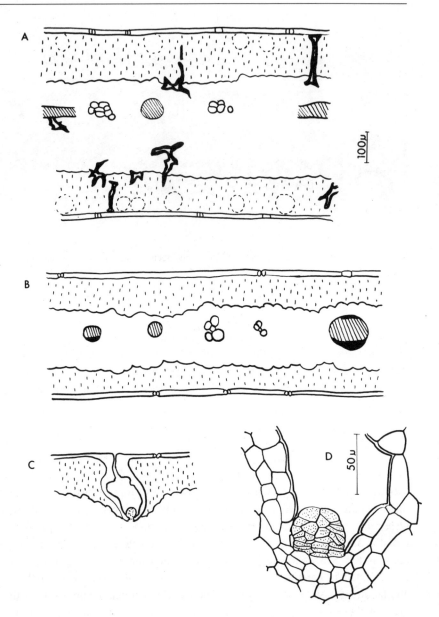

Figure 4.16. Leaf anatomy, diagrammatic isolateral leaves. (A) *Sonneratia alba*, T.S. (B–D) *Laguncularia racemosa*: (B) T.S.; (C) gland in pit on adaxial surface; (D) detail of presumed salt-secreting gland at base of pit.

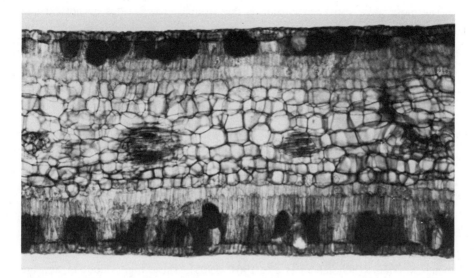

Figure 4.17. *Sonneratia alba*. T.S. lamina (× 10) to show isolateral structure. Note enlarged hypodermal mucous cells.

Wood anatomy

The anatomy of mangrove woods has been studied extensively, in part because of their economic importance and in part to correlate wood structure with physiological specialization toward the mangrove habitat (e.g. Jansonnius 1950; Panshin 1932; and for the Rhizophoraceae alone, van Vliet 1976). The following generalizations seem appropriate:

Mangroves usually lack growth rings, which might allow age to be determined. An exception is the wood of *Diospyros ferrea*, reported by Duke et al. (1981), which produces microscopically visible growth rings at an average rate of seven rings every four years. This unusual frequency seems determined by irregular trends in seasonal rainfall. Superficial examination of other mangrove woods often suggests the presence of growth rings, but these seem related to variations in the density of chemical wall deposits and not to microscopic discontinuities, which are the result of variations in cambial activity. These discontinuities probably account for the claim by Davis (1940) that *Rhizophora mangle* produces annual growth rings in Florida. *Avicennia* has conspicuous growth rings which relate to the anomalous alternation of bands of xylem and phloem tissue (Fig. B.12), but again these are not produced seasonally (Gill 1971). *Aegialitis* also has successive cambia, producing alternate rings of xylem and phloem.

The most conservative conclusion is that datable growth rings do not exist in mangrove woods. This puts severe constraints on ecological analysis because no indirect measure of the age of a tree is possible, a general problem in tropical ecology (see Bormann and Berlyn 1981).

Mangrove woods have narrow and densely distributed vessels. Panshin found that the tangential diameter of vessels were in the range of small (i.e., narrow) to very small [usually less than 100 μm (e.g., Fig. 4.18), rarely exceeding 150 μm] when compared with the total range of vessels in woods (range 50-250+ μm diameter). These results were confirmed by Jansonnius, who suggested that vessel density was in proportion to the frequency of inundation. The extreme concentration is found in *Aegiceras* with vessels all narrower than 50 μm and up to 150 per square millimeter. *Pelliciera* also has narrow vessels.

The high density of narrow vessels can be related on theoretical grounds to the fact that high tensions (negative pressures) have been measured in the xylem of mangrove stems (Scholander *et al.*, 1964, 1965). These high values in turn are related to the high osmotic potential of sea-water, which mangroves must overcome initially to absorb water, and to the high temperatures which promote transpiration. Frequent high tensions in the xylem increase the likelihood of cavitation within vessels during extreme stressful conditions. Since such embolisms are contained within a single unit (the vessel), an element of safety is introduced if there are a large number of conducting units, since it is reasonable to assume that embolism is a random event, independent of the number of units, but directly dependent on vessel diameter (Zimmermann, 1983). However, since the conductivity of a capillary is proportional to the fourth power of its radius, from the Haagen-Poiseuille equation

$$\frac{dV}{dt} = L_p \cdot \frac{dP}{dl} \quad \text{and} \quad L_p = \frac{r^4}{8\eta}$$

where dV/dt = flow rate

dP/dl = applied pressure gradient

L_p = hydraulic conductivity

r = radius of capillary

η = viscosity of liquid

A slight decrease in diameter therefore brings about a proportionally much greater increase in resistance to flow (the reciprocal of conductivity). In other words, if one substitutes, for reasons of safety, narrow vessels for wide ones, maintaining a uniform flow rate requires the replacement of each capillary of unit diameter with

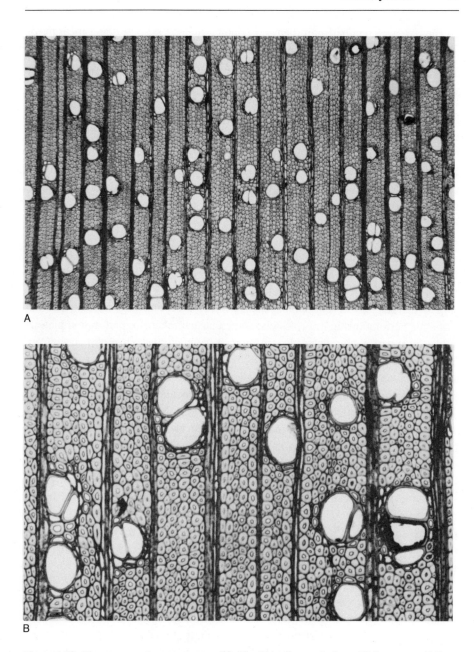

Figure 4.18. Mangrove wood anatomy exemplified by *Rhizophora apiculata*. (A) Low-power T.S.; (B) detail at higher power. Vessels average 85 μm in diameter.

16 capillaries of one-half unit diameter (Zimmermann 1983). A direct measure of relative conductive capacities can be obtained by comparing the sum of the radius raised to the fourth power of all conducting elements. The narrowness of vessels in mangrove woods (for safety, since it reduces the loss of conducting units by embolism) is therefore understandably compensated by an increase in their numbers, hence their density. Jansonnius gives vessel densities in mangrove woods (vessel number per unit cross-sectional area) from 2 to 10 times the average value for a wide range of woods, which is expected by theoretical considerations. The reverse trend towards few wide vessels is found in lianes, which have relatively narrow stems because they are not self-supporting, and which cannot accommodate large numbers of vessels. However, this is not a safe system; lianes do not occur in the strict mangal.

Vessel elements in mangrove woods have simple perforation plates, with the notable exception of the mangrove Rhizophoraceae (tribe Rhizophoreae) which have scalariform perforation plates with few (2-7) broad thickening bars (Fig. 4.19). (*Dolichandrone spathaceae* is a minor exception since it has in addition to simple perforations in most vessels, some unusual reticulate perforation plates.) Scalariform perforation plates also distinguish the mangrove Rhizophoraceae from their terrestrial relatives (van Vliet 1976). Since mangroves with and without scalariform perforation plates in their wood grow close together, it is not easy to provide an adaptive explanation for this structural peculiarity of the Rhizophoreae. The simplest explanation, but not a very satisfactory one because it is not a functional explanation, is that this tribe is monophyletic and evolved from an ancestor that had scalariform perforation plates in its wood. Scalariform perforations were retained in descendants because they are adaptively neutral. An alternative explanation might be suggested if a function for perforation plates could be discovered. However, speculation about the functional attributes of wood features is dangerous in the absence of experimental investigation because the water economy of a tree is complex and requires a comprehensive appreciation of all the factors involved: efficiency of absorption, control of transpiration, extent of conducting channels, rates of flow, and hydraulic architecture.

Apart from vessel dimensions, other wood structural variation relates mostly to the systematic position of a species. This suggests that there are no very precise adaptive constraints that relate anatomy to habit; that is, the structure of the mangrove wood says as much about the systematic position of a species as does its ecological location. This allows Panshin, for example, to construct keys for the identification of the mangrove woods he studied, often distinguishing closely related species. Some caution should be exercised in using these keys, since

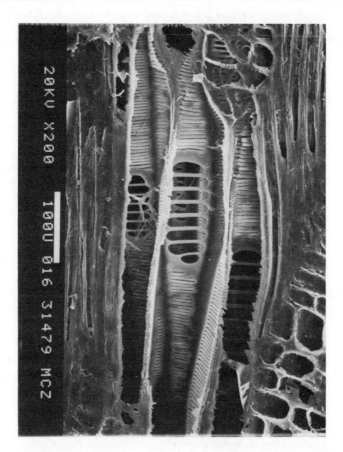

Figure 4.19. *Bruguiera gymnorrhiza.* Scanning electron micrograph of radial longitudinal surface to show scalariform perforation plates that characterize wood of the mangrove Rhizophoraceae.

quantitative differences are often used and might refer only to the rather limited range of samples studied by Panshin. A more comprehensive evaluation using qualitative characters might be attempted and would be useful to ecologists and paleobotanists.

Synoptic key to common mangrove genera based on leaf anatomy

1A. Leaves dorsiventral, stomata restricted to lower surface; colorless tissue, if present, hypodermal 2

1B. Leaves isobilateral, stomata equally abundant on both surfaces; colorless tissue, if present, never hypodermal 14

2. (*Leaves dorsiventral*); all have *enlarged terminal tracheids* except *Bruguiera* and *Kandelia*) . . . .

2A. Hypodermis of compact colorless (nonchlorophyllous) cells below one or both surfaces; sclereids and salt glands present or absent 3

2B. Hypodermis of colorless cells not developed; branched stellate sclereids abundant, salt glands conspicuous. .... *Aegialitis* (Fig. 6.4A,B)

3A. Hypodermis below both surfaces; abaxial hypodermis always least conspicuous 4

3B. Hypodermis restricted to adaxial surface 10

4A. Indumentum on lower epidermis of three- or four-celled uniseriate capitate hairs forming a dense palisade .... *Avicennia* (Figs 4.15, 6.4G)

4B. Indumentum usually absent, if present not palisade-like 5

5A. Adaxial hypodermis up to seven cells deep, differentiated into outer two or three layers of tabular cells, inner layer(s) of anticlinally extended cells; branched sclereids usually well-developed; cork warts present....... ....*Rhizophora* (Fig. 4.14B)

5B. Adaxial hypodermis not differentiated into distinct layers; sclereids present or absent; cork warts absent 6

6A. Sclereids or fiber sclereids well developed, conspicuous 7

6B. Sclereids absent or inconspicuously developed 8

7A. Sclereids stellately branched, thick-walled salt glands present; secretory cavities in mesophyll.. *Aegiceras*

7B. Sclereids fiberlike, thin walled; salt glands and secretory cavities absent ....*Pelliciera*

8A. Terminal tracheids well developed .... *Ceriops*

8B. Terminal tracheids not well developed............................. 9

9A. Mesophyll somewhat isodiametric, with assimilating palisade-like mesophyll layers below each hypodermis .... *Kandelia*

9B. Mesophyll dorsiventral, mesophyll palisade tissue present only adaxially .... *Bruguiera*

10A. Fibrous bundle sheath of most veins extended to one or both surfaces; indumentum of conspicuous peltate scales on abaxial surface............ ....*Heritiera* (Fig. 4.14A)

10B. Bundle sheath not forming fibrous extensions to surface; scaly indumentum absent 11

11A. Salt glands present, especially on upper surface .... *Acanthus*

11B. Salt glands absent... 12

12A. Mesophyll somewhat bifacial with palisade-like layers toward each surface; laticifers present but not conspicuous in spongy mesophyll .... *Excoecaria*

12B. Mesophyll dorsiventral; laticifers absent 13

13A. Terminal tracheids enlarged, conspicuous; bundle sheath not fibrous, not crystalliferous; adaxial hypodermis not enlarged............. *Scyphiphora*

13B. Terminal tracheids not enlarged, bundle sheath fibers well developed, often associated with crystals; adaxial hypodermis anticlinally enlarged in succulent leaves ....*Xylocarpus* (Fig. 6.6)

14. (*Leaves isolateral; all have enlarged terminal tracheids except Pemphis*) 15

15A. Enlarged spherical multicellular oil cavities, with lysigenous development, conspicuous and abundant below each surface.................. *Osbornia*

15B. Enlarged oil cavities absent, hypodermal secretory structures, if present (*Sonneratia*), unicellular and not developing lysigenously 16

16A. Leaf mesophyll relatively uniform, not conspicuously differentiated into central and peripheral layers; capitate hairs on both surfaces...*Conocarpus*

16B. Leaf mesophyll conspicuously differentiated into central colorless and peripheral assimilating layers; capitate hairs absent 17

17A. Leaves with conspicuous slender, unbranched unicellular hairs ...*Pemphis*

17B. Leaves glabrous (or unicellular hairs only in early development)....... 18

18A. Sclereids conspicuous, abundant; hypodermal mucilage cells common and conspicuous....................................*Sonneratia* (Fig. 4.16A, 4.17)

18B. Sclereids absent or infrequent, hypodermal tannin or mucilage cells absent 19

19A. "Salt glands" in epidermal depressions often present, but sparse ....
....*Laguncularia* (Fig. 4.16B)

19B. Salt glands absent.. *Lumnitzera*

5   Root systems

Aerial roots

A feature of more highly specialized mangroves is that they develop some part of the root system exposed to the atmosphere, at least at low tide. This unusual example of the root system of trees in tidal forests is built into the definition of "aerial root" given by Gill and Tomlinson (1975): "a root which is exposed to the atmosphere at least part of the day." This qualification is, of course, not necessary for most aerial roots in tropical plants, which are continually exposed. Ascending or above-ground parts of aerial roots are common in trees of tropical swamp forests where they are continually exposed. In the more seaward mangroves, however, aerial roots are covered at high tide. This turns out to be important to their functioning. Physiologically the need for the development of an above-ground part of the root system in trees in the tropics is related to the anaerobic nature of substrates in swamps and the need for atmospheric oxygenation of the absorbing system rather than oxygenation via the absorbing surface itself. The presumed mechanism is discussed in further detail later. Descriptions and illustrations of root systems of the trees of fresh-water swamps in the tropics are by Ogura (1940), Jeník (1978), and Corner (1978). A survey of root systems in mangroves is provided by Troll (1943), based largely on his own extensive field studies.

There are several types of aerial roots in mangroves. Only general features that demonstrate common characteristics are described.

Stilt-roots

This refers to the branched, looping aerial roots that arise from the trunk and lower branches of *Rhizophora*. The description of them functionally as "stilt" roots applies to them only in older trees where they function as flying buttresses, since the trunk at the base of the tree characteristically becomes obconical with age; most secondary thickening at the base of the tree occurs in the aerial roots rather than the trunk (Fig. 5.1). Stilt roots of the same type are developed to a limited extent in *Bruguiera* and *Ceriops* in the sapling stage close to the stem base, becoming shallow buttresses in old trees. Stilt roots occur sporadically in other

Figure 5.1. *Rhizophora harrisonii.* Tivives, Costa Rica. Aerial roots form flying buttresses supporting the trunk, which tapers below.

Figure 5.2. *Avicennia officinalis* (Avicenniaceae). Aerial stilt roots, a common feature of this species. Ponggol, Singapore.

97

mangroves and are a common feature, for example, of *Avicennia* species (notably *A. alba* and *A. officinalis*, Fig. 5.2). In other *Avicennia* species (e.g., *A. germinans*) aerial roots growing from the trunk seem in part to be a response to wounding.

The structure and development of the root system of *Rhizophora* have been investigated in detail by Gill and Tomlinson (1971a, 1977). They emphasize (following Docters van Leeuwen 1911) the opportunistic nature of the spread of the roots, since new roots arise only where the apex is damaged or where the axis becomes rooted distally.

Pneumatophores

Erect roots that are some form of direct upward extension or appendage of the subterranean root system are called "pneumatophores" in a descriptive sense; the term has no precise morphological connotation. Quite different developmental types can be recognized, each characteristic of a genus or group of species. The essential features are shown in Figure 5.3.

Pneumatophores in *Avicennia* and *Sonneratia*. The pneumatophores are erect lateral branches of the horizontal roots, which are themselves buried in the substrate (Fig. 5.4). The first-order horizontal roots originate at the base of the trunk and extend horizontally for long distances through the substrate: the visible parts of the root system are pencil-like erect branches spaced at regular intervals along the first-order roots. In *Avicennia* the roots are of limited height (usually less than 30 cm) and develop little secondary thickening. In *Sonneratia* the roots have a much longer period of development and undergo secondary thickening so they become quite tall–exceptionally up to 3m (Fig. 5.5). The pneumatophores usually remain unbranched, but are capable of branching, seemingly only when damaged. The surface texture in these two types differs: in *Avicennia* the surface remains smooth and spongy, and in *Sonneratia* a characteristically flaky bark develops in younger roots; in older roots the bark is smoother. In both genera the roots include chlorophyll in the subsurface layers.

Pneumatophores in *Laguncularia*. This genus develops erect, blunt-tipped pneumatophores under certain circumstances, rarely exceeding a height of 20 cm (Jeník 1970). They are more frequently branched than in *Avicennia* and *Sonneratia*. The pneumatophores seem facultative in their development so that populations in some localities lack them. Consequently, the existence of pneumatophores in *Laguncularia* sometimes goes unrecorded (e.g., Allen 1956).

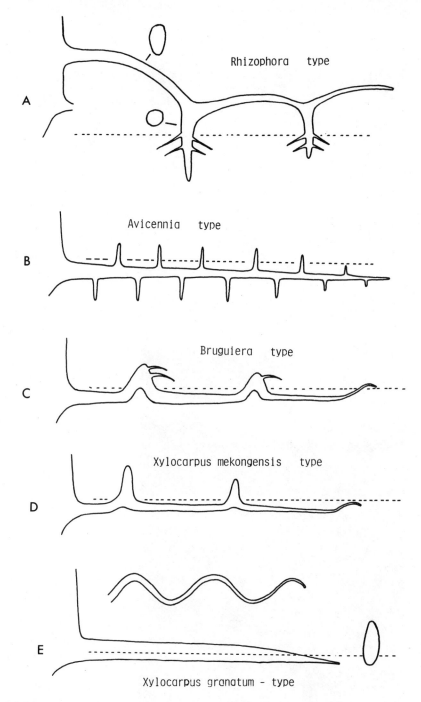

Figure 5.3. Schematic demonstration of types of aerial roots in mangroves. All have developed from left to right. Dotted line, substrate level.

Figure 5.4. *Avicennia* species (Avicenniaceae). Pneumatophores extending beyond the tree canopy in a mud flat. Townsville, Queensland. (From a color transparency by A. M. Gill)

Figure 5.5. *Sonneratia* species. A single pneumatophore suitable for the *Guinness Book of Records*, held by N. C. Duke, Australian Institute of Marine Science. (From a color transparency by A. M. Juncosa)

Root knees

Root knees in *Bruguiera* and *Ceriops*. In these genera the horizontal roots periodically reorient as they grow through the substrate away from the parent tree so that the tip forms a pronounced loop before continuing its horizontal growth (Fig. 5.6). At the site of the loop, eccentric secondary thickening occurs mainly on the upper side so that a blunt, knoblike structure is raised above the substrate surface. Branching of the root system is largely restricted to these "knees"; the branches both proliferate new horizontal roots and form anchoring roots. By the repeated development of loops, a single horizontal root develops a series of outgrowths at regular intervals. The size, shape, and frequency of root knees are characteristic for each species, though somewhat variable according to substrate condition.

Pneumatophores in *Lumnitzera*. The root knees of *Lumnitzera* are rather inconspicuous and seem mainly to be derived from laterals that execute the looping growth and develop little secondary growth. They may be thought of as structures intermediate between pneumatophores and knees.

Root knees in *Xylocarpus*. Localized erect outgrowths of the upper surface of the horizontal root system are a feature of *Xylocarpus mekongensis*. Here the structures are secondary and produced entirely by localized cambial activity on the upper surface of the horizontal root (Figs. 5.7, B.41). This should be contrasted with *Bruguiera* and *Ceriops*, where the knee is initiated as the result of primary growth. Root knees in *Xylocarpus* may grow to a height of 50 cm. The knobby aerial structures of *Camptostemon* roots correspond closely to this type (Troll 1933a).

Plank roots

In *Xylocarpus granatum* the horizontal root becomes extended vertically by eccentric cambial activity throughout its length. These roots are laterally sinuous in their course, so the result in the mature tree is a series of wavy, planklike structures growing away from the base of the trunk (Fig. B.39B).

Smaller plank buttresses occur in species of *Heritiera* (e.g., *H. fomes* and *H. littoralis*, Fig. B.82) as narrow surface outgrowths of the base of the trunk. These are transitional to the plank buttresses of the trunks of many rain forest trees.

Aerial roots of *Pelliciera*

Fluted buttresses are formed by enclosed aerial roots, (See Fig. B.55).

A

B

Figure 5.6. *Bruguiera gymnorrhiza* (Rhizophoraceae) root knees. (A) Exposed at low tide; (B) system exposed by erosion of substrate; (C) single pneumatophores with lenticels and descending branches. Hinchinbrook Island, Queensland.

102

C

Figure 5.7. *Xylocarpus mekongensis* (Meliaceae). Root knees in L.S. to show that they are secondary extensions of the horizontal cable roots. Townsville, Queensland.

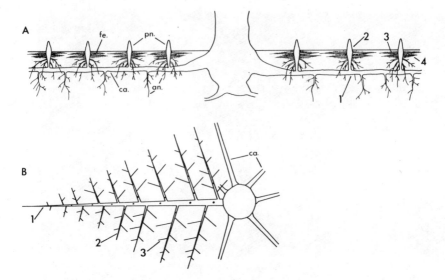

Figure 5.8. Root system of *Sonneratia*. (A) Profile schematic view. (B) Plan schematic view. Numbers refer to branch orders; the ultimate (5) order branches are too fine to appear in part A. Specialized root structures are referred to as cable roots (ca.), pneumatophores (pn.), anchoring roots (an.), and feeding roots (fe.). Order number does not determine the type of root (e.g., pneumatophores can be second- or third-order erect branches; second-order branches may be pneumatophores, anchoring roots, or high-order cable roots). Note that second-order cable roots must intersect extensively close to the trunk if all roots developed according to this scheme. (After Troll and Dragendorff 1931)

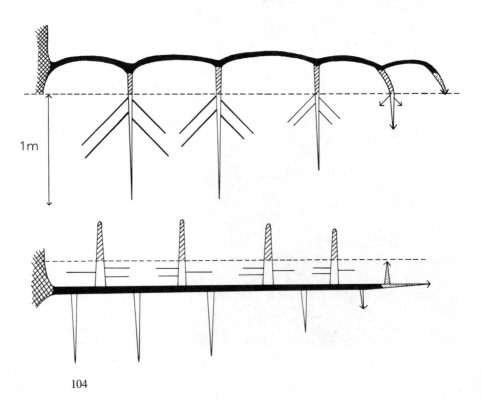

Mangroves without aerial roots

The following mangroves normally lack any elaborated aerial part to the root system: *Aegiceras, Aegialitis, Excoecaria, Kandelia, Osbornia, Scyphiphora*, and *Nypa*. However, certain structures may be developed to aerate the root system. The base of the stem in *Aegialitis* is enlarged and fluted, with a very spongy texture. *Kandelia* tends to develop aerial roots in some limiting environments (e.g., Hosakawa et al. 1977). In *Nypa*, the individual leaves function like a giant pneumatophore growing close to the stem apex, facilitated by the somewhat spongy texture of its ground tissue and the spongy ground tissue of the rhizome itself. This function may continue in the leaf base, which persists after the distal portion breaks off. Breakage does not seem to be the result of anatomical specialization, although its height is quite consistent.

Mangal associates do not develop any aerial roots, with the exception of the palms *Phoenix, Raphia*, and *Oncosperma*, which usually develop slender pneumatophores. This condition is characteristic of many palms that grow in wet sites (e.g., *Metroxylon* and *Mauritia*).

Root architecture

The preceding description refers to only the aerial portion of the total root system, which has received much emphasis because it forms such a conspicuous element of mangrove forests. If one examines in toto the root system of the more specialized mangroves, three basic structural components can be recognized, even though these components may have a different morphological origin in different species (Figs. 5.8, 5.9). Each component has a different function. It is the existence of these three components that is the most distinctive feature of mangroves, rather than just the aerial portion. The functional correlations between the various parts of the root system need emphasis.

Aerating component

The erect part of the root system, which in all specialized mangroves projects above the root system, relates to gas exchange. As we have seen, this part is produced differently in different root types: the columns of the looping root system of *Rhizophora*; the pneumatophores of *Avicennia, Laguncularia*, and *Son-*

Figure 5.9. Simplified representation of two major contrasted types of root system in mangroves (upper, *Rhizophora* type; lower, *Avicennia* type) to show common structural features. Key to symbols: Cross hatched area, trunk; black area, cable root; hatched area, pneumatophore; white area, anchoring roots; fine lines, absorbing roots; stippled and arrows, growing regions; dashed line, substrate level. (After Gill and Tomlinson 1977)

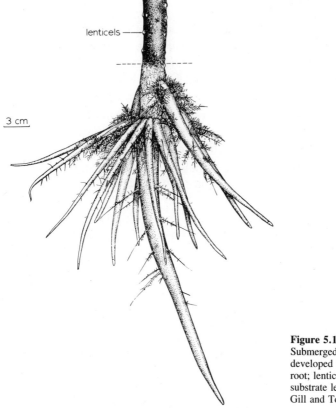

lenticels

3 cm

Figure 5.10. *Rhizophora mangle.* Submerged portion of root system developed from anchored aerial root; lenticels on "column" above substrate level (dotted line). (From Gill and Tomlinson 1977)

neratia; the root knees of *Bruguiera, Ceriops*, and *Xylocarpus mekogensis*; and the plank extension of the roots of *Xylocarpus granatum*.

Absorbing/anchoring component

This component has a dual function: It forms the basis of anchorage for the whole system and is the site of absorption. In most mangroves this is the result of the development of appendages that grow downward from the horizontal system. They are direct first-order laterals in *Avicennia* and *Sonneratia*, for example, but second-order laterals from the region of the root knee in *Bruguiera* and *Ceriops*. In *Rhizophora* each unit is the direct extension of one loop, which penetrates the substrate (Fig. 5.10).

The absorbing and anchoring components can be morphologically segregated because the more massive downward-directed root appendages provide the main

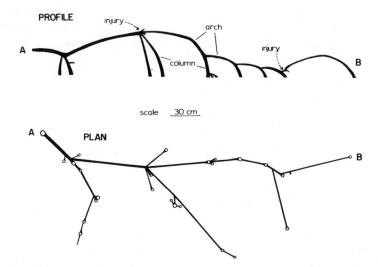

Figure 5.11. *Rhizophora mangle.* Sector of the aerial root system in profile and plan, growing from the trunk (A) to the most recently attached but still unbranched segment (B). Open circles, insertion of column into substrate. Branching of cable roots may be due to injury or subsequent to rooting. The branch insertion distinguishes between two regions referred to as "column" and "arch." (From Gill and Tomlinson 1977)

anchorage; the finer ultimate orders of the branching system are the immediate sites of absorption, as with the capillary rootlets of *Rhizophora*.

Cable component

This is the extended horizontal part of the root system that unifies the aerating and absorbing/anchoring part of the root system. In *Rhizophora* this system is above ground and is a direct result of extension of the primary aerial roots (Fig. 5.11). In other mangroves the cable system is underground.

It should be emphasized that the absorbing or anchoring and the cable components exist in root systems of all plants; the aerating component is the frequent additional component in swamp or mangrove forest species.

Integrated functions of roots

Disregarding the finer details of root architecture, which have diverse morphological origins, recognition of the three basic components allows an overview of the root architecture in functional terms (Figs. 5.8, 5.9).

The sites of absorption are the ultimate branches of the descending roots; their immediate parent roots, the more massive portion of the descending system, have the primary mechanical function of attaching the whole system to the substrate. In mangroves this absorptive system is characteristically shallow, its depth presumably limited by the lack of available oxygen for deep-growing roots. Localized sites of aeration are provided by that portion of the root system that extends into the atmosphere. Commonly there is a close structural association between ascending and descending components (e.g., *Rhizophora, Bruguiera, Sonneratia*) such that the pathway for oxygen diffusion is minimized. The ascending and descending components are linked by the cable system, initially simply a developmental necessity, but finally providing the transport system so that water and nutrients can move from distal sites of absorption to the trunk (in the xylem), while assimilates needed for metabolic processes are moved from the trunk to the distal parts of the root system (in the phloem). Viewed in this way the root system of mangroves can be appreciated as an integrated whole. The special requirements for existence in an anaerobic environment are met. The further specialization for absorbing fresh water from sea water by the exclusion of salt is a micromorphological mechanism at or close to the absorbing surface that does not necessitate major morphological modification.

Gas exchange in roots

General theory

The general concept that the above-ground component of the root system in mangroves ventilates the buried portion for respiratory purposes continues to be supported by experimental methods. That the composite root system is permeable to the mass flow of gases, with atmospheric exchange occurring through lenticels in the aerial vents or "chimneys," is evident from the large internal gas space system that develops largely by the enlargement of intercellular spaces supplemented by cell breakdown. The gas space can be measured in a cut length of root by weighing it before and after artificial waterlogging under a partial vacuum. The weight difference can be converted to a volume difference, representing the volume of displaced gas, provided there is no significant shrinkage or swelling of tissue. Values of about 40 percent of total root volume have been measured (Gill and Tomlinson 1971a) and may be fairly representative. Continuity through the gas space system can readily be demonstrated by blowing air through the root. Resistance can be accounted for, in part, by known anatomical constrictions, as at the level of attachment of branch roots to the main axis in *Avicennia* (Chapman 1940).

Morphological analysis has demonstrated a uniform structural system of vents, absorbing roots, and connecting roots, even though there are several contrasted developmental procedures for achieving this structural set. The spacing of vents can be highly deterministic, as in *Avicennia*, where pneumatophores are spaced regularly and are 15 to 30 cm apart (Fig. 5.4). The total extent of these vents is enormous. Scholander et al. (1955) calculated that a single tree of *Avicennia* only 2 to 3 m tall had well over 10,000 pneumatophores. As they point out, such an extensive structural specialization requires some functional explanation.

Because diffusion is inadequate to account for the extent of gas exchange of the requisite volumes and over the distances involved, some mechanism for mass flow has been sought. Westermaeir (1900) suggested that alternate compression and relaxation of roots, caused by tidal fluctuation, could work a simple pumping or exhalation and inhalation mechanism. However, Scholander et al. (1955) considered the pneumatophores insufficiently compressible. Furthermore, measurements showed that the gas pressure in the roots was, for the most part, *reduced* at high tide, rather than increased as would be required by Westermaeir's compression hypothesis. Scholander and his colleagues suggested, and to a large extent substantiated, an alternate theory dependent on changes in the partial pressure of different gases in the internal gas space system during the tidal cycle. These authors showed that the reduced pressure occurred at high tides when the pneumatophores were completely covered with water. Reduced pressure is the result of the respiratory consumption of oxygen when the lenticels are closed and the relatively high solubility in water of released carbon dioxide. The maximum suction (negative pressure) generated was equivalent to about a 55-cm water column. The vents are not flooded by the hydrostatic pressure of the water above them because the suction is insufficient to overcome the surface tension of the intercellular spaces at the lenticel surface. The suction developed is proportional to the length of time the pneumatophores remain submerged because respiration is continuous. When the pneumatophores eventually become uncovered as the tide falls, the pressure imbalance between the external and internal atmosphere is restored as air is drawn into the root. The proposed mechanism operates only in aerial roots that are regularly inundated. In the higher reaches of the mangal where the tidal influence is minimal, the mechanism cannot operate and roots are presumably aerated (or oxygenated) by diffusion. Nevertheless Scholander et al. showed a slight fluctuation in the gas composition of two roots of *Avicennia* above high tide. Aeration of above-tide roots may be less critical if the substrate itself is better aerated. Some mangroves (e.g., *Laguncularia* and possibly *Lumnitzera*) are facultative in their development of pneumatophores; this may be controlled by the extent to which the substrate is anaerobic.

Strong supporting evidence for the tidal suction mechanism comes from the observation that the relative concentrations of carbon dioxide and oxygen change in the manner required ("molar gas reduction"). Oxygen concentrations can change as much as 10 percent during the tidal cycle in pneumatophores that become in-undated. Artificial and prolonged sealing of the lenticels by vaseline or wax reduces the oxygen concentration to very low values; the roots become asphyxiated. This may cause widespread mangrove death after storm tides, where high water levels may be impounded for abnormally long times.

Scholander et al. (1955) emphasize that ventilation of the root system through the lenticels of the pneumatophores is essentially a continuous process and is mod-ified only by the tide in the significant ways they measured. Simple diffusion remains an important component of the process, as is suggested by the observation that the anchoring and absorbing roots are relatively shallow and never penetrate the sub-strate much beyond a meter.

The mechanism of root ventilation cannot be discussed in terms of an isolated root segment; the root system as a whole is highly integrated. Further experiments by Scholander and his associates show that a single pneumatophore can serve as a port of gas entry for an extended portion of the system from which other pneu-matophores have been removed. The unrelated observations of Dacey (1980, 1981) of what were termed ''internal winds'' in the yellow water lily (*Nuphar luteum*) are noteworthy. An internal cycling or mass flow of gas was demonstrated from young, newly emergent leaves down into the rhizome and back to the older leaves via the long petioles. The mechanism depends on the development of a positive pressure in young leaves as they are heated by the sun, and the relative imperme-ability of the leaf surface to gas exchange. The pressure generated is vented via the internal gas space through older leaves, which have much lower resistance to gas exchange. Apart from eliminating respiratory carbon dioxide, the mechanism circulates methane (CH_4) absorbed by the roots from the sediment. Since the internal wind is a simple consequence of the physical properties of the system, though of benefit to the plant, it could operate in mangrove roots if they developed some differential permeability. However, emergent roots do not present a large surface area to the sun. Nevertheless, somewhat similar conditions could develop where submerged and emergent pneumatophores are attached to different ends of the same cable root.

Technical methods

Because the apparatus used by Scholander et al. is simple and easily reproduced, diagrams are included here. Gas is sampled from the substrate portion of the root with a hypodermic needle, and a distant plunger is inserted into a root that is first excavated and then reburied (Fig. 5.12). Samples can be drawn off at regular intervals. For measuring changes in gas pressure, a simple water manometer is attached tightly to the stump of a pneumatophore whose woody cone is greased

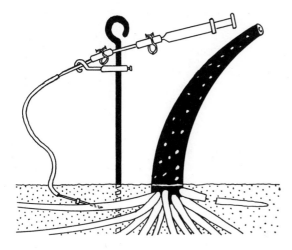

Figure 5.12. Apparatus used by Scholander et al. (1955) to extract gas samples from intact roots of mangroves.

to make it airtight, that is, so that gas exchange occurs only through the aerenchymatous cortex (Fig. 5.13 left). For measurements of volume changes, the apparatus shown on the right in Figure 5.13 is attached to an entire pneumatophore in such a way that the collar is airtight but the enclosed lenticels are not obstructed. The horizontal calibrated tube has a kerosene indicator drop.

Alternative theories

The general conclusion that aerial roots have a demonstrable ventilatory function seems well founded but has not always been supported. Troll and Dragendorff (1931) present views that differ with this standard interpretation. Working primarily with *Sonneratia*, they made extensive observations on the respiratory mechanism and concluded that the major function of the root system is to adjust the level of the absorbing and anchoring system to changes in the substrate level, especially those due to progressive sedimentation. The method of development of the root framework undoubtedly permits the necessary degree of plasticity, and instances of adjustment to depth are easily seen. However, the root construction seems overelaborate for the adjustments necessary. Most mangrove root systems are not subjected to regular or extensive changes in substrate levels. Similar pneumatophores occur in fresh-water swamp-forest species that are rarely subjected to sedimentation. Terrestrial trees adjust to changes in substrate levels quite frequently, but without developing specialized tiering mechanisms. It may be simplest to recognize that structures that develop primarily in relation to root aeration are also beneficial in allowing depth adjustment.

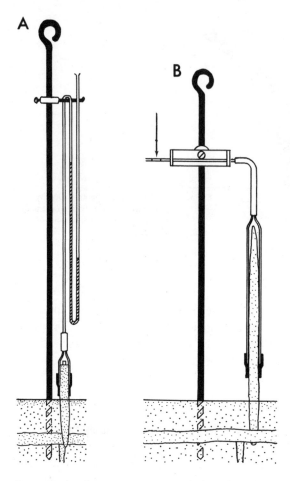

Figure 5.13. Apparatus used by Scholander et al. (1955) to measure gas pressures in pneumatophores of *Avicennia*: (A) manometer attached tightly to cut pneumatophore records pressure changes compared to atmosphere; (B) sleeve fitted tightly around intact pneumatophore measures changes in rate of gas exchange via recording bubble (arrow).

Root anatomy

The anatomy of mangrove roots has been extensively investigated in relation to development and function. Summaries are provided by Chapman (1976) and Troll (1943); individual articles of particular relevance are by Baylis (1950), Brenner (1902), Chapman (1940, 1947), Goebel (1886), Liebau (1914), and Schenk (1889). Most information is available for the above-ground portions of root systems, since these are relatively accessible.

In all roots a basic pattern of rootlike characters can be recognized: root cap, endogenous origin of lateral roots, exarch protoxylem, and alternate primary phloem and xylem strands. A common modification is an enlarged, polyarch stele with a wide parenchymatous medulla (Figs. 5.14, 5.15). Sometimes some normal root features are obscured and have been misinterpreted.

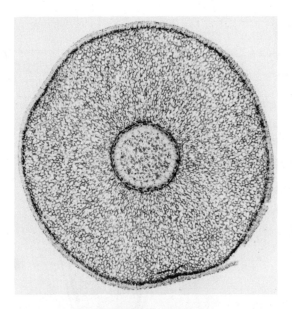

Figure 5.14. *Rhizophora mangle.* T.S. subterranean root to show broad lacunose cortex, narrow medullated stele ($\times$ 12). No trichosclereids are developed.

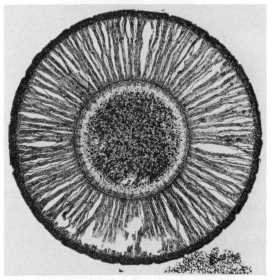

Figure 5.15. *Bruguiera gymnorrhiza.* T.S. underground anchoring root to show well-developed aerenchyma in cortex and wide, medullated stele ($\times$ 10).

In *Rhizophora* superficial observation suggested that the aerial root had bicollateral primary xylem and phloem strands and endarch protoxylem, leading to the suggestion that these organs combined features of stem and root or were even novel organs, comparable to the "rhizophore" of *Selaginella* (Pitot 1958). Developmental study shows that the aerial root of *Rhizophora* is entirely normal in the early stages of vascular differentiation, but protoxylem appears mesarch and even endarch as

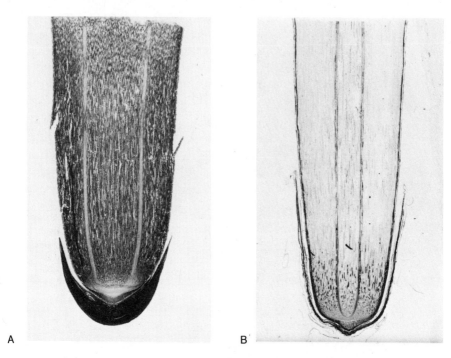

Figure 5.16. *Rhizophora mangle*. (A) Apex of aerial root in longitudinal section. Note pronounced root cap ($\times$ 8). (B) The same root after it becomes submerged. Note the marked change in shape of the root cap and the absence of tannin and trichosclereids.

development proceeds. This unusual development is correlated with an extended region of elongation behind the apical meristem proper (Gill and Tomlinson 1971a); underground roots have a short zone of apical elongation. The suite of characters developed by the aerial roots of *Rhizophora* is possible because (1) they are not mechanically constrained by the substrate and (2) they are never directly absorbing organs. The initial exarch organization in this sense is almost atavistic. Aerial roots therefore are appropriately modified for life above ground. They have a well-developed periderm, trichosclereids are abundant, and secondary thickening is pronounced (Fig. 5.16A).

When the same roots reach the ground (Fig. 5.16B) they undergo a remarkable transformation that adapts them to a subterranean existence: The extension zone becomes short, primary vascular differentiation is normal, the phellogen is little developed, trichosclereids are absent, and secondary thickening is limited. In addition the ground tissue becomes decidedly spongy (Fig. 5.14).

The contrasted suite of characters developed by the aerial root of *Rhizophora* before and after it becomes buried develops in varying combinations in the roots of other mangroves where there is a more consistent differentiation between roots of different orders. A phellogen-producing cork is present in above-ground root parts. In young pneumatophores of *Sonneratia* it exfoliates in a characteristic way. The cortex is often lacunose; in *Avicennia* the cells of the middle cortex become lobed, creating enlarged intercellular spaces, and they are supported by bands of wall thickenings. Similar thickening bands occur in some Rhizophoreae. Secondary thickening in the roots of mangroves is roughly proportional to the order number, so that cable roots have most and ultimate orders none. There are many exceptions to this generalization. Pneumatophores of *Avicennia* have limited secondary growth; those of *Sonneratia* are made up of largely secondary tissue. Root knees in *Bruguiera, Camptostemon, Ceriops*, and *Xylocarpus granatum* are composed of mostly secondary wood.

Lenticels are important components of the aerial roots of mangroves for aeration. They are especially conspicuous in Rhizophoreae; on the columns of the root system of *Rhizophora* (Fig. 5.10) and the root knees of *Bruguiera* (Fig. 5.6C) and *Ceriops*.

The surface layers of the aerial roots of mangroves often undergo extensive secondary changes with age. These relate to their unusual milieu, since they may undergo alternate submersion and exposure, adjust to internal enlargement, and effect gas exchange when exposed. Apart from the normal cork, extensive secondary cortical tissues are often produced internally by the phellogen.

Less is known about the anatomy of higher-order roots. In *Rhizophora*, there is a progressive reduction in diameter and complexity with each order of submerged roots; the ultimate orders or "capillary rootlet" has a very simple organization. It has a diarch stele and a narrow cortex and the epidermis develops no root hairs (Attims and Cremer 1967). The absence of root hairs seems to characterize all strict mangroves; it is a feature common in all aquatic plants. It is surprising that the ultrastructure of the epidermis and endodermis of capillary rootlets has not been investigated, since these are the most likely sites for the ultrafiltration mechanism that makes life in sea water possible for mangroves. In view of the absence of root hairs, the endodermis is the most likely controlling layer because it would become the effective absorbing layer.

6   Water relations and salt balance

Sap ascent

The water relations of mangroves can be discussed in terms of the water potential (ψ) of the tissues of the plant and of the environment, where in any system, the water potential is the algebraic sum of the pressure potential and the osmotic potential ($\psi = \psi_p + \psi_\pi$). In a plant cell, the water potential is determined by the osmotic potential of the cell sap and the turgor pressure of the cytoplasmic membranes against the cell wall. In effect, the plant rooted in sea water has to generate an additional hydrostatic pressure of about 25 bars (the negative water potential of sea water) as compared with a plant rooted in fresh water. Mangroves also raise water against a hydrostatic gradient to heights of up to 30 m. The process can be understood in fairly straightforward physical terms, although the detailed structural features that provide the physical mechanism are not well understood. By convention, the water potential for pure water is established as zero; sea water, at a concentration of 3 percent sodium chloride (NaCl) has a negative pressure (tension) of about 25. Values expressed in "atmospheres" or "bars" in the older literature can be directly transcribed into SI (Standard International) units (MPa) now used by plant physiologists by the appropriate conversion factors.* Simple integers are used in the examples for easy comparison.

Observations and experiments by Scholander (1968) demonstrate that the root system effectively functions as a partial to almost complete ultrafiltration system and that the water potentials of the leaf cells are always lower than that of sea water. Under these conditions, if leaf tissue was in direct contact with sea water, it would absorb water to an extent determined by the water potential difference. In the living tree, under conditions of no transpiration (stomata closed), that is, no xylem flow of water to the leaves to replace water lost by transpiration, there is an additional hydrostatic pressure in the column of water in the xylem equal to -1 bar per 10 m to be overcome. This hydrostatic pressure occurs, of course, in all trees. Figure 6.1 shows the essential relationships and representative values for a mangrove rooted in sea water.

Under conditions of transpiration (stomata open), work is done to move the

---

* 1 bar = 0.1 MPa; 1 atmosphere = 0.1013 MPa.

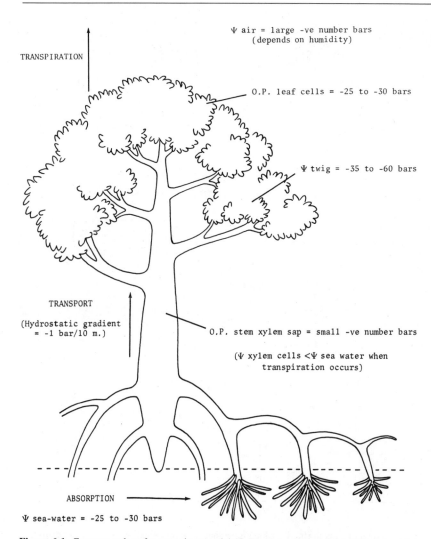

Ψ air = large -ve number bars
(depends on humidity)

TRANSPIRATION

O.P. leaf cells = -25 to -30 bars

Ψ twig = -35 to -60 bars

TRANSPORT

(Hydrostatic gradient
= -1 bar/10 m.)

O.P. stem xylem sap = small -ve number bars

(Ψ xylem cells <Ψ sea water when
transpiration occurs)

ABSORPTION

Ψ sea-water = -25 to -30 bars

Figure 6.1. Common values for osmotic potential (O.P.) and water potential (ψ) involved in water balance of a mangrove rooted in sea water. All values are either calculated or measured. For water to move from sites of absorption to sites of transpiration a negative hydrostatic gradient must be maintained. If ψ sea water exceeds ψ xylem cells, water should flow back into the sea.

continuous water column and requires a pressure gradient. The pressure in the crown is decreased by transpiration. Water molecules move along a steep gradient from the leaf (at a low negative water potential determined by the osmotic potential and turgor pressure of the leaf cells) to the atmosphere (at a very low negative water potential for relatively dry air) via the open stomata. Transpiration is controlled collectively by the stomata. These interpretations and generalizations are supported

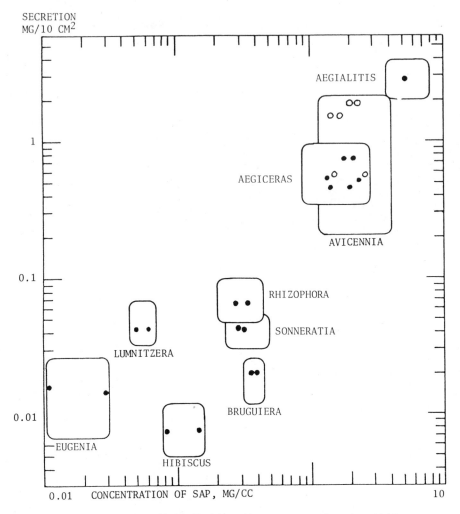

Figure 6.2. Secretion of sodium chloride (dots) in various mangroves and two terrestrial trees over nine daylight hours, as related to salt concentration in the xylem sap (range of values within the rectangles). (After Scholander et al. 1962)

by our knowledge of the morphology and structure of mangroves and the measurements of Scholander and his colleagues.

Structural modifications and transpiration

Water loss is most immediately controlled by stomatal closure. There is some controversy as to whether mangroves have high or low rates of transpiration. High

Table 6.1. *Rate of transpiration from rate of salt secretion*

Species	Rate (mg/dm²/min)
Salt secreters	
Aegiceras	2.5
Aegialitis	5
Avicennia	6.5
Nonsecreters	
Sonneratia	1.5
Rhizophora	2.5
Lumnitzera	6.5
Hibiscus	6.5
Eugenia	7.5

Source: From Scholander et al. (1962).

rates would reflect the high insolation of the mangal environment and the continual availability of water, even though it may have to be absorbed against a considerable negative substrate water potential. Low rates would reflect the general physiognomy of mangrove plants, which suggests that they conserve water as "physiological xerophytes"; that is, they grow in an environment of low water availability entirely because the substrate water potential would tend to extract water from root tissues. Such a designation is artificial; the mangal environment is unique and mangroves have unique physiological mechanisms to adapt to it. The controversy about the rate of transpiration is genuine, since it depends on the true direct measurement of a functional process. Early estimates of low transpiration rates in mangroves (e.g., Faber 1913, 1923) were somewhat contradicted by later workers (e.g., Walter and Steiner 1936). Measurements involving detached shoots in potometers are artificial, however, because hydraulic relations are disturbed. This is not entirely eliminated by alternative methods that measure water loss at the surface of attached leaves by porometry. Modern psychrometric methods measure water potential from the equilibrium value of the relative humidity of air in a closed chamber enclosing a leaf disc.

Scholander et al. (1962) measured transpiration rates indirectly by the ingenious method of comparing the amount of salt excreted at the leaf surface with the concentration of salt in the xylem sap (Fig. 6.2). This gave values (shown in Table 6.1) significantly lower than those that typically occur in terrestrial trees (ranging from 10 to 55 mg/dm²/min.). Scholander suggested that the effect of permanently sequestering salt by transferring it to leaf or other tissue would constitute a larger error for nonsecreters than for secreters, so the figures for the latter group are likely to be on the low side.

Structural modifications for the leaves of mangroves that seem to be related to these presumed low values include a thick cutinized outer epidermal wall, somewhat elaborated stomata, well-developed colorless mesophyll or hypodermal layers, terminal tracheid clusters, and the frequent presence of sclereids. The general succulent or coriaceous texture of mangrove leaves has led to their comparison to the leaves of xerophytes, perhaps unfortunately because it predisposes observers to functional explanations without supporting experimental evidence.

Hydrostatic pressure measurements

As we have seen, even in a steady, nontranspiring state, mangroves must maintain a higher water potential in their xylem elements if water is not to be lost to the substrate through the roots by "back filtration." The development and application of the pressure bomb technique, primarily by Scholander (e.g., Scholander et al, 1965), greatly facilitated measurements of hydrostatic pressure in mangrove xylem. Only indirect methods can be used because xylem water under negative pressure is in a metastable state in the undisturbed tree. In this technique a detached shoot is enclosed within a sealed pressure bomb and a measured positive pressure is progressively applied until leakage from the cut xylem begins. This critical positive pressure is assumed to represent the negative pressure that was in the xylem at the time the shoot was detached, since it restores water to the previous volume state. The method is best used comparatively. Scholander found very low values for mangroves, -30 to -60 bars, but the values always exceeded the negative water potential of normal sea water. Under these circumstances, back filtration would not occur. In progressively more saline conditions (when ψ for sea water becomes increasingly more negative), however, the ability of the mangrove to absorb water decreases. The upper limit seems to be a value of 90 0/00 NaCl, since even the most efficient mangroves are then excluded, as a number of distribution/salinity correlations have shown.

Despite the physical nature of most mechanisms involved in tree growth in tidal waters, a combination of factors is necessary for successful intertidal colonization. The floristic poverty of mangroves and their taxonomic diversity show that the combination has rarely been developed by higher plants, but also independently in many groups. It is misleading to categorize mangroves as either halophytes or xerophytes or to refer to them as being subject to physiological drought (Schimper 1891) because the mangrove environment is characterized above all by its variability: in salinity, substrate texture, and degree of waterlogging. The only constant factor is a fairly uniform high temperature.

Coping with excess salt

Scholander (1968) has discussed the physiological status of plants that grow in mangal in an article titled "How mangroves desalinate water," which describes the necessary process in direct terms. Salt in high concentrations in plant tissues is seemingly toxic (although the physiological reason is not clear) and must be largely excluded. All mangroves exclude most of the salt in sea water. They can be further divided into two groups (Figs. 6.2, 6.3): those that secrete salt (secreters) and those that do not secrete salt (nonsecreters). In *salt secreters*, such as *Aegialitis, Aegiceras,* and *Avicennia*, the NaCl concentration of xylem sap is relatively high, but still about one-tenth of the concentration of salt in sea water. Salt is only partially excluded at the roots. The absorbed salt is primarily excreted metabolically via salt glands, which are described later. The voided salt in solution can crystallize by evaporation, can be blown away, or is otherwise washed off. In *nonsecreters*, such as *Bruguiera, Lumnitzera, Rhizophora*, and *Sonneratia*, xylem sap has a salt concentration less than one-hundredth of that of sea water, but still almost 10 times more concentrated than that of nonmangrove plants such as *Eugenia* and *Hibiscus*. Salt exclusion is here more efficient, but the small amount absorbed must accumulate and has to be disposed of.

Although nonsecreters have no specialized mechanism for actively secreting salt, they lose some salt through the leaf surface, possibly by cuticular transpiration. Scholander was able to leach a salt solution from the leaves of plants in this category with mineral elements in the same proportion as they occur in the xylem sap. An additional mechanism for the elimination of salt in all mangroves is simply by loss of parts, notably leaves. There is no indication that salt accumulates preferentially in older leaves, however; that is, there is no active transport mechanism toward senescing leaves. This would be expected from simple physical considerations because any localized imbalance of the water potential by unequal osmotic potential would generate pressure gradients.

Scholander demonstrated experimentally that the salt separation process must occur at or near the root surface and is mediated by physical processes alone, since it is not inhibited by poisons or high temperatures like a metabolic process. Most convincing of all was Scholander's ability to reverse the process, that is, to move water out of the root into the substrate by simply applying a high hydrostatic pressure to a cut root. No detailed information about the structure of the ultrafiltration mechanism is known. Presumably it occurs either at the absorbing root surface (epidermis) or at the root endodermis; the latter region might be the most likely because the ultimate absorbing roots (e.g., the capillary rootlets of *Rhizophora*) lack root hairs. This indicates that the absorbing area is reduced in comparison with a terrestrial plant. However, compensation for this loss of absorbing ability may

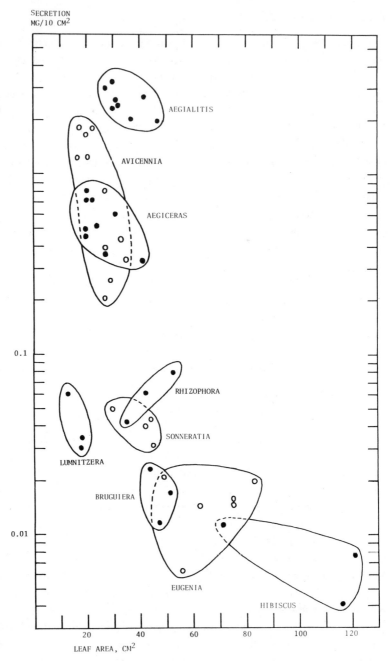

SECRETION
MG/10 CM^2

AEGIALITIS

AVICENNIA

AEGICERAS

0.1

RHIZOPHORA

SONNERATIA

LUMNITZERA

BRUGUIERA

0.01

EUGENIA

HIBISCUS

20   40   60   80   100   120

LEAF AREA, CM^2

Figure 6.3. Secretion of sodium chloride from leaves of several mangroves and two terrestrial species (*Eugenia* and *Hibiscus*) over nine daylight hours. (After Scholander et al. 1962)

be possible if absorbing rootlets live a long time or are produced in especially large numbers. No quantitative estimates are available.

Salt glands and salt secretion

The most distinctive trichome developed in certain mangrove leaves is the structure that secretes solutions of certain ions, mainly Na^+ and Cl^-. These form a general class of secretory structures called "salt glands" by Fahn (1979), which also occurs in other halophytes. The particular kind found in mangroves is referred to as "multicellular glands" by Fahn. Salt glands occur in *Acanthus, Aegiceras, Aegialitis,* and *Avicennia* (Fig. 6.4), all of which are mangroves that control their salt balance by secreting sodium chloride. The salt evaporates and crystallizes in a conspicuous manner (Fig. 6.5). Salt glands are abundant on leaves of these plants but are not necessarily equally frequent on upper and lower surfaces. Other mangroves may have epidermal structures that somewhat resemble salt glands, where a secretory ability has not been precisely demonstrated, like *Laguncularia* and *Conocarpus.*

The salt glands of *Avicennia* have been much investigated; recent accounts by Fahn and Shimony (1977) and Shimony et al. (1973) summarize the relevant literature and present detailed structural, developmental, and physiological information. These authors describe the hairs of *A. marina,* but the information undoubtedly applies to all species of the anatomically uniform genus *Avicennia.* Salt glands in *Avicennia* are scattered in individual shallow pits on the upper leaf surface (Fig. 6.4H) and much more densely within the abaxial indumentum (Fig. 6.4G), where they are not sunken but still obscured by the palisade of three- to four-celled nonglandular hairs. Each gland consists of two to four basal vacuolated cells at the level of the epidermis, a single stalk cell with an almost completely cutinized wall, and a radiating series of at least eight terminal cells that have a thin, minutely perforated cuticle separated apically from the noncutinized cell wall to leave a shallow cavity. The stalk and terminal cells are nonvacuolate, and the dense cytoplasm is rich in ribosomes and has a large nucleus and numerous organelles such as mitochondria, all features that suggest high metabolic activity. Plasmodesmata intimately connect the cells of the gland.

Developmentally, glandular and nonglandular hairs are similar until the three-celled stage, when the appearance of the short middle (stalk) cell distinguishes the future gland. This early similarity led Fahn and Shimony to homologize the two kinds of hair and suggest that the glandular hair is phyletically derived from the nonglandular type. The homology is supported by the thick cuticle of the stalk cells in both types of hair.

The precise mechanism of salt secretion is not understood, but it does require

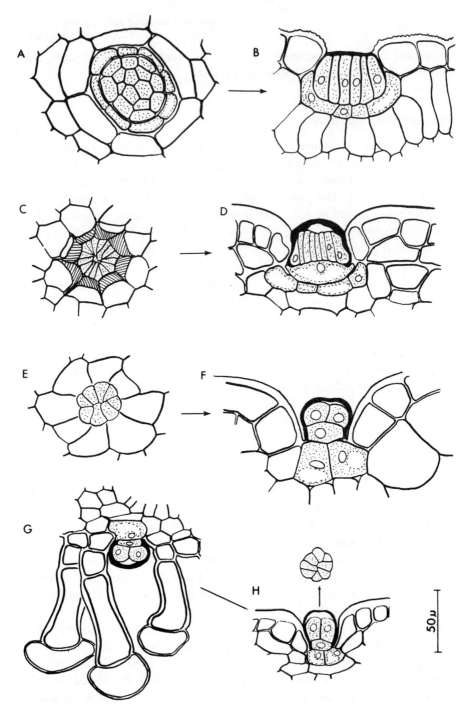

Figure 6.4. Salt glands in mangroves. (A,B) *Aegialitis annulata*; (C,D) *Aegiceras corniculata*; (E,F) *Acanthus ilicifolius*; (G,H) *Avicennia marina*. All are from T.S. adaxial epidermis except G, which is from abaxial epidermis.

Figure 6.5. *Aegiceras corniculatum* (Myrsinaceae) flowers and foliage with salt crystals. (From a color transparency by A. M. Juncosa)

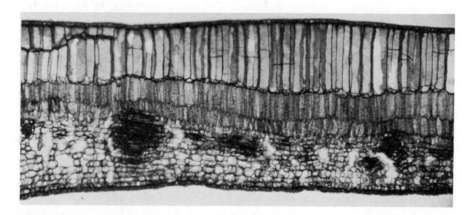

Figure 6.6. *Xylocarpus mekongensis*. T.S. succulent leaf ($\times$ 45) to show single enlarged hypodermal layer.

energy and can be stopped by metabolic inhibitors. Salts move symplastically into and along the hair down a decreasing ionic gradient (Shimony et al. 1973) and eventually into the terminal cells, where they are voided into the subcuticular cavity and hence, via the cuticular pores, to the leaf surface where they visibly accumulate in solution (Scholander 1968). Reabsorption via the apoplast is prevented by the leaf cuticle and especially the cutinized wall of the basal cell.

The structure of salt glands in other salt-secreting mangroves is surprisingly similar in view of the fairly remote systematic affinity of the several families involved; it is good evidence for evolutionary convergence. Salt glands all have basal (collecting) cells, a single cutinized stalk cell, and a capitate group of terminal cells. The ultrastructure, insofar as it is known, seems similar. Comparable experimental evidence to that found for *Avicennia*, supporting a secretory function, is available for *Aegialitis* (Atkinson et al. 1967).

Looking at other organs in mangroves, one discovers that glandular hairs with similar structure may have different secretory functions. Evidence for this is found in mangroves that have glands on the leaves that secrete salt and identical glands in the corolla that are presumably responsible for nectar secretion. The flowers of *Avicennia, Acanthus, Aegialitis*, and *Aegiceras* are all pollinated by insects and offer nectar as the main floral reward. There are very similar epidermal leaf glands in the back mangrove *Dolichandrone*, but these are not known to secrete salt. On the other hand, capitate glandular hairs are common in the Bignoniaceae; the association here therefore seems to be systematic.

In other mangroves salt secretion is less clearly established. In *Laguncularia racemosa*, for example, Biebl and Kinzel (1965) described three types of secretory structures on leaves. The smallest are simply depressions of the leaf surface with a rather irregular group of densely cytoplasmic cells prominent at the base of the pit (See Fig. 4.16C,D). These structures occur in rather infrequent but variable numbers. Salt crystals are extruded from them. In *Pelliciera* glands on the inner margin of the inrolled leaf are secretory, but contrary to the report of Collins et al. (1977), the exudate is not salty.

Salt balance

High salt concentrations are inimical to the growth of most plants (glycophytes) either because they increase the water potential of the soil to levels beyond which the plant can absorb water through the roots, or more usually because absorbed salts are toxic. This is a major problem in irrigated soils where salts may accumulate. Knowledge of the salt-tolerance mechanism is therefore of commercial significance (Greenway and Munns 1980). Halophytes, including mangroves, develop a tolerance to soil salinity because they both maintain a high cellular water potential and are relatively insensitive to salt toxicity. The mechanisms of salt tolerance are complex and variable and involve factors such as ionic potentials across membranes, osmotic relationships, enzyme activation, and protein synthesis. There is no simple explanation for the toxicity of high anion or cation concentrations. An extended discussion of the subject is provided by Waisel (1972) and Flowers et al. (1977).

Photosynthetic rates and stomatal responses to changes in both salinity (short term or long term) and humidity were measured by Ball and Farquhar (1984a,b) in *Avicennia* and *Aegiceras* to quantify the effect of increased salinity in depressing carbon assimilation. *Aegiceras* was shown to be more sensitive to high salinity than *Avicennia*. The short-term salinity experiments show that there is some ''mesophyll resistance'' to small increases in salinity; that is, the species does not respond immediately. An important conclusion in these experiments was that even when leaf metabolism was affected, water loss in relation to carbon gain was always minimized so that in one sense photosynthesis was not less efficient in high salinities even though the carbon gain was reduced. These authors provide a useful discussion of how salinity affects photosynthetic capacity.

It has been suggested that some mangroves have a limited requirement for sodium chloride; they have been described as ''obligate halophytes.'' This question was discussed rather inconclusively by Benecke and Arnold (1931). Flowers et al. (1977) cite evidence that enzymes in vitro are inhibited by salt concentrations only half of that in the leaf cells of halophytes. This is evidence for a nonobligatory relationship between plant and salt, since it shows that enzyme systems neither require salt nor are particularly resistant to it. Stern and Voigt (1959) obtained better growth of *Rhizophora* seedlings in salt solutions of low concentrations compared with those in fresh water. Connor (1969) found that optimum growth of *Avicennia marina* occurred in culture solutions with a sodium concentration about half that of sea water. Higher than normal concentrations of calcium and potassium depressed growth. This was specifically an ion effect and not due to osmotic potential alone. Because of the complexity of the cell mechanisms involving salts, which are not primary plant nutrients, however, the physiological basis for this observation remains obscure. Most mangroves grow very well in fresh water and some penetrate considerable distances inland along river banks where water is permanently fresh and tidal fluctuations are small or absent. It is usually assumed that mangroves are excluded from terrestrial communities by competition with other kinds of plants that do not carry the burden of features that are adaptive only in stressed environments. Mangroves toward their ecological and geographical limits become stunted, for example in the transition from salt to fresh-water marsh communities in the Florida Everglades. We do not know the reason for this stunting. More obviously the decline in the stature of mangroves along gradients of increasing salinity (e.g., Soto and Jiménez 1982) shows the increasing stress of high salt concentrations.

The simplest conclusion is that mangroves tolerate salt to a limited degree. The usual figure quoted is 90 0/00 of NaCl on the basis of the maximum salinity of soils in which mangroves will grow. Soto and Jiménez (1982) suggest values up to 155 0/00 with annual averages of 100 0/00. Evidence for adaption to a saline environment comes from the progressive development of leaf succulence in most mangroves. Nevertheless, mangroves have a fairly low limit to their salt tolerance compared with strict, herbaceous halophytes, which are usually much more succulent but never very tall. A landward transition to salt marsh is common next to mangal. Where salt persists at very high concentrations, no plants grow and a "salt desert" results (Fig. 1.3). The interrelation between mangal and salt marsh in the tropics is discussed extensively in Chapman (1977a).

It seems appropriate to mention at this point the experiments by Walsh et al. (1979) on the toxicity to heavy metal ions of *Rhizophora mangle* seedlings. At the concentrations applied (up to 500 μg/g soil, 250 μg for lead) seedlings were more tolerant of lead and cadmium than of mercury. The authors suggest several mechanisms – formation of insoluble sulfides at the root surface, ion exclusion at the root surface, and internal detoxification – but favor the first.

Leaf succulence

A fleshy texture and a high water content usually develop in mangrove leaves with an increase in leaf thickness with age. In *Laguncularia racemosa*, for example, Biebl and Kinzel (1965) measured a fourfold increase in leaf thickness from the youngest to oldest leaves along a shoot. Such an increase is probably representative for many species. Anatomical changes resulting in increased succulence always involve an enlargement of existing colorless cells, not cell division. Cell enlargement may be localized or generalized; Walter and Steiner (1936) provide illustrations of the changes. In dorsiventral leaves, like *Rhizophora*, the hypodermis alone expands; in isolateral leaves the central mesophyll layers enlarge, as in *Laguncularia* and *Sonneratia*, or the whole leaf tissue becomes succulent, as in *Pemphis*. In *Xylocarpus* a single palisade layer expands considerably (Fig. 6.6). Biebl and Kinzel (1965) measured a decrease in the number of stomata per unit area with increasing succulence, presumably because of the increased surface area of the epidermis.

Leaf succulence in mangroves seemingly has a simple explanation in terms of salt balance. The osmotic potential of the leaf cells of mangroves is high, as measured in the epidermis by plasmolytic methods (Walter and Steiner 1936) or as measured cryoscopically in expressed sap (Scholander et al. 1964). A high osmotic potential is essential if mangroves are to draw water from the sea with its high negative water potential, as has been mentioned. However, these and several other authors note that the salt concentration of mangrove leaves remains constant and independent of leaf age. We also have mentioned that measurements of the salt content of xylem sap demonstrate incomplete salt exclusion at the roots. Therefore, mangroves accumulate salt. Accumulation is in part compensated by salt secretion via salt glands in the less efficient salt excluders. Otherwise the salt is sequestered and can be voided only when the leaves eventually fall. Since salt concentration is constant and independent of leaf age, salt must accumulate by an increase in the volume of the leaf cells, inducing succulence. This has also been mentioned as understandable in terms of water relations because different salt concentrations would complicate the balance of water potential in adjacent leaves.

An exception to this generalization is found in developing seedlings in viviparous mangroves in which a salt concentration lower than that in leaves has been measured (see Kipp-Goller 1940, Pannier 1962). This has led to the development of theories that explain the viviparous condition in terms of differential salt tolerance. Nevertheless, the permeability barrier that allows this differential salinity has not yet been detected.

Leaf succulence in mangroves, in these simple terms, may therefore be accepted as a consequence of life in an environment that provides ample water at the expense of some compensation for that water's high salinity. Further evidence comes from the observations of Camilleri and Ribi (1983) that leaf thickness (presumably also succulence) in *Rhizophora mangle* is correlated with soil salinity, since leaves were thicker in sites of constant high salinity. These authors deny that succulence (which they term "formation of water storage tissue") is a simple function of leaf age, but they also fail to cite Biebl and Kinzel's results.

These simplistic conclusions avoid the developmental question of how succulence is induced in conditions of high salinity, that is, how cell volume is increased. The answer is remote from our present level of understanding of the control of developmental mechanisms in plants.

Selective ion absorption

Of the mineral cations essential to the nutrition of plants, potassium is required in the largest amounts. The salt exclusion mechanism of mangroves therefore must be selective, it must have a sufficient discriminatory capacity to absorb ions in appropriate concentrations from external solutions of lower potassium concentration, and there must be a preferential selection of potassium in competition with high concentrations of sodium. Rains and Epstein (1967), studying the leaf tissue of *Avicennia*, provide evidence for two mechanisms; one operating at low and the other at high potassium concentrations but resistant to interference by sodium. The net result is a preferential absorption of potassium, and it is suggested that the mechanism occurs in roots as well as leaves.

Resistance to high salt concentrations is not complete, however, because Ball and Farquhar (1984a) have measured decreased potassium concentrations at high salinities in *Avicennia* and *Aegiceras*. This is significant to their studies of photosynthetic responses to high salinities, since potassium ions are involved in the stomatal mechanism and indirectly the potassium concentration could affect gas exchange.

7 Flowering

Floral biology and breeding mechanisms

As a consequence of vivipary and other seedling peculiarities, interest in the reproductive biology of mangroves has been centered on seed biology, which will be discussed later. Less is known about floral biology, pollination, and breeding mechanisms, although a knowledge of the effectiveness of floral mechanics and genetic isolating mechanisms is an important prerequisite to the study of successful dispersal and establishment. Limited studies have been devoted to special floral and pollination mechanisms; there are some records of flower visitors and potential pollinators but most are anecdotal. For this reason I use the words "visitor" and "pollinator" interchangeably. Little experimental work has been done and there is virtually no information about sterility barriers and types of incompatability mechanisms. This account is therefore a generalized overview based on information presented in greater detail in the separate descriptions.

Breeding mechanisms

The simple analysis of sexual differentiation and floral mechanisms in Table 7.1 shows a trend toward outbreeding mechanisms based on the analysis by Primack and Tomlinson (1980). These authors contrast mangroves with trees in terrestrial lowland forests and relate differences to the pioneering propensities of mangrove taxa. In the strict mangroves dioecy is rare (4 percent) and monoecy uncommon (11 percent) and most taxa are hermaphroditic. In dioecious organisms outbreeding is obligate. It may be promoted in monoecious organisms especially where sexual segregation is reinforced by lack of synchrony in the maturation of male and female flowers. This occurs in *Nypa* where the inflorescence is strikingly protogynous. Each vegetative shoot supports only one functional inflorescence at a time, with the terminal female flowers becoming exposed and receptive well before pollen is released on lateral spikes by the numerous male flowers. *Nypa* is an example of extreme dimorphism between male and female flowers. On the other hand, in *Xylocarpus* the difference between male and female flowers is slight and the functional difference between them only recently appreciated; the two types of flower are almost identical. Female flowers produce sterile pollen, however, and the ovules

131

Table 7.1. *Breeding mechanisms and distribution of sexual types in common mangroves*

Genus	Perfect	Monoecious	Dioecious	Outbreeding mechanisms
Rhizophora	+			Possibly weak protandry, self-compatible (?)
Bruguiera	+			Pollination may favor outcrossing
Ceriops	+			Pollination may favor outcrossing
Kandelia	+			
Avicennia	+			Protandrous
Sonneratia	+			
Lumnitzera	+			
Osbornia	+			
Pemphis	+			Heterostyly plus self-incompatibility
Aegiceras	+			
Aegialitis	+			
Scyphiphora	+			Protandrous
Pelliciera	+			
Camptostemon	+			
Nypa		+		Protogynous
Heritiera		+		
Xylocarpus		+		
Laguncularia[a]			+	Outbreeding obligate
Excoecaria[a]			+	Outbreeding obligate

Generic summary: Hermaphrodite 74%, monoecious 16%, dioecious 10%.
Specific summary: Hermaphrodite 85%, monoecious 11%, dioecious 4%.

[a] Probably not consistently dioecious.

in male flowers are nonfunctional. This morphological similarity presumably facilitates pollen transfer, since visitors (mainly bees) make no discrimination between the two sexes; the reward in both flower types is nectar.

Among mangroves with perfect flowers there is clear evidence for an outbreeding mechanism in *Pemphis*, which has floral dimorphism (heterostyly) associated with self-incompatibility. Otherwise there is evidence for varying degrees of dichogamy, usually protandry, as in *Avicennia* and *Scyphiphora*, whereas weak protandry is recorded for the mangrove Rhizophoraceae.

In many of these species there is limited evidence for self-compatibility based either on a few controlled pollinations or on observation of the fruiting of isolated individuals or of heavy fruit set in natural populations. Theoretical arguments would support the idea that mangroves, if they are primarily colonizing species, would retain the need for self-fertility if they are to establish populations in isolated localities, as argued by Primack and Tomlinson (1980). However, more direct confirmation of this hypothesis is needed; for example, abundant fruit set could be evidence of efficient pollination as well as of self-compatibility.

Pollination biology

This subject refers to the actual mechanism of pollen transfer from one flower to another. Mangroves are almost exclusively pollinated by animals, and the classes of flower visitors are remarkably diverse. Mangroves must interact with each other or even compete with each other for pollinators, and the evidence, particularly for the mangrove Rhizophoraceae, which have been studied in comparative detail, suggests that the interaction minimizes the competition for pollinators so that the "pollinator resource" (the total available population of pollinating agents) is used in the most efficient manner possible. First we must deal with one conspicuous exception to our generalization; *Rhizophora*, which seems to be mainly pollinated by wind. Subsequently, pollination is best described in terms of flower visitors, that is, presumed pollinators.

Wind pollination

The chief evidence for wind pollination in *Rhizophora* is a knowledge of the floral mechanism in comparison with that of related genera in the tribe Rhizophoreae and prolific pollen production. Evidence is presented more completely in the detailed description of the group. Here it is only summarized.

1. Method of pollen release: dispersed by petal hairs after stamens dehisce in bud (most species) or by a "pepper pot" mechanism (*R. apiculata*)
2. High pollen/ovule ratio* as much as an order of magnitude higher than related genera
3. Light powdery pollen (but also characteristic of other Rhizophoreae!)
4. Absence of an attractive odor
5. Absence of abundant pollinator reward (other than pollen)
6. Inverted flowers
7. Ephemeral pollen presentation time (stamens and petals fall within 12 hours)
8. Screened flowers set fruit

Arguments against *Rhizophora* being pollinated by wind are the absence of an elaborated stigma suited to catching wind-borne pollen and the frequent records of flower visitors (bees) that may visit the flower for pollen. Also, there is some

* P/O ratio, the ratio of pollen produced per ovule, is characteristically high in wind-pollinated plants and an indication of a low efficiency of pollination. *Nypa*, which is pollinated by insects, probably has a very high pollen/ovule ratio, suggesting inefficient pollination but facilitating outcrossing in an organism that spreads clonally by rhizomes.

indication that flowers produce an exudate (nectar?) after the floral organs have been lost. Chai (1982) has produced evidence that *Rhizophora* sets fruit when animal visitors are denied access to flowers enclosed by a fine mesh bag that does not exclude wind-borne pollen. *Rhizophora* appears to be self-compatible, however, and there is therefore the possibility of self-pollination. Again, stigmas may not become receptive until after pollen is released (weak protandry). Much more experimental work is needed.

Animal pollinators: nocturnal

Bats. *Sonneratia* is a well-known example of a flower visited by bats. The flowers open in the early evening, when the calyx expands to expose the attractive mass of extended stamens with powdery pollen. Flowers produce nectar in some quantity from a basal disc. The stamens mostly fall from the flower by the following morning, so that the flowers are ephemeral. In west Malaysia bats are known to fly as far as 50 km from inland roosting sites to feed on *Sonneratia*. Bats are not the only flower visitors, however; there are also records of hawk moths, another important group of mangrove pollinators.

Hawk moths. (Sphingidae). Primack et al. (1981) have recorded hawk moths visiting *Sonneratia* in Queensland. These animals are important alternative pollinators for the genus because they may occur in areas where there are few or no bats. That *Sonneratia* is pollinated exclusively by bats is implied in the literature, but for many mangroves flexibility in the pollinating agent is very important in view of their wide distribution. *Dolichandrone spathacea* is also probably pollinated by sphingid moths. Its flowers open at night and are ephemeral. The floral tube is very long (up to 20 cm) and the nectar can be reached only by an animal with a long tongue. No visitor records exist, however.

Moths. Small moths seem to be the most likely pollinators of *Ceriops tagal*; its flowers are small and inconspicuous but have white petals, and a sweet scent develops in the early evening. Tomlinson et al. (1979) record moths as flower visitors in Queensland.

Animal pollinators: diurnal

Birds. The large-flowered species of *Bruguiera* (e.g., *B. gymnorrhiza*) are well adapted to bird visitors and there are several independent records of such visitors (Davey 1975, Tomlinson et al. 1979). The flowers are recurved and typically point backwards into the crown of the tree (Fig. 7.1); this facilitates an approach

Figure 7.1. *Bruguiera exaristata* (Rhizophoraceae). The solitary flowers are strongly recurved, a characteristic of the large-flowered species of *Bruguiera* and apparently suited to the needs of nectarivorous birds. Townsville, Queensland.

by a perching bird. Nectar is produced in abundance and held in the deep floral cup. In *B. gymnorrhiza* the calyx is red, a color attractive to birds. The bird probes the base of the flower, exploding the petal pouches, and pollen is thrown onto the head of the animal, which is likely to transfer pollen to the stigma of a subsequently visited flower. Cross-pollination may be favored by weak protandry because the stigma may not become sensitive until two or three days after the flower opens. Pollen is not usually discharged in the absence of a flower visitor. Birds as a class are flower visitors to *Bruguiera*, since there are records of sun birds in some parts of its range (e.g., East Africa and Sarawak) and of honey eaters in Queensland.

Lumnitzera littorea (see Fig. B.22) is described as bird pollinated in contrast to *L. racemosa*, which is visited by small insects (see Fig. B.23).

Bees. These are the most common flower visitors in the mangroves and have been observed on flowers of *Acanthus, Aegiceras, Avicennia, Excoecaria, Rhizophora, Scyphiphora,* and *Xylocarpus*. In South Florida "mangrove honey" is made, a major constituent of which is said to be the nectar from *Avicennia germinans*. Most bee flowers are rather small, although they are commonly aggregated, with a short corolla tube; the petals are typically white, pink, or lilac; nectar is produced in small quantity; the flowers typically have an attractive scent; and they may last for several days. *Excoecaria* is exceptional but has elaborate glands on its spicate unisexual inflorescence. The kinds and sizes of bees vary considerably. *Nypa* is visited by small trigonid bees. A large bee is required to extract nectar from the flower of *Acanthus*, and the pollination mechanism depends in part on the forceful separation of the flower parts (Primack et al. 1981).

Butterflies. Records of butterflies visiting *Bruguiera parviflora* have been provided by Tomlinson et al. (1979), and the flowers in this and other small-flowered *Bruguiera* species seem well adapted to this kind of visitor. The flowers are directed outward, facilitating visits by an insect approaching the shoot. Crab spiders may predate such visitors.

Generalized small insects. Some mangrove flowers seem to be pollinated by a generalized population of flying insects of suitable size, like *Ceriops decandra* and *Kandelia candel*. Visitors of this kind, apart from bees and butterflies, are wasps and flies. There seem to be no reliable records of mangrove flowers being visited by beetles, which are considered to be "primitive" pollinators and typically associated with large flowers or inflorescences of very generalized type. Such "beetle" flowers are absent from mangroves.

Pollinator resource and interaction

The diversity and generalized nature of the pollinators that are effective in mangroves allow two basic conclusions: First, the spectrum of pollinators is broad so that no plant is highly dependent on one specific pollinator; and second, plants are specialized only to the extent of being associated with a given class of pollinator. These conclusions have a reasonable explanation in the wide geographical range of mangroves, so that plants are not constrained by a dependence on a specific pollinating agent with a limited geographic range. Furthermore, since each mangrove adapts primarily to a generalized type of pollinator, competition for the available pollinator resource is reduced. The primary division is between nocturnal and diurnal flower visitors; secondary divisions involve adaptation to different classes of pollinators or pollinators distinguished by size as well as by behavior. This is most clearly seen in the mangrove Rhizophoraceae, where each genus or group of species within a genus is adapted to a given class, with *Rhizophora* "opting out" of competitive interaction because of its primary dependence on wind as the

Table 7.2. *Pollinators and pollen-discharge mechanisms in Rhizophoreae*

Genus or subgroup	Flower visitor or vector	Pollen discharge
Rhizophora	Wind	Nonexplosive
Bruguiera		
Large-flowered species	Birds	Explosive
Small-flowered species	Butterflies, etc.	Explosive
Ceriops		
C. tagal	Moths	Explosive
C. decandra	Small insects	Nonexplosive
Kandelia	Small insects	Nonexplosive

pollen vector (Table 7.2). A particular feature of this group of related species is that in some of them (*Bruguiera* and *Ceriops tagal*) the mechanism of pollen discharge is explosive and highly specialized, yet is related to visits from three classes of pollinators: birds, butterflies, and moths. *Ceriops decandra* and *Kandelia candel* are much more generalized in their method of pollination and seem to lack a highly elaborated mechanism. The subject is presented in greater detail in the description of the group but is summarized here. A feature common to all mangrove Rhizophoraceae is the precocious dehiscence of the anther so that pollen is released within the unopened flower bud. This feature may say something about the common ancestor of these diverse flower types.

Vector dependence

So far we have dealt with pollination in mangroves primarily from the point of view of the needs of the plant, and we have stressed their partitioning of the pollinator resource. It is equally important, however, to consider the system from the point of view of the pollinators. We can disregard *Rhizophora*, in which pollen transfer seems to be primarily by physical and not biological agents. A feature of the plant–pollinator interdependence in mangroves is that most of the pollinators live outside the community and the mangroves share them, or even compete for them, with plants of adjacent terrestrial communities. One can argue that the pollinators are less dependent on the plants than the plants are on the pollinators. Close scrutiny shows that there may be seasonal interdependence in certain areas that works to the advantage of the mangroves.

Sonneratia visited by *Macroglossus minimus* (Fig. 7.2A) in the Malay Peninsula provides a well-investigated example (Start and Marshall 1976). One of its major pollinators in the Klang area is the bat *Eonycteris spelea* (the cave fruit bat), which

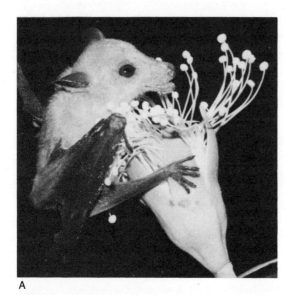

A

Figure 7.2. (A) The bat *Macro-glossus minimus* feeding on the flower of *Sonneratia caseolaris*. (B) The bat *Eonycteris spelaea* feeding on the flowers of the durian, *Durio zibethinus*. (Photos by A. N. Start, from Marshall 1983)

B

eats the nectar and pollen of the flowers. Each flower in *Sonneratia* functions for just one night. The bat roosts inland in large caves such as those in the limestone area known as Batu near Kuala Lumpur. The bats have a range of up to 50 km in their feeding flight, and *Sonneratia* (especially *S. alba*) is one of a number of favored flowers, which also include the commercially important durian (*Durio zibethinus*, Fig. 7.2B). We do not know what proportion of the bats' feeding is on *Sonneratia*, but it is clear that this mangrove species forms a link in the chain that ensures a good durian crop, because the durian has a limited flowering (and hence, fruiting) season and the bats need alternative food resources. In return, *Sonneratia* is indirectly dependent for its successful pollination on the ability of bats to sustain themselves elsewhere during seasons when *Sonneratia* overlaps little with durian, so that again competition is minimized. The bat–mangrove–durian interrelationship represents an interdependence between mangroves and terrestrial plants that may be quite common.

Another example is provided in parts of north Queensland where species of *Bruguiera* that are pollinated by birds (*B. gymnorrhiza* and *B. exaristata*) flower most abundantly in the dry season, at which time terrestrial vegetation may support few flowers that are suited to nectar-eating birds (e.g., honey eaters). Under these circumstances, the mangal may become the main food source for birds that otherwise feed extensively in terrestrial plant communities. *Bruguiera* is less directly dependent than terrestrial plants on rainfall for growth, since it is rooted in tidally inundated areas; it can take maximum advantage of the availability of bird visitors, so a mutual dependence results. If this interpretation is correct, the mechanism of control of flowering is presumably selected for in relation to some climatic trigger, since flowering obviously cannot be stimulated by flower visitors.

Where several species of *Avicennia* grow together, there is evidence of nonsynchrony in flowering times, which might minimize the competition for pollinators (probably bees) and at the same time spread the availability of nectar over a more extended period. In Malaysia, for example, Watson comments (in a letter in the Herbarium of the Singapore Botanic Garden) on a visit to an area where the three common Malayan species (*A. alba, A. marina*, and *A. officinalis*) grew together that each species was at a different stage of the reproductive cycle. Mature flowers in abundance were present in only one species at one time. This may be part of the reproductive isolating mechanism for these species, which have very similar flowers and may well be served by the same class, if not the same species, of pollinator.

Bees and wasps represent a group of pollinators that nest in mangroves, and some populations are therefore more completely dependent on mangal for their existence than is usual in plant–animal interaction in mangroves.

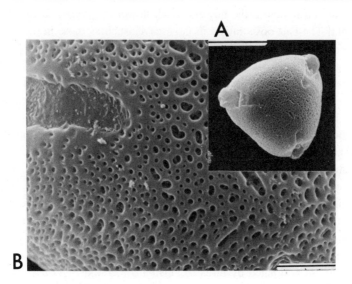

Figure 7.3. *Rhizophora apiculata* (Rhizophoraceae). Pollen morphology via scanning electron microscopy. (A) Single grain ($\times$ 200, scale $=$ 10 μm); (B) detail of sculpturing ($\times$ 10,000, scale $=$ 2 μm). Pollen in the mangrove Rhizophoraceae is quite constant in size and morphology, yet the group includes both wind-pollinated and animal-pollinated species.

Pollen

The study of the pollen grains of mangroves may appear to be a somewhat arbitrarily selected topic, since it deals with pollen from an unrelated group of plants, but it has several important applications. First, fossil pollen is the most extensive and systematically reliable source of information about the ancestry of individual mangroves. Available information has been reviewed by Muller (1981). Because extant mangroves are only in coastal areas, their pollen record can be interpreted as a record of the existence of mangrove vegetation, provided it has not been widely redistributed before fossilization. The dispersibility of mangrove pollen is therefore relevant to the study of fossil mangal. Graham (1977) compares the distribution of *Pelliciera* pollen in Tertiary deposits with the present distribution of the genus to demonstrate that the range has contracted in recent geological time (see Fig. 3.2).

Second, pollen structure is related to the floral mechanism, and its analysis may be an important adjunct to a study of floral biology and may give some indication of the pollen vectors. The large grains of *Scyphiphora* and the sticky grains of *Nypa*, for example, clearly can be transported only by animal agents. The abundant dry, light pollen of *Rhizophora* (together with other evidence) suggests pollination by wind, and the likelihood of wide dispersal in marine sediments has to be considered when fossil *Rhizophora* pollen is examined (Fig. 7.3). On the other hand, a study of the floral biology of *Bruguiera, Ceriops,* and *Kandelia,* which have pollen structurally similar to that of *Rhizophora* but in less quantities, is necessary to show that they are not likely to be pollinated by wind. Nevertheless, their pollen may have considerable dispersibility not associated with the pollination mechanism.

Third, pollen abortion can give evidence of hybridization, an application found most extensively in the work of Muller and Hou Liu (1966) on *Sonneratia* (see also Wright 1977).

Fourth, there is limited application of information about pollen structure in the recognition of the source of honey, which is sometimes derived from the nectar of mangroves (notably *Avicennia*).

Fossil pollen

In his summary of the appearance of angiospermous families in the geological record based on their identifiable pollen, Muller (1981) documents the existence of the genus *Nypa* in the very late Cretaceous (Maestrichtian, 69×10^6 years B.P.). He interprets this as an invasion of the mangrove environment by the palm followed by rapid sea-borne dispersal, rather than the origin of the genus. His evidence seems to be primarily the simultaneous occurrence of *Nypa* pollen in South America, Africa, India, and Borneo. Dicotyledonous mangrove pollen (Rhizophoraceae, Pellicieraceae, Combretaceae) is recorded for the upper Eocene (44×10^6 years B.P.) and Sonneratiaceae (*Sonneratia,* Fig. 7.4) for the Lower Miocene (22.5×10^6 years B.P.). Muller suggests that the present species of *Rhizophora* had already differentiated by the Miocene. This supports the idea that the present range restriction of eastern and western species is a result of the isolation of earlier undifferentiated populations by tectonic movements. The records of *Sonneratia* are of interest because they are the result of a gradual change from types referable to Lythraceae, and they strongly support the evolutionary divergence of Sonneratiaceae from a lythraceous ancestor, which comparative morphology independently suggests. Because the earliest identifiable angiosperm fossil pollens are much older, for example, Chloranthaceae, Lower Cretaceous (114×10^6 years B.P.), these records suggest that the origin of modern mangroves is at a relatively late stage of angiosperm diversification but still early enough to render their geographic redistribution complex and difficult to interpret.

Pollen structure

A detailed discussion of pollen micromorphology in mangroves is not particularly relevant to the scope of this book, even though reasons have been given for its importance. Most variation in mangrove pollens is of systematic significance

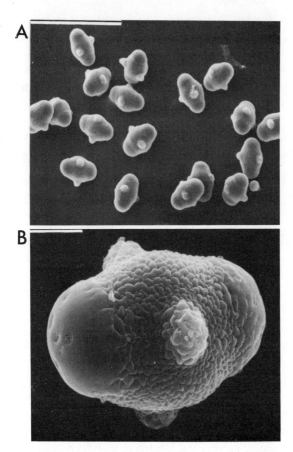

Figure 7.4. *Sonneratia caseolaris* (Sonneratiaceae). Pollen morphology via scanning electron microscopy. (A) Grains at low power (× 350, scale = 100 μm); (B) single grain (× 2000, scale = 10 μm). Grains are large and appropriate for an animal-pollinated species.

and relates to the taxonomic status of each group. Information tends to support the principle established earlier: that the most specialized mangroves are taxonomically isolated. A useful survey of the north Queensland mangrove pollen flora by Wright (1977), available only as an honours thesis submitted to James Cook University in Townsville, is helpful because it deals with the common genera of the Old World mangroves. The following key to genera, adapted from this thesis, gives some indication of the structural range and includes comments taken from its summaries.

Key to major pollen groups represented by eastern mangrove genera

1A. Grains annulocolpate, spinose .... *Nypa* (Fig. 3.3)
    (Pollen of *Nypa* is distinct from that of other palms)
1B. Grains otherwise.. 2
2A. Grains tetracolporate .... *Xylocarpus*
2B. Grains otherwise.. 3

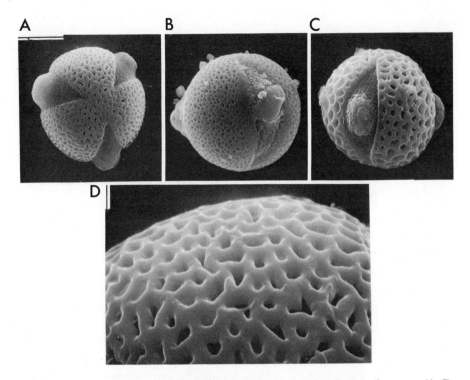

Figure 7.5. *Avicennia* (Avicenniaceae). Pollen morphology via scanning electron microscopy. (A–C) Whole grains (× 2000, scale = 10 μm); (D) detail of wall sculpturing from part A (× 10,000, scale = 1 μm). (A) *A. marina* var *resinifera* (New Zealand); (B) *A. germinans* (Florida); (D) *A. eucalyptifolia* (Queensland). Size and morphology seem quite constant in the genus, but wall sculpturing is variable.

3A. Grains with pseudocolpi................................... *Lumnitzera*
3B. Grains otherwise... 4
4A. Grains with arci over poles .... *Osbornia*
4B. Grains otherwise... 5
5A. Grains colporate, colpi long and narrow .... *Acanthus*
5B. Grains otherwise... 6
6A. Grains triporate................................*Sonneratia* (Fig. 7.4)
6B. Grains otherwise... 7
7A. Grains tricolporate, ornamentation finely reticulate, the reticulum
     becoming progressively less distinct from pole to equator ....mangrove
                                  Rhizophoraceae (Fig. 7.3)
     (Although the genera are similar, they can be distinguished by size and
     details of wall structure)
7B. Grains otherwise... 8
8A. Grains with operculum... 9
8B. Grains otherwise... 10

9A. Grains with circular operculum; colpi broad and long with a granular
membrane. . . . . *Scyphiphora*
(Pollen characters support the separation of this genus in the family
Rubiaceae)

9B. Grains with circular operculum; colpi broad and short, without a
membrane. . . . . *Excoecaria*

10A. Grains with colpi characterized by prominent central wartlike sculptures . .
. . . . *Aegialitis*
(Pollen evidence supports the isolation of this genus in the order
Primulales)

10B. Grains otherwise. 11

11A. Grains tricolporate with distinct annulus . . . . *Aegiceras*
(The grains are not very distinctive among mangroves but are remarkably
uniform)

11B. Grains tricolporate, reticulate; muri broad, flat, thick; lumina small
irregularly shaped; colpi deeply intruding . . . . *Avicennia* (Fig. 7.5)

Other information obtained from these pollen studies is: palynological evidence
to support the hybrid status of *R*. x *lamarckii* (a large proportion of collapsed and
irregular grains); evidence to support the hybrid status of *Lumnitzera rosea* (Tomlinson et al. 1978) (reduced viability of its pollen, structurally intermediate between
that of its two parents); and populational complexity of *Sonneratia* in the north
Queensland region.

Within individual genera such as *Rhizophora* and *Lumnitzera*, it is said to be
possible to distinguish species on the basis of pollen morphology. My own observations suggest that *Avicennia* species can also be distinguished by pollen; this
information would be helpful in clarifying the subgeneric classification of this genus.
Pollen size varies considerably; the largest grains are those of *Aegialitis*, up to 90
μm in the longest dimension. *Sonneratia* grains are also relatively large. Rhizophoraceae, on the other hand, have small grains ranging from 6 to 20 μm in diameter.

In view of the wide range of mangroves and the known variability of pollen
structures in even small samples, a complete description of mangrove pollen will
require a very extensive survey.

8   Seedlings and seeds

Germination and establishment

Establishment of the seedling is critical in the life cycle of all seed plants, but it is rendered difficult for mangroves by the unstable, variable substrates and the tidal influence within mangal (Fig. 8.1). The range of germination types shows a simple correlation with seed size in Table 8.1, where a distinction is made between hypogeal germination (cotyledons not expanded and not exposed), epigeal germination (cotyledons expanded and exposed), and the specialized type of germination known as the ''*Rhizophora* type.'' Propagule length is only a crude measure of size, but it is easily obtained. Propagule weight and length are not closely correlated: *Pelliciera* has seeds about 8 cm long that weigh 60 to 90 g; *Rhizophora mangle* has seedlings 20 to 25 cm long that weigh about 15 g (Rabinowitz 1978a). All propagules float on release, however. The values are nevertheless sufficient to show that seeds with epigeal germination are generally small, and those with hypogeal germination are large. Overlap obscures this generalization, since large seeds with cryptoviviparous seedlings in which the seed expands beyond the testa within the developing fruit have epigeal germination (e.g., *Avicennia*). This information is relevant to the later discussion of vivipary.

Dispersal

All mangroves are dispersed by water, and the propagule (fruit, seed, or seedling) has some initial ability to float, even if for only a limited time. In the most highly specialized types, represented by the mangrove Rhizophoraceae, the unit of dispersal is the viviparously developed seedling (Fig. 8.2). In *Ceriops, Kandelia*, and *Rhizophora* the seedling is abscised from the cotyledons, which remain together with the fruit on the tree. In *Bruguiera* the fruit is abscised with the seedling, so the unit of dispersal is more complex.

In a number of mangroves and mangrove associates the fruit is a capsule or like a capsule, and the individual seeds become the units of dispersal after they are released, as in *Acanthus, Aegialitis, Conocarpus, Dolichandrone, Excoecaria*, and *Xylocarpus* (Table 8.2). In *Acanthus* each capsule usually contains four seeds that are discharged explosively in the manner usual for the family Acanthaceae; the

145

Figure 8.1. *Avicennia alba* (Avicenniaceae). Seedlings at low tide. Ponggol, Singapore (Straits of Johore behind).

seeds are flattened and their discus shape facilitates this initial discharge to a distance of up to 2 m. Seeds have a thin testa and continue their dispersal by floating. The large woody capsules of *Xylocarpus* may split on the parent tree to release the angular seeds, or may shatter when they fall. The initial dispersal of *Sonneratia* is imprecise. The fruit falls green from the tree, but the calyx soon separates so that the numerous small angular seeds are exposed and readily lost, especially by further decay of the fruit wall. Fruits initially may be carried some distance before releasing the seeds, but the agent for final establishment is always the seed.

Flotation

Where the seed or fruit is the dispersal agent, some part of this propagule is modified to facilitate flotation. This is usally a fibrous portion of the fruit wall, typically the mesocarp, as in *Cerbera, Cynometra, Heritiera, Laguncularia, Lumnitzera, Nypa, Pelliciera, Scyphiphora,* and *Terminalia.* In *Xylocarpus* the testa is thick and corky. Special adaptation is particularly notable where the method of dispersal is different between mangrove and terrestrial relatives. In *Heritiera* and *Terminalia*, for example, other members of the genus have winged seeds suited for wind-assisted dispersal. In *Dolichandrone* the corky seed wing differs from the papery wings

Table 8.1. *Germination type and seed size in major and minor mangroves*

Genus	Type of propagule	Vivipary[++] or cryptovivipary[+]	Seed length (cm)
"Rhizophora-type" germination (modified epigeal)[a]			
Rhizophora	Seedling	+ +	70–15
Kandelia	Seedling	+ +	40–30
Bruguiera	Seedling	+ +	30–15
Ceriops	Seedling	+ +	20–15
Aegiceras	One-seeded fruit	+	7–5
Hypogeal or modified hypogeal germination			
Nypa[b]	One-seeded fruit	+	10
Xylocarpus	Seed	–	7–6
Epigeal or modified epigeal germination			
Pelliciera[c]	One-seeded fruit	+	8–7
Aegialitis	Seed	+	4–3
Avicennia[d]	One-seeded fruit and seed	+	3–2
Lumnitzera	One-seeded fruit	–	2–1
Laguncularia	One-seeded fruit	–	2
Scyphiphora	One-seeded fruit	–	2–1
Osbornia	Seed	–	0.5
Excoecaria	Seed	–	0.3
Sonneratia[e]	Seed	–	0.2–0.1

[a] "*Rhizophora* type" of De Vogel (1980); "Durian type" of Ng (1978).
[b] In *Nypa* the plumule protrudes; the radicle remains enclosed (Fig. B.51).
[c] "Semi-hypogeal type" of De Vogel (1980).
[d] "*Sloanea* type" of De Vogel (1980).
[e] "*Macaranga* type" of De Vogel (1980).

suited for wind dispersal, which are characteristic of the seeds of many Bignoniaceae. The genus *Osbornia* is difficult to place in the Myrtaceae because its fruits are neither capsular nor fleshy, whereas the family as a whole can be subdivided on the basis of this character. Where the seedling is the propagule, the cortex of the hypocotyl is aerenchymatous in varying degrees (mangrove Rhizophoraceae, e.g., Fig. 8.7).

Vivipary

The development of the embryo in seed plants is a continual process, normally interrupted by a longer or shorter period of seed dormancy, with tropical species tending to have the briefest dormancy. Thus, there is a potential continuum of

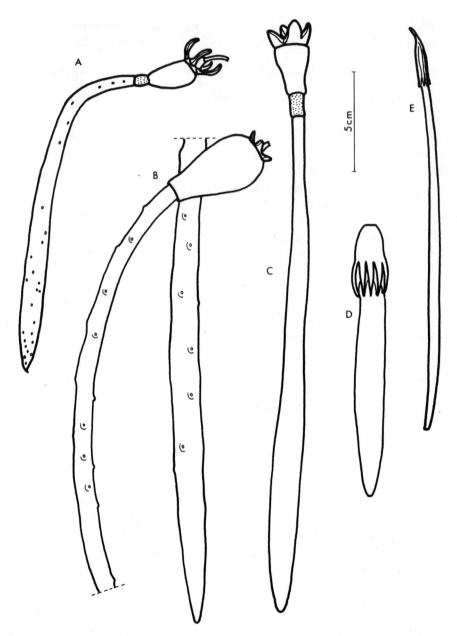

Figure 8.2. Range of size of viviparous seedlings in mangrove Rhizophoraceae. (A) *Rhizophora mangle*; (B) *R. mucronata*; (C) *R. apiculata*; (D) *Bruguiera gymnorrhiza*; (E) *B. parviflora*. In *Bruguiera* the fruit falls with the seedling. In *Rhizophora* the seedling is detached at the cotyledonary collar (stippled in A and C).

Table 8.2. *Germination type and seed size in mangrove associates*

Genus	Type of propagule	Vivipary[++] or cryptovivipary[+]	Seed length (cm)
Hypogeal germination			
Cerbera	One-seeded fruit	−	7–5
Terminalia	One-seeded fruit	−	7–5
Heritiera	One-seeded fruit	−	6–5
Calophyllum	One-seeded fruit	−	4–3
Epigeal germination			
Cynometra	One-seeded fruit	−	3–2
Dolichandrone	Seed	−	1
Acanthus	Seed	−	1–0.5
Conocarpus	Seed	−	0.4–0.2

developmental possibilities. In the extreme condition of "true vivipary" (Fig. 8.3), the embryo that results from normal sexual reproduction has no dormancy but grows first out of the seed coat and then out of the fruit while still attached to the parent plant (Juncosa 1982); the propagating organ (propagule) is thus the seedling and not the seed. Development of the seedling on the parent tree in this way is a feature of several mangroves and is found in the whole tribe Rhizophoreae of the family Rhizophoraceae. Why vivipary should be so common in mangroves is a topic that has fostered plenty of discussion but no generally accepted explanation.

Tan and Rao (1981) describe what they refer to as vivipary in *Opiorrhiza* (Rubiaceae), a plant found in wet places. Seedlings simply germinate within the undetached infructescence and are not specialized. Premature germination is known in some cultivated fruits, as in *Citrus* where it is usually associated with some form of cleavage polyembryony, whereas adventitious vegetative buds can develop on inflorescences in such plants as *Agave* and *Poa alpina* without involving sexual processes. These nonsexual or parasexual processes are often referred to as "false vivipary" and are not uncommon in flowering plants.

Vivipary needs initially to be studied as part of the continuum of normal seed development and not as an isolated phenomenon, since explanations often involve a comparison with terrestrial species. The absence of a dormant phase in the development of the embryo of *Rhizophora mangle* has been documented by Sussex (1975), who uses the example to describe the processes that govern embryo development in seed plants generally. The reviews and classifications by Ng (1978) and De Vogel (1980) of germination processes in tropical woody plants are valuable for comparative information. These authors confirm that rapid germination of the

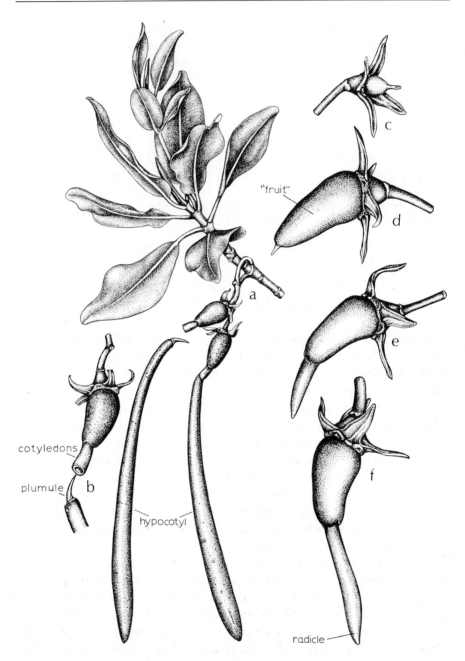

Figure 8.3. *Rhizophora mangle* (Rhizophoraceae) shoot, fruit, and seedling. (a) Shoot with mature seedlings, one just detached ($\times$ 1/3); (b) fruit ($\times$ 1/2), with plumule of seedling detached from cotyledonary collar; (c–f) fruit ($\times$ 1) with stages in its development and protrusion of seedling hypocotyl. (From Tomlinson 1980)

seed (i.e., with limited dormancy) is common in tropical rain forest trees and that the seed life span is usually short (a few weeks). Rapid germination of this type does not, however, normally involve exposure of the seedling before the fruit falls. In a more advanced state found in some mangroves, "cryptovivipary" exists, in which the embryo emerges from the seed coat but not the fruit before it abscises (Carey 1934). Examples are *Aegiceras, Avicennia, Nypa*, and *Pelliciera*. In the palm *Nypa*, the plumule emerges from the fruit at the time this is released and may even assist abscission. In the mangrove Rhizophoraceae, although true vivipary represents the most advanced condition, there are significant differences between the genera so that comparative study is possible. In other marine ecosystems, true vivipary occurs only in the seagrass *Amphibolis*. Vivipary is not known in halophytes.

The association between vivipary and the littoral habitat has occasioned much comment, with many authors attempting to account for the correlation in adaptive terms. Morphological, ecological, and physiological explanations have been developed, but only recently have comparative studies begun to explain the phenomenon in evolutionary terms (Juncosa 1982, 1984a,b). This is a rather surprising omission in view of the diversity of systematic characters revealed by embryological study (e.g., Tobe and Raven 1983). Inland members of Rhizophoraceae need to be studied because they have a range of germination types from epigeal (e.g., *Cassipourea*) to hypogeal (e.g., *Anisophyllea*), either of which could represent the ancestral state, since it is reasonable to assume that the mangrove Rhizophoraceae had a terrestrial ancestor among the same complex that gave rise to the present-day family. Most of the evidence, summarized later, suggests that the Rhizophoreae have modified epigeal germination.

An explanation of vivipary might therefore account for its origin among other types of germination, its high incidence in mangroves (but not in other halophytic habitats), its convergent development in unrelated groups, and its functional or adaptive significance. The structural relation between flower and fruit is shown in Figure 8.4. Figure 8.5 shows details of embryonic development in one species; Figure 8.6 shows growth regions in embryos of different ages.

Morphological features

Rhizophora species

A study of *Rhizophora mangle* by Juncosa (1982b) can illustrate the distinctive features of embryo development and germination in the genus (Fig. 8.5); see also Cook (1907). In *Rhizophora* the cotyledons develop as a cylindrical body, which remains in the fruit at the time the seedling is released. Of the four ovules in each ovary, normally only one becomes an embryo (see Fig. 8.4F–H), but there

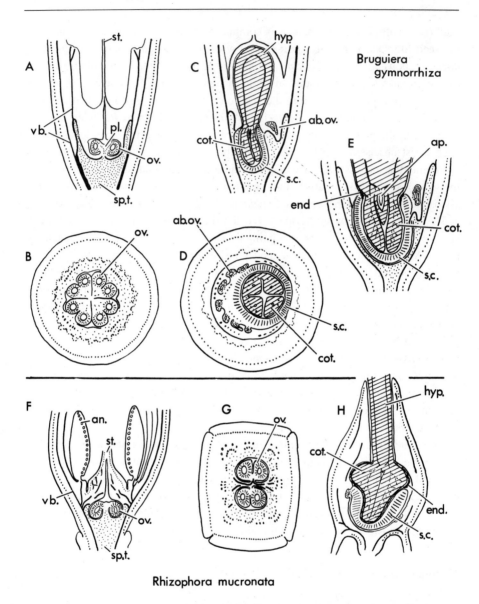

Bruguiera
gymnorrhiza

Rhizophora mucronata

Figure 8.4. Structural features of vivipary in mangrove Rhizophoraceae. (A–E) *Bruguiera gymnorrhiza*: (A) L.S. flower; (B) T.S. ovary at level of placenta; (C) L.S. fruit with developing seedling still enclosed within ovary; (D) T.S. fruit at level of cotyledons of an almost mature seedling; (E) the same with details of seedling apex. (F–H) *Rhizophora mucronata*: (F) L.S. flower; (G) T.S. flower at level of placenta; (H) L.S. fruit with apical part of emerged seedling. Key to symbols and abbreviations: dotted line, layer of secretory cells (laticifers); hatched area, embryo; ab. ov., aborted ovule; an., anther; ap., plumule; cot., cotyledon; end., remains of endosperm; hyp., hypocotyl of seedling; ov., ovule; pl., placenta; s.c., seed coat; sp.t., spongy tissue; st., stigma; vb., vascular bundle. The remains of the endosperm in *Bruguiera* scarcely protrude from the seed, whereas in *Rhizophora* they are extruded well beyond the seed. (Redrawn from Kipp-Goller 1940)

is a low incidence of two- or even three-seeded fruits, where two or three ovules develop. Each ovule has two integuments. Endosperm development is initially free-nuclear and then cellular, but normal for this type in angiosperms. The embryo is initially attached to the integument at the micropylar end by a long multicellular suspensor whose basal cells disintegrate (Fig. 8.5A). This allows the embryo to move deeper into the endosperm. The latter features are considered by Juncosa to be unique among the angiosperms. Transfer cells are conspicuous at the interface between embryo and endosperm. Intrusive growth of the endosperm envelops the embryo, forces the micropyle open, and may carry the embryo out of the integument (Fig. 8.5B–D). This is curious because if ''germination'' refers to the extrusion of the embryo from the seed, it is growth of the endosperm, rather than of the embryo, that initiates the process.

The cotyledons are initiated on the top-shaped embryo as a single torus that only later becomes lobed so that there is at first virtually no indication of a pair of cotyledons. Early growth occurs apically in the cotyledonary body (Fig. 8.6A), but the most active growth is soon restricted to a diffuse intercalary hypocotylary meristem below, but separated from the cotyledon by a region of differentiated cells (Fig. 8.6B). Growth of the hypocotyl thus extends the seedling well beyond the fruit. A final stage is the further elongation of the cotyledons, whose basal abscission zone becomes exposed from the fruits (Figs. 8.3, 8.6C). The radicle apex is enclosed by embryonic tissue and its own root cap; it remains almost wholly inactive and usually plays no part in establishing the seedling. Lateral roots, visible on older seedlings as apical protuberances, arise early in seedling development and delineate the morphological root at the end of the hypocotyl.

Three pairs of leaves and associated stipule pairs are produced while the seedling is still on the tree; the first pair of leaves (after the cotyledons) aborts and persists only as minute vestiges on the detached seedling so that the plumule is actually protected by well-developed stipules.

Other species of *Rhizophora* have not been studied in the same detail, but the seedling size at maturity varies and is somewhat diagnostic. *Rhizophora mangle* has short seedlings (20 to 30 cm long), whereas in *R. mucronata* seedlings 70 cm and longer are common. Juncosa emphasizes that the *Rhizophora* seedling is distinctive in the timing of meristematic activity within the embryo, rather than in any major structural modification. It seems appropriate to view the *Rhizophora* seedling as having modified epigeal germination, since both types are characterized by an elongated hypocotyl.

Histologically the *Rhizophora* seedling is well protected, which seems appropriate in view of its long exposure both on the tree and during dispersal. There is an incipient phellogen toward the radicle end, usually sufficiently developed to mask the superficial chlorophyllous tissue. The aerenchymatous ground tissue is abundantly tanniniferous and there are numerous trichosclereids in the cortex and med-

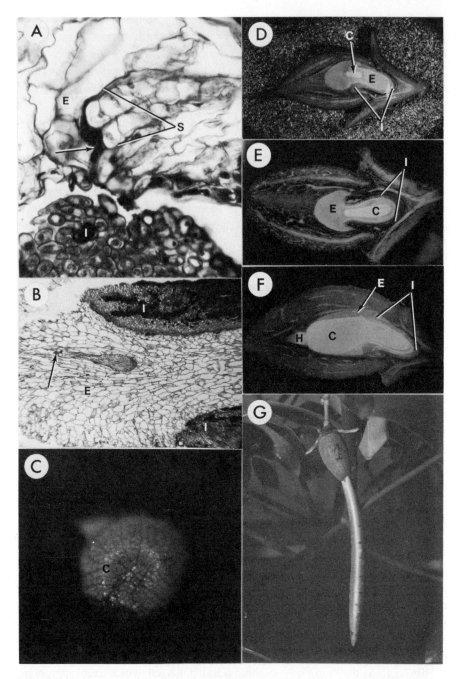

Figure 8.5. Developmental morphology of the embryo and seedling of *Rhizophora mangle*. (A)
Early development of embryo, separating from the integument (I) by degeneration of basal cells of
the suspensor (arrow) but still surrounded by endosperm (E) (× 420). (B) L.S. of ovule showing
"germination" by expansion of endosperm (E) out of the micropyle, between the integuments (I);
the remains of the detached suspensor are represented by the tanniniferous cells shown by the arrow

154

ulla. Clusters of stone cells occur in the outer cortex. A peculiar feature of the epidermis is the absence of stomata so that gas exchange may occur through the numerous lenticels, whose development has been described by Roth (1965). In *R. mucronata* the lenticels are warty and prominent.

Bruguiera exaristata

Juncosa (1984b), in his study of *Bruguiera*, placed it between *Cassipourea*, a terrestrial member of the Rhizophoraceae with epigeal germination (Juncosa 1984a), and *Rhizophora* in terms of the degree of specialization in embryo development. The same growth phases as in *Rhizophora* are recognized, that is, apical growth of the cotyledons followed by hypocotylary extension. The ovule has a very narrow nucellus. The suspensor is wide and short to a degree considered aberrant in the family, but it does not become detached from the integument as in *Rhizophora*. Transfer cells are less well developed than in *Rhizophora*. Two or even three discrete cotyledons are recognizable as the embryo differentiates, and their apical growth lengthens the embryo. The radicle apex is protected by a specialized tissue referred to as the coleorhiza, which begins to develop before the tip emerges from the fruit. The second growth phase, mainly localized in the upper hypocotyl, causes "germination" in the sense that the radicular end of the embryo is extruded from the seed and ultimately the fruit. Juncosa emphasizes that this process corresponds exactly to that in plants with epigeal germination except that the seed is still within the attached fruit. Unlike *Rhizophora* there is no extrusion of the endosperm. The first plumular leaves are poorly developed. A distinctive feature of this species is the development of buds in the axils of the cotyledons. The histology of the seedling is less elaborate than that of *Rhizophora*; only the inner cortex is aerenchymatous and sclereids and trichosclereids are absent (Fig. 8.7).

Bruguiera differs from all other Rhizophoraceae in that the fruit remains attached to the seedling and is dispersed with it (See Fig. 8.2D,E). Less obvious is that the cotyledons of *Bruguiera* early develop flattened, fundamentally laminar structures and only later become solid and fleshy. In *Bruguiera parviflora* the cotyledons remain laminar and provide further evidence for an epigeal ancestry.

Caption to Figure 8.5. (*cont.*)
(× 50). (C) Surface view of the shoot end of an embryo showing the initiation of the toroidal cotyledonary primordium (C) (× 150). (D) Fruit split longitudinally, "germinated" embryo during the first phase of its growth; C, cotyledonary body; E, endosperm; I, integument (× 3). (E) The same at a later stage, with the cotyledonary body (C) more completely replacing the endosperm (E) within the integuments (I); the embryo is less completely emerged than in D (× 3.4). (F) The same at a yet later stage, with initial growth in the cotyledonary body (C) complete; endosperm almost consumed; rapid growth of the hypocotyl (H) beginning; and the radicle penetrating the fruit wall (× 3.3). (G) Attached fruit and germinated seedling; ink marks demonstrate intercalary growth of the hypocotyl (× 0.7). (Plate prepared by A. M. Juncosa, from Juncosa 1982b)

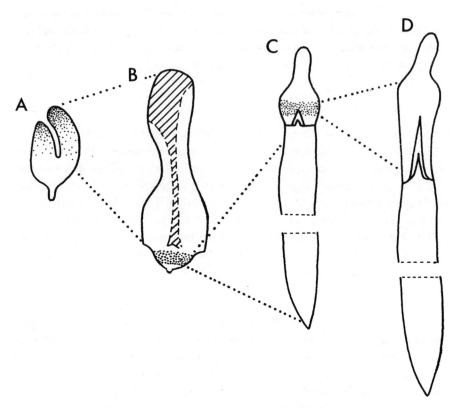

Figure 8.6. Embryo development in *Rhizophora*, schematic; steps are about half an order of magnitude different in size. Stippling represents region of maximum growth (cell division and expansion). (A) Growth in cotyledons; (B) growth in hypocotyl; (C) second growth phase in cotyledons; (D) mature embryo. Hatched area in part B represents the median sinus of cotyledons.

Additional evidence for placing *Bruguiera* between *Cassipourea* and *Rhizophora* is the intermediate complexity of the cotyledonary vasculature, with *Cassipourea* the least and *Rhizophora* the most specialized. In this respect, a feature of the Rhizophoraceae that Juncosa emphasizes as occuring in few other angiosperms is the late differentiation of provascular tissue, after cotyledonary initiation. This might be a preadaptation for vivipary. *Rhizophora* can be interpreted as the most specialized in its vivipary compared with terrestrial relatives because modifications associated with vivipary occur much earlier in development.

These considerations are an indication of the approaches that are needed to place mangrove vivipary in a comparative context. They show that the range of morphological variation in viviparous embryos is considerable so that trends of increasing specialization can be established. Further extensive comparison with extant

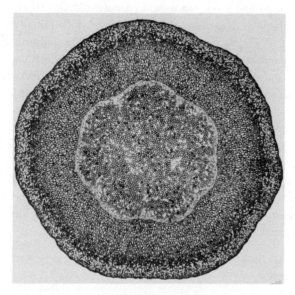

Figure 8.7. *Bruguiera cylindrica.* T.S. seedling hypocotyl with lacunose cortex; wide medullated stele with limited peripheral vascular tissue ($\times$ 10).

terrestrial relatives might well allow the process of evolutionary specialization to be reconstructed in detail.

Comparison with nonviviparous seedlings

Ng (1978) distinguished four main seedling types in his study of Malayan forest trees. The least common (only 8 percent of his total sample) he describes as the "Durian type," in which the hypocotyl is extended, thus raising the cotyledons above the ground. These remain unexpanded within the seed coats, however, and are shed as a separate unit. *Rhizophora* is included in this category because of morphological similarities. We should not assume that the Durian type represents a stage in the evolution of the *Rhizophora* seedling; Ng (1978) emphasizes that the Durian type is not very efficient, since the cotyledons are raised by the hypocotyl to no apparent advantage as they do not become photosynthetic. The common feature here seems to be the elongated hypocotyl. Vivipary in the mangrove Rhizophoraceae therefore most likely evolved directly from an ancestor with epigeal germination, since this is the usual condition in terrestrial Rhizophoraceae.

The more elaborate subdivision of seedling types by De Vogel (1980) not only emphasizes the degree of specialization of mangrove Rhizophoraceae, but also contrasts nonviviparous mangrove and mangrove-associate species. Mangrove Rhizophoraceae are segregated as a separate type (*"Rhizophora* type"), which also includes *Aegiceras*. *Cynometra ramiflora* exemplifies a type with fleshy, food-storing cotyledons that persist at ground level, since the hypocotyl remains short. A peculiarity is the extended plumular axis *above* the cotyledons, which initially produces a series of scale leaves. *Pelliciera* of the New World apparently conforms to this type; semiepigeal and semihypogeal seem equally appropriate designations (Ng 1978). *Barringtonia* is also made a distinct type by De Vogel; either it lacks cotyledons or they are vestigial and the hypocotyl becomes fleshy as a food-storage organ *within* the seed so that only radicle and plumule eventually protrude. The plumule is again elongated and initially bears only scale leaves. This peculiar morphology seems primarily an adaptation that allows the seed to germinate when buried quite deeply. The long plumular axis grows upward until it finds the light. Most other mangroves with hypogeal germination correspond to less specialized types in De Vogel's classification. There is some minor subdivision of epigeal types; mangroves like *Avicennia* (*Sloanea* type) and *Sonneratia* (*Macaranga* type) are contrasted on the basis of thickness of the cotyledons.

Dispersibility and establishment

There is much discussion in botanical literature about the ability of mangroves to disperse and subsequently establish their propagules (e.g., Guppy 1906, 1917; Ridley 1930). Clearly, mangrove propagules are adapted to float, at least initially. What then prevents them from occupying all suitable pantropical sites? Is it lack of a suitable substrate rather than an inability to disperse sufficiently broadly that prevents mangroves from reaching some remote islands? *Pemphis acidula*, for example, seems to be the only mangrove on the more remote atolls of southeastern Polynesia (Brown 1935) because it behaves more like a beach or strand plant. Here lack of suitable growing conditions seems to exclude other (strict mangrove) species. On the other hand, even though the Hawaiian Islands have no native mangrove flora, artificially imported seedlings of *Rhizophora* have flourished and built syn-thetic mangal. Dispersibility seems to be the factor that limits natural establishment on this isolated island group. The question is of phytogeographic as well as eco-logical interest because the present distribution of mangroves (See Chapter 3) sug-gests that ancestrally mangroves may have crossed the Pacific at least once in geological time, but can no longer do so. The marked contrast between the eastern and western groups of mangroves shows that dispersal and subsequent establishment are not random.

A number of investigators have experimentally studied dispersibility and the ability for establishment of mangrove propagules, chiefly by recording the length of time they float in salt and fresh water. Some "mark and capture" experiments have been carried out. Rabinowitz (1978a), for example, established "dispersal" parameters for the common mangroves of Central America. These include longevity (how long the propagule floats while retaining viability), period of floating, period for establishment (permanent rooting), and period of "obligate dispersal" (how long a seedling floats before it can initiate the root system that will allow it to anchor). Clearly, the last is an absolute minimum value for time available for dispersal. She gives values from 35 days to more than a year for longevity, from 8 to 40 days for obligate dispersal, and a flotation time from as little as 1 day for *Pelliciera* in fresh water (but 6 days in salt water) to an "unlimited" period for *Avicennia*. It takes most mangroves 5 to 15 days to develop an anchoring root system. Differences between values for experiments carried out in sea water and fresh water are small and do not really affect performance. Some seedlings regularly regain buoyancy after they have sunk, increasing their dispersibility. These observations are later built into the hypothesis that propagule weight affects ecological distribution, since larger units like those in *Avicennia* and *Pelliciera* are supposed to be more resistant to tidal buffeting. Hence seed weight determines zonation, according to Rabinowitz (1978b).

The frequently expressed notion that *Rhizophora* propagules are pointed to promote self-planting as the seedling falls from the trees deserves little serious consideration, even though it is not difficult to watch it happen. Yamashiro (1961), working with *Kandelia* in southern Japan (the northernmost mangrove "forest" in the world), demonstrated with painted seedlings that most seedlings are transported within 30 days at least 50 m away from the mother tree. Out of 1854 marked seedlings, 1627 (87.76 percent) were "lost," that is, were carried beyond the 50-m study plot. Only 31 (1.67 percent) were collected under the mother tree. Figures for trees in different tidal regimes were comparable.

The establishment of propagules, once dispersed, is a more critical problem because the figures presented by Rabinowitz (1978a) and Chai (1982) show that establishment cannot occur during a single low tide. *Rhizophora* seedlings in still water seem to be able to take root while completely submerged. The radicle (embryonic root) itself is often not involved in the process; in *Rhizophora*, for example, it aborts and anchorage is the responsibility of the lateral (adventitious) root primordia developed subapically by the short root segment of the propagule. In *Bruguiera*, on the other hand, the radicle is developed. Another problem is how the elongated hypocotyl of the Rhizophoreae, which is usually stranded in a horizontal position, becomes erect. Curvature, presumably by unequal expansion of cortical cells, is common; it is a characteristic feature of *Bruguiera parviflora* (Fig. 8.8). No other mechanical process has been detected. Clearly the mangroves, like other water-dispersed plants, have developed a "stranding strategy." Whether vivipary is part of this strategy has been discussed, although it seems unlikely.

Figure 8.8. Seedlings of *Bruguiera parviflora* (Rhizophoraceae) stranded at low tide. They show the characteristic curvature that erects the axis and is part of the establishment strategy. Klang, Malaysia.

Accounting for vivipary

The viviparous condition, which is pronounced in mangrove Rhizophoraceae and approached in the cryptoviviparous genera *Aegiceras, Avicennia, Pelliciera*, and *Nypa*, is so strongly associated with mangal that it is suggested to have adaptive significance in an intertidal environment. *Amphibolis* is the only reported instance of vivipary among the seagrasses. No fully acceptable explanation for the correlation has been offered, however. The mangal environment is unique, not in any single feature, but as a complex of factors, so that it has proved difficult to assign an adaptation for the viviparous seedling to a single factor. Halophytes and tropical swamp forest species do not show vivipary so the condition is not a response to either salinity or wet soil alone. On the other hand, mangrove propagules are widely distributed and must become established in slightly to extremely unstable substrate conditions. Rabinowitz (1978b) suggested that the length of the seedling promoted establishment in mangal environments to the extent that the zonation could be a simple correlation of propagule sorting by wave action. A longer propagule is most easy to produce by hypocotyl extension before the embryo is detached, since it has been shown that vivipary in Rhizophoraceae involves an exaggeration of normal development. On the other hand, the diversity of seed and fruit size, especially in Indo-Pacific mangroves, shows little correlation between propagule size and a species position in mangal zonation. In contrast to these mechanical explanations, other authors have turned to aspects of the physiology of viviparous seedlings and have investigated variations in salt concentration, respiration activity, and enzyme distribution (notably the series of articles by Pannier 1959, 1962, 1965, and Pannier et al. 1967).

It was early established (e.g., Walter and Steiner 1936) that developing viviparous seedlings in Rhizophoraceae retain lower salt concentrations than the parent shoot and especially leaves; these observations have been verified recently (e.g., Lotschert and Liemann 1967). There is a two-stage drop from the pedicel to the fruit and then from the testa to the embryo. The same condition is found in the crypto-viviparous *Aegiceras* (Bhosale and Shinde 1983). Localization of a physiological barrier in the floral pedicel or testa (until its rupture) is uncertain, even though there is a specialized "endothelium" or endothelium-like layer, whose cells have some of the properties of transfer cells, at the limit of the endosperm (Juncosa 1982b). No barrier in terms of ionic restriction has been identified. A physiological barrier is not itself necessary to account for different salt concentrations, since the differences could result from a diminished passive accumulation of salts in organs that do not transpire or that transpire slowly. Structural evidence for this is the absence of stomata from the hypocotyls of *Ceriops* and *Rhizophora* seedlings (Wilkinson 1981); stomata are present in the seedlings of *Bruguiera* (Rhizophoraceae, e.g., Kipp-Goller 1940) although they may not function. On the other hand, seedlings develop numerous and often conspicuous lenticels (Roth 1965).

This lowered salt concentration, whatever its origin, is interpreted as a mechanism to "protect" the embryo from the deleterious effects of high salt concentrations until maturity. Sudden submergence in sea water is presumably not detrimental, and because most propagules spend some time in sea water before they become established, there is a presumed gradual adjustment to saline conditions (Joshi et al. 1972). Comparable studies for other halophytes that germinate in fresh water but where vivipary does not occur could be informative.

The seedling is clearly nutritionally dependent on its parent shoot (Pannier and Pannier 1975, Bhosale and Shinde 1983) but makes a contribution to its own energy requirements via photosynthesis after the hypocotyl is exposed. Studies have shown that species have different rates of seedling respiration (Chapman 1962a,b). Brown et al. (1969) studied the capability of seedlings to respire anaerobically in the dark using thick slices of hypocotyl tissue, but the methods seem somewhat artificial in relation to normal seedling biology.

Establishment of the elongated seedlings of Rhizophoraceae has occasioned discussion and experimentation, since observers have had difficulty accepting that seemingly rigid axes can become erect after rooting (Bowman 1916, 1917; Egler 1948; Larue and Muzik 1951). Field observation as well as experiments show that appropriate growth curvatures are possible; in *Bruguiera parviflora* this curvature is a characteristic of established seedlings (See Fig. 8.8). A developed radicle is present only in *Bruguiera* (a presumed primitive feature); in other genera rooting is initiated by lateral roots at the radicular end of the seedling, with the radicle itself aborted. Establishment in most species is rapid; Chai (1982), for example, gives values of as little as one to two days for the initiation of root development in mangrove Rhizophoraceae, with plants firmly rooted after 10 days. Root development is most rapid in seedlings that are planted erect.

In reviewing literature on mangrove ecology, one is made aware of inconsistencies between different accounts because the basic biological features of seedling morphology and physiology are not well appreciated. The approach of Juncosa (1982b, 1984a,b) is exemplary in this respect, since the objective is to study viviparity in relation to biological mechanisms in tropical woody species generally. Until this background information is more fully developed, ecological explanations for the viviparous condition are likely to be premature. Juncosa (1982a) makes the interesting suggestion that vivipary is a simple consequence of the amplification of the normal ''torpedo'' stage of embryo development by the extended growth of the hypocotyl, the absence of temporary cessation of intercalary growth that occurs in seed dormancy in nonviviparous plants, and the advantage of immediate germination in tidal habitats. The prevalence of elongated, pointed, or curved propagules aids in anchorage, which is further facilitated in soft substrates by the rapid development of many lateral roots. Self-erection becomes an important attribute in such seedlings (Lawrence 1949). This hypothesis does not really account for vivipary in direct adaptive terms, but it answers the question Why vivipary? with the response Why not!

9   Utilization and exploitation

Mangroves are an entrepreneur's dream: They produce raw material (lignocellulose) from sea water by using renewable energy sources. Most energy is supplied by sunlight. Tidal energy is also used, since nutrients are replenished by tidal action and estuarine runoff and detritus is conveniently flushed away. Furthermore, if this energy is to be harvested, immediate transport by water in and out of mangal can be facilitated by the proximity of river mouths and the sea. If the products are to be exported, they are already on the coast. Despite these benefits and extensive exploitation, the question has been asked: "Mangroves – what are they worth"? (Christensen 1983) The answer depends on whether one sees mangal as a resource either to be harvested once as a realizable asset or to be sustained on a renewable basis as a dividend from a fixed investment, or as land to be converted into a more profitable resource by a conversion of capital. The question is also phrased in terms of a direct economic assessment of mangroves per se, so that some monetary value can be put on both capital and its yield. The question is not exactly an academic one and litigation has, in one instance, established a value of $751,368.30 per hectare for the restitution of mangrove in Puerto Rico damaged by an oil spill (Lewis 1983).

In contrast, one may express a concern for the long-term socioeconomic implications of mangrove exploitation at various levels: from the extremes of a national economy to a local resource important only in a peasant economy. It may be possible to give reasonably accurate dollar values at the regional level, but it is much more difficult to do so at the local level, even though the greatest total utilization occurs locally and probably benefits most individuals. The total number of people who derive a livelihood from mangrove plants may be more significant in social terms than total profits to some industrial complex.

The covert uncertainty of who benefits most underlies any discussion of mangrove utilization no matter how precisely assessments have been made. Decisions about appropriate policies for mangrove use are thus made difficult.

Economic complexity

A major problem in economic analysis is the difficulty of recognizing or distinguishing indirect and direct benefits. Apart from problems of establishing economic

163

benefits where either there is no cash flow or it cannot be measured very accurately, the fact that mangal represents an interphase community (which makes it such an attractive ecosystem for study by biologists) creates administrative complexity, since competing economic interests are at work. Major industries, such as forestry, fisheries, and agriculture, may be in conflict because each can claim mangroves as their administrative domain, and the policy that is best for one industry may be detrimental to another (Christensen 1983). A forestry department will emphasize utilization that may degrade the resource, a fisheries department will emphasize conservation with a minimum of disturbance, and an agricultural department may advocate conversion and replacement by some putatively more valuable resource. This conflict is the background to mangrove management as well as much of the economic justification for extended research on mangrove communities. In addition to recent summaries by Christensen (1983) and Walsh (1977), which cover much of the modern literature, there are extensive older accounts, such as Foxworthy (1910), Watson (1928), Brown and Fischer (1920), and Brown and Merrill (1920).

Forestry

Mangroves constitute a minor part of the forest resources of most tropical countries, since they can occupy only a limited area. For example, 2 percent of all forest types in Papua New Guinea is designated mangrove, according to Percival and Womersley (1975). Nevertheless, some forestry departments have emphasized mangrove management and utilization, notably in Malaysia. The classic study by Watson (1928) was an attempt to produce information from which a management program was to be developed. The main objective in the program was to produce biomass on a sustained yield basis. Changing technologies tend to move faster than recommended rotation cycles so that a growth rate designed to support a sawn timber industry might be ineffective if directed toward a pulpwood or a pole timber resource. A short-term rotation might be suitable for pulpwood, but allows no thinning for firewood. In terms of actual harvesting, it is usually agreed that clear-felling is the preferred procedure; discussion then centers on whether subsequent regeneration should be either natural or artificial and to what extent selective culling is to be allowed during the regeneration cycle.

Sawn timber. A number of major mangrove constituents produce timbers with desirable qualities (e.g., high density, termite and marine borer resistance) so that woods may have special uses, such as in boatbuilding and fishtraps. Low floristic diversity is an asset when extraction is necessary and can be used as an argument for clear-felling. Large-scale exploitation using heavy equipment is very disruptive of the community. Mangroves are a decreasing source of merchantable timber, however, because of their decreasing average maturity as they are exploited. A few species of the back-mangrove community, for example, *Heritiera* and *Xylocarpus*, produce high-quality timber but their scarcity and difficulty of access have made them unattractive for harvesting. Perhaps *Heritiera fomes* in the Sundarbans is the most actively exploited species.

Poles. Unsawn poles are now the most common extraction product, especially at the local level, because they can be harvested readily by simple manual methods and because of short rotations. This produces a cyclical effect because poles are also suited to the short rotation often practiced in managed forests (often by necessity where population pressure is great). These practices, although they do not allow the regeneration of tall forest, are probably minimally disruptive of the community. Regeneration from seed is not inhibited because most mangroves flower as saplings, but there could be selection for slow-growing species.

Fuel wood. This may be used directly or after conversion to charcoal and is probably the main biomass utilization of mangroves at the local level. Mangrove management as a fuel resource is also suited to minimally disruptive short-term rotations, since only small trees are needed and these are most easily harvested by manual woodcutting, which minimally disrupts the substrate.

Tannins and dyes. Use of mangroves for tan bark has at times been extensive; the entries under "tannin" in the index by Rollet (1981) amount to eight columns. Industrial development of synthetic or more accessible alternatives has tended to shut down most commercial operations (e.g., in Papua New Guinea, as described by Percival and Womersley 1975). The high tannin content of most mangroves, especially in the mangrove Rhizophoraceae, presumably increases their resistance to herbivores, so that this biological attribute has had a direct industrial spinoff. Local use of mangrove tannins may still be common. Mangrove sap is the source of a black dye used in the manufacture of Polynesian tapa cloth.

Minor forest products. Because of their low floristic diversity and absence of a specialized understorey, mangroves do not offer much resource diversity. They have limited local use in medicine (Burkill 1935), but many of their properties are assumed, if not magical. On the beneficial side, there are few poisonous plants in the mangroves. *Gluta* (Anacardiaceae) has an irritant sap that causes skin rashes; *Excoecaria* and *Hippomane* (Euphorbiaceae) have an irritant sap that causes skin ulcers. The seeds of *Cerbera* are said to be poisonous. Where plants contain saponins they have been used as fish poisons (e.g., *Barringtonia* and *Derris*). None of these plants is likely to be eaten by humans.

The Nypa palm. The common nypa palm, which often dominates quieter estuarine and lagoon areas in Southeast Asia, is exceptional in that it provides a diversity of minor products. Industrial ethanol is made by distilling the fermented sugary phloem sap, which is obtained by tapping the lateral inflorescence and has at times been produced on a large scale. An early account of the industry in the Philippines is given by Brown and Merrill (1920), who comment that the source at that time was the cheapest known form of industrial ethanol, a statement hardly likely to apply in modern times. A modern account of the tapping process is provided by Päivöke (1984) with some information on yield. The physiology of sap production is not understood; it is presumably similar to the process in other palms tapped for sugar (notably *Arenga, Borassus, Cocos, Phoenix*, and *Raphia*). *Nypa* ethanol represents a renewable fuel resource, but it remains questionable whether it could become a large-scale, successful commercial proposition. *Nypa* provides other minor commodities: attap or thatch from its leaves (which are rated superior to any other palm thatch), cigarette papers from the stripped surface layers of the leaflets, an edible jelly from the unripe endosperm, and a sort of salt from the ashed leaflets (see Burkill 1935).

Fisheries

The indirect relationship between mangroves and some inshore commercial fishing and shrimping is well recognized. First, mangroves supply nutrition (mainly via a detritus chain) to marine communities. Second, they may provide a habitat for some commercially exploited marine organisms at critical phases of their life cycle; they may function as "nurseries" in certain instances. Third, they may directly support some organisms, notably oysters and other shellfish that usually grow in mud. Promotional literature about South Florida emphasizes it as a place where oysters grow on trees (actually the stilt roots of *Rhizophora*). Finally, mangroves stabilize shorelines and protect inshore fish habitats from sediment pollution. Since figures for the cash value of commercial fish catches are relatively easily assembled, they can be used to demonstrate the indirect commercial value of mangroves (e.g., Christensen 1983).

Agriculture

Mangrove plants are demonstrably unpalatable and provide little direct human food. This may be related to the lack of close plant–animal interactions within the community. Although pollination by animals is common, these pollinators are very generalized. Mangrove fruits, seeds, and seedlings are all dispersed by water. Fleshy, edible fruits are not produced; the most conspicuous exceptions are the mistletoes (e.g., *Amyemena*) and epiphytes (e.g., *Hydnophytum*), where bird dispersal of fruits is essential to their establishment. In general mangroves are designed to resist herbivores. This is reflected in their high tannin content and leathery texture, which may even involve extensive development of ideoblastic sclereids (notably *Rhizophora, Sonneratia,* and *Aegialitis*). Mangroves are thus unlike pioneer species in their poor leaf palatability. Nevertheless, there are reports of young viviparous seedlings of *Bruguiera* or of *Avicennia* seeds being boiled and eaten, but seemingly only as food during a famine. On the other hand, seeds of a few mangrove associates are edible (*Barringtonia edulis,* and *Inocarpus fagifer*). Animal-pollinated flowers in most instances have nectar as a pollinator reward, and in some areas mangroves are a commercial source of honey.

Agricultural use of mangal is directed mostly toward its conversion, and salt-resistant rice varieties have been successfully cultivated, notably in Sierra Leone (Walsh 1977) in converted, impounded mangal. Otherwise conversion is usually difficult either because the texture of the mangrove substrate is unsuited to the subsequent use of standard agricultural equipment or because the previously anaerobic soils, when oxygenated, become highly acidic. Conversion of mangal to rice fields in Costa Rica has been legislated against as a conservation measure.

More successful conversion for mariculture and aquaculture has been achieved, notably in the Philippines and Indochina, yielding fish, shrimp, and shellfish. Oysters have been successfully cultivated with little or no mangrove conversion. This simply uses artificial supports for the crop, such as stakes at the edge of estuarine creeks. In mariculture of this type, the mangal simply coincides with a suitable substrate and medium for fish culture. Nutrient resources are made available via the mangal so that enterprises of this kind again depend on a minimally disturbed plant community.

Other industrial and quasi-industrial uses

Raw materials. Mention has been made of industries that use the mangrove biomass with little or no subsequent industrial conversion. In contrast, modern technology may demand little more than a supply of lignocellulose as a manufacturing substrate. Recent development toward the total chipping of mangroves for pulpwood and cheap synthetics (notably rayon) has been extensive in Southeast Asia, especially Malaysia, the Philippines, and Papua New Guinea. The material is exported for conversion to industrially developed countries, such as Japan. This is probably the worst kind of exploitation. The disturbance to the community is maximal because all standing biomass is removed. Soil textures are disrupted and there may be little subsequent regeneration. Furthermore, there may be little direct benefit to local communities, since the foreign currencies that are generated do not percolate to the local level. The employment benefit may be short term.

Salt conversion. In drier areas, mangroves can be converted to salt pans; there may even be an alternation between salt and shrimp production within the same enclosure in dry and wet seasons, respectively. The product may not compete commercially with the large-scale production in the drier subtropics (e.g., the Bahamas and Western Australia), where tides, topography, and climate are optimally favorable. Ashing of mangroves is a local alternative. Some mangrove plants secrete salt, presumably very pure, but no attempt seems to have been made to capitalize on this natural process.

Sewage treatment. Mangrove swamps close to major centers of populations are familiar to travelers as unofficial garbage dumps and local sewage-disposal units. The detritus of the consumer economy is often not readily dispersed because of the restricting aerial root systems of mangroves. A common (but discrete) sight around Asian coastal villages is the "thunder box" accessible by a boardwalk and built into the mangal. Tidal scour makes the unit self-flushing. A more serious consideration for the use of mangroves as natural sewage-treatment plants has been presented by Australian ecologists, usually with scrubby mangal in mind (e.g., at Westernport Bay, Victoria). The effectiveness of this proposal depends on the knowledge of nutrient cycling in these communities. An extended evaluation is provided by Clough et al. (1983).

Coastal protection

There is much discussion in the ecological literature on the extent to which mangroves trap sediment and so contribute to land building. The general conclusion seems to be that they do so only in geomorphologically prograding regions, that is, areas that would undergo sedimentation even in the absence of shoreline vegetation. Mangroves nevertheless stabilize these prograding shores and prevent excessive shifting of coastlines. Consequently they are particularly important on coasts that are subject to major tropical storms, and they buffer the destructiveness of wind and storm tides during hurricanes and typhoons. Deforestation of deltaic regions in Bangladesh (the Sundarbans) has been widely accepted as a cause of increased property damage and loss of human life during tropical storms.

Direct economic measurement of the benefit of the protection of undisturbed mangal to coastal regions is again difficult, except where one can estimate property damage in adjacent unprotected regions. The buffering effects of mangal are comparable to the benefits of terrestrial watershed forests in preserving water quality and preventing erosion. In some areas real estate development of mangroves may be economically attractive in temporarily increasing land value (or by dredging operations, actually creating salable building sites.) The "developer" may adopt a Janus-like attitude, expressing concern for the quality of the environment and even seeking limited reestablishment of natural communities, such as seagrass meadows and mangal, on sites that would have retained their economic and esthetic status in the absence of "development." This ethical dilemma has been acute in areas like South Florida and Singapore, where affluence and population density demand the ready conversion of a resource.

Wildlife management

Christensen (1983) and MacNae (1968) provide information about the relationship between faunal communities in mangals and the need for preservation measures, listing species of mammals, birds, and reptiles. There is not a high degree of interdependence, as we have seen, because few animals are exclusive mangrove dwellers. It is of interest, however, that the few "dangerous animals" in mangroves are largely themselves "endangered," for example, crocodiles and tigers. Mangroves have been a refuge for the "Key deer" in South Florida.

Productivity

The economic utilization of mangroves depends ultimately on biological productivity, usually measured in the amount of biomass produced per unit weight of standing biomass. I make no attempt here to summarize or review the extensive literature on the subject because it is not entirely relevant to the scope of the book and it is difficult to make generalizations based on studies that have been carried out in different parts of the world and in communities of different ages that may have very different canopy heights and very diversified nutrient patterns. Good reviews are provided by Christensen (1978, 1983), Lugo and Snedaker (1974), and Chai (1982). One notes in this literature the wide range of values (from 2 to 16 m³/ha/year of wood mean annual increment). The higher values are comparable to those that can be produced for both temperate and tropical forests. Bearing in mind that wood is not necessarily the only commercially utilizable product of mangroves, this suggests that carefully managed mangal is indeed an excellent economic resource. A modern, comparative assessment of mangal productivity at the hands of skilled production ecologists is urgently needed.

Conclusions

Preservation and utilization of mangal directly or indirectly for agriculture, fisheries, and forestry at both the local and industrial level are seen as competing influences for the same limited and often sensitive resource. The mangrove community is almost a microcosm of the socioeconomic complications attendant on the human use of a natural resource. Undoubtedly mangal is useful, although ''what they are worth'' is not a very specific question until it is qualified by ''to whom'' and ''in what terms.'' Mangal may be a simple ecological community in terms of species diversity, but it is an exceedingly complex ecosystem to evaluate. The biological basis for sustained use and management is still deficient; the safest policy minimizes direct utilization and especially disturbance. The woodcutter puttering his way through mangrove creeks with a load of firewood in a shallow-draft boat may not be a very dramatic figure in entrepreneurial terms, but this is probably the best image that can be associated with a mangrove forest treated as a resource to be sustained in perpetuity.

Section B

Detailed description by family

Family: Acanthaceae

A family of mainly tropical herbs, shrubs, or even small trees with zygomorphic, sympetalous, usually conspicuous flowers with (2)–4(–5) stamens. Ovary bilocular; fruits usually capsular with the 2 or more seeds compressed and surrounded by the hardened funiculus, which functions as a jaculator when the capsule itself bursts. *Acanthus* is the only genus of the family that has representatives in mangrove communities.

Acanthus L. 1753
[*Sp. Pl.*:639]

A genus of some 30 species in tropical Asia and Africa but with a center of diversity in the Mediterranean region (including southern Europe). It is distinguished from related genera that also have 1-celled anthers and only 4 calyx segments (e.g., *Blepharis* and *Crossandra*) by its usually spiny leaves, spicate terminal inflorescences, 2 bracteoles, and uniform anthers. It is distinctive within the family, which lacks stipules, in often having a pair of spines on each side of the stem in the stipular position at most nodes (Fig. B.1d). These persist, and the stem itself is commonly described as spiny. Sometimes the spines are double.

The species described are characteristic associates of mangroves with a total range from India to the Western Pacific (New Caledonia), tropical Australia, and the Philippines. Three species are circumscribed (van Steenis 1937) as indicated in the accompanying key, but there has been a tendency to treat them as one single variable species. They do not seem to differ in any consistent vegetative feature.

1A. Open flower 3.5–4 cm long; corolla in part light blue or violet (rarely white). Bracteoles persistent, in fruit up to 1 cm long. Ripe fruit 2.5–3 cm long or longer; seed approximately 10 mm in diameter. Inflorescence usually longer than 10 cm. Plants typically robust with spiny to very spiny leaves. Common in the back mangal throughout the Asian tropics from India to Polynesia and Australia; rarely inland............. *A. ilicifolius* L.

1B. Open flower 2–2.5 cm; all parts smaller than in previous species. Corolla always white. Bracteoles small, inconspicuous, early deciduous, or absent. Ripe fruit shorter than 2.0 cm; seed 5–7 mm in diameter. Inflorescence variable. Plants sometimes delicate 2

2A. Plant usually spiny with thick stems. Leaves usually widest below middle. Bract distinctly shorter than the calyx; deciduous before flowering. Bracteoles usually present but early deciduous. Common, often sympatric

with the previous species, and with a similar geographic range . . . .
. . . . *A. ebracteatus* Vahl.

2B. Plant usually unarmed, with slender delicate sprawling stems. Leaves
usually widest above the middle. Bract always longer than the calyx;
deciduous during flowering. Bracteoles never present. Uncommon, locally
present in Indonesia and New Guinea. . . . . *A. volubilis* Wall.

Acanthus ilicifolius L. 1753   (Figs. B.1, B.2)
[*Sp. Pl.*:639]

Acanthus neo-guineensis Engl. 1886 [*Bot. Jahrb.* 7:474]

A low sprawling or somewhat viny herb, scarcely woody, to a height of
2 m. Hermaphroditic. Axes initially erect but reclining with age; branching infrequent and commonly from older parts. Aerial roots from lower surface of reclining
stems (Fig. B.1b). Leaves decussate, usually with a pair of spines at the insertion
of each leaf. Leaves glabrous; petiole short (1 to 1.5 cm); blade up to 20 cm long,
gradually tapered below, either broadly lanceolate with an *entire margin* (Fig. B.1a),
the apex rounded or mucronate, or more usually with a sinuous, *spiny margin* (Fig.
B.1c), the apex broadly tridentate including an apical spine; spines sometimes also
present above and below on major veins. Major spines at end of leaf lobes horizontal
or erect; minor spines between major lobes erect. Inflorescences terminal, forming
bracteate spikes 10 to 20 cm long, the spike extending with age. Flowers in 4 ranks,
up to 20 pairs, the bract below each flower 5 mm or shorter, often caducous; lateral
bracteoles 2, conspicuous and persistent. Flowers perfect; calyx 4-lobed, the upper
lobe conspicuous and enclosing the flower in bud, the lower lobe somewhat smaller;
lateral calyx lobes narrow, wholly enclosed by upper and lower sepal. Corolla
zygomorphic, at least 3 cm long with a short tube closed by basal hairs; abaxial
lip broadly 3-lobed to entire, adaxial lobes absent. Stamens 4, subequal, with thick
hairy connectives, anthers medifixed, each with 2 cells, aggregated around the style.
Ovary bilocular, with 2 superposed ovules in each loculus, style enclosed by stamens, the capitate to pointed stigma exposed. Fruit a capsule 2 to 3 cm long and
1 cm wide, usually with 4 rugose angular seeds about 1 cm long; testa delicate,
wrinkled whitish green. Jaculator conspicuous. Germination hypogeal.

Growth

Acanthus ilicifolius is typically a low woody herb that owes its ability for
vegetative spread to its reclining stems so that it forms large patches by vegetative
means. Growth is continuous in the sense that there are neither resting terminal
buds nor obvious articulations; in unfavorable seasons, shoot growth may be inactive.

The distal, leafy shoots are at first erect, to a height of as much as 2 m. They
are determinate by flowering, and the old terminal spikes are substituted by 1 branch
(sometimes 2) that originates from buds in the axils of leaves just below the inflorescence. Axes eventually recline under their own weight, apparently for mechanical

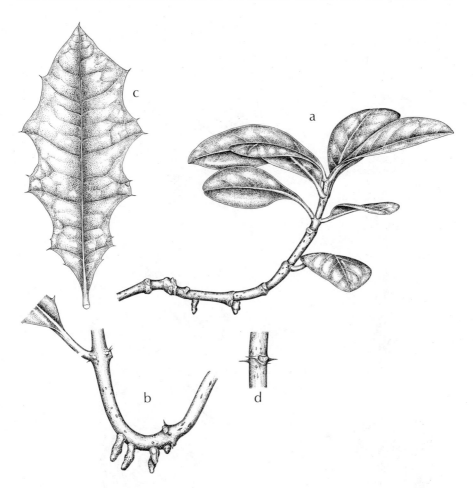

Figure B.1. *Acanthus ilicifolius* (Acanthaceae). Vegetative morphology. (a) Distal end of reclining stem (× ½), leaves not spiny; (b) adventitious roots on lower surface of bent axis (× ½); (c) distal leaf with spines (× ½); (d) node with spine pairs in stipular position (× ½). (Material cultivated at Fairchild Tropical Garden, Miami, Florida)

reasons. This reclining growth in itself is not proliferative and is in part a consequence of the limited production of secondary tissue. In addition proleptic branches develop from the older horizontal axes at infrequent and seemingly irregular intervals. This lack of self-support coupled with infrequent branching accounts for the common description of the plants as viny, since they may scramble over any adjacent vegetation that will provide support. The development of adventitious aerial roots is important in the sprawling habit because it anticipates the need for secondary

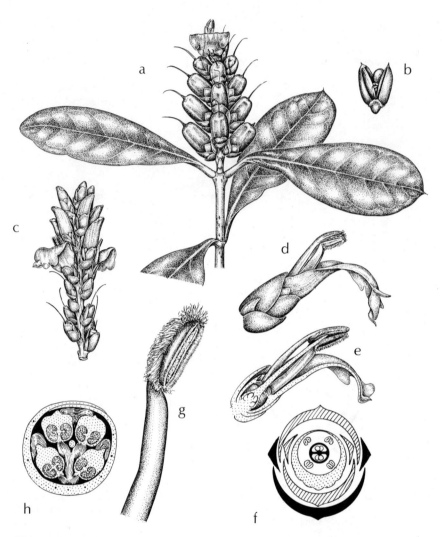

Figure B.2. *Acanthus ilicifolius* (Acanthaceae) flowers and fruit. (a) Old flowering head with maturing capsules (× ½); (b) dehisced but unexploded capsule (× ½); (c) flowering head (× ½); (d) flower in side view (× ⅔); (e) flower in L.S. (× ⅔); (f) floral diagram; (g) single abaxial stamen (× 3); (h) flower bud in T.S. (× 3) to show interlocking stamen hairs. (Material cultivated at Fairchild Tropical Garden, Miami, Florida)

rooting of shoots after they become horizontal. The aerial roots are initially thick, somewhat fleshy, and with a well-developed periderm.

Leaf morphology and variation

There is considerable variation in leaf form, mainly in relation to the degree of spininess. Leaves are sometimes entirely spineless, ovate-lanceolate with a rounded apex and entire margins, but more commonly they have a sinuous dentate margin with spines on the margin and both surfaces, as is common in the terrestrial (non-mangrove) species of the genus. The variation is part genotypic and part phenotypic. Lack of spines seems to be a juvenile character, but one that may recur on the leaves immediately below the inflorescence or on proleptic branches. Leaf spininess also seems to be accentuated by the degree of water stress, which may be related to salinity, seasonality, and light intensity. Some populations characteristically have a high proportion of spineless leaves, however, and have been considered of potential horticultural value because of their salt tolerance.

Reproductive biology

Flowers are probably pollinated by both birds and insects. As the corolla expands, the lip is formed by marginal recurving, which physically tears the corolla on each side at its base. This allows the distal part of the corolla to separate. The 4 stamens are in 2 slightly unequal pairs with their anthers pressed together under tension facing the lip of the corolla. The anthers dehisce longitudinally by a single slit, with a thick line of hairs bordering the split so that pollen is presented on the lower side of the stamens. The stigma rests on top of the anthers but is prevented by the hairs from coming near the pollen. There is an additional ring of dense hairs at the base of the stamens where the floral tube narrows. All these hairs point upward and outward, preventing small insects from crawling into the floral tube.

A large animal visiting the flower probes into the channel, about 12 mm long, formed between the stamens and the large corolla lobe below. When the bases of the stamens are forced apart by the visitor, they diverge in two lateral pairs and the style and stigma descend. When pressure on the stamen bases is released, the stigma lifts up and the stamens return to their original appressed position. The result is that the 2 divergent lobes of the stigma touch the back of the flower visitor first and pick up any pollen it already may be carrying, whereas pollen is deposited by the flower as the stamens come together and the visitor withdraws. As a result of this functional morphology, the stigma and the dehisced anthers rarely contact each other. Self-pollination can occur if the stigma does not return to its original position sufficiently rapidly.

The stigma does not become receptive until the second day of anthesis. Flowers usually last 2 days (exceptionally 3 or 4). This weak protandry is likely to restrict self-pollination.

Figure B.3. *Acanthus ebracteatus* (Acanthaceae), the mangrove "thistle." Semetan, Sarawak.

Observations by Primack et al. (1981) in Queensland show that flowers are visited by sunbirds (e.g., *Nectarina jugularis*); large bees (*Xylocopa*) were suspected of being pollinators, although no visits were actually observed. The size of the flower and its mechanics indicate that a relatively large pollinator is needed for effective pollen transfer. From the abundance of fruits in the population studied by these authors it seems that pollination is normally very effective.

The seeds, with a wrinkled whitish testa that tears easily as the seed matures, consist of the 2 flattened green cotyledons enclosing the minute seedling axis. Usually all 4 ovules form seeds. Release is explosive, with the capsule splitting violently in the dorsiventral plane and the seeds propelled away with a spinning action like a discus. They are dispersed up to about 2 m.

Acanthus ebracteatus Vahl. 1791   (Fig. B.3)
[*Symb. Bot.* 2, 75, t. 40]

Where this species grows with *A. ilicifolius* the two seem distinct in the characters used in the key, but they are often confused.

Acanthus volubilis Wall. 1831
[*Pl. As. Rar.* 2, 56, t. 172]

This species is much less robust in its vegetative parts than the other 2 species and is described as unarmed. The inflorescence usually has fewer flowers and is shorter.

Taxonomy and nomenclature

That *A. ilicifolius* and *A. ebracteatus* are conspecific has been suggested by earlier authors and accepted by some recent ones (e.g., Percival and Womersley 1975). The situation is confusing because Vahl, in his original description of *A. ebracteatus*, emphasized differences in leaf shape (which are not consistent) as diagnostic and, as pointed out by C. B. Clark (1884, *Hook. Fl. Br.* 4:481), his description of a 4-fold difference in flower size between the two species is certainly incorrect. His diagnosis also emphasizes *A. ebracteatus* as having the largest flowers, a difference reversed in modern treatments.

Anatomy

The epidermal glands of *A. ilicifolius* (see Fig. 6.4E,F) have been described (e.g., Areschoug 1902, Van Steenis 1937). These apparently are the source of secreted salt, which gives the upper leaf surface a greasy feel on occasions.

Family: Anacardiaceae

A major tropical family of some 80 genera and 600 species. It includes some commercially important timber species as well as the mango and cashew nut. Many species have active alkenyl phenols, which produce a contact dermatitis [e.g., poison ivy, *Toxicodendron radicans* (L.) O. Ktze (syn. *Rhus radicans* L).] The following genus represents the occurrence of the family as mangrove associates.

Gluta L. 1771
[*Mant. Pl.* 2: 293]

A genus of about 20 species with a range from Madagascar (1 species) to New Guinea, but with a concentration in Indo-Malaya. The following species, with a wide distribution from Burma and Thailand, the Malay Peninsula, and Malesia to Sumatra, northern Borneo, and West Java, is common along the edge of tidal rivers and is associated with mangroves, most characteristically behind *Nypa* swamps. It is included in this account because it has an irritant sap.

Gluta velutina Blume 1850
[*Mus. Bot. Lugd. Bot.* 1:183]

There is a full account in Ding Hou (1977).

A large shrub or small tree growing to 7 m high with numerous branched aerial stilt roots, developing an extended fluted stem base. Leaves simple, spirally arranged, elliptic, 12 to 30 by 5 to 8 cm, cuneate at the base, blunt apically, somewhat resembling those of mango. Inflorescence a terminal panicle up to 12 cm long with numerous small white or pink flowers with velutinous calyx and pedicels. Flowers each with 5 stamens surrounding the ovary raised on a central stalk (torus). Fruit usually solitary, a 1-seeded drupe up to 7 cm in diameter but with a distinctive scurfy brown tuberculate surface and ridged toward the base.

The best field characters are the black-spotted older leaves and the clear sap, which exudes readily and rapidly turns black so that the trunk becomes black stained. Any tree with this character should be avoided because it usually indicates a member of the family and the sap is often slightly to extremely irritant and toxic.

Family: Apocynaceae

A mainly tropical family of herbs, shrubs, lianes, and trees, typically exuding a white latex from cut surfaces. Two genera are treated here; *Cerbera* occurs as trees in the Old World, and *Rhabdadenia* as vines in the New World.

Cerbera L. 1753
[*Sp. Pl.*:208]

A genus of 6 species, of which at least 3 have a coastal distribution and may sometimes be constituents of back-mangal communities. They are readily recognized by their stout twigs with broadly oblanceolate leaves in dense terminal spirals, copious latex, and terminal cymes of conspicuous white, fragrant flowers. The genus superficially resembles the cultivated species of *Plumeria* (frangipani) but is distinguished by the large calyx segments, the stamens inserted toward or at the top of the corolla tube, the well-developed style with a thick disciform stigma, and particularly the fleshy, drupaceous (not capsular) fruit, which is dispersed by water. One species, *C. odollam*, is sometimes cultivated, and its range may have been extended artificially. It is, for example, a common street tree in Singapore. The fruits are described as poisonous, however, and used as a means of committing suicide by the Marquesans (Brown, 1935).

Two closely related species (*C. odollam* and *C. manghas*) are occasional elements of coastal forest and back-mangal communities but with only limited salt tolerance. They have been much confused, even though they were clearly differentiated by Valeton (1895). (See also Backer and Bakhuizen van den Brink 1964, 2:233.) Commenting on the confusion, Corner (1939) says: "How backward is botany in Tropial Asia, this genus will illustrate." They occur most typically in sandy soils, not regularly inundated tidally. A third species, *C. floribunda*, restricted to New Guinea, may also be encountered. The three are very similar vegetatively but may be distinguished as follows:

1A. Stamens inserted below middle of corolla tube; flowers numerous (40 or more) per cyme, each with a short inflated corolla tube, scarcely 1 cm long .... *C. floribunda* Schum.
1B. Stamens inserted at or above middle of corolla tube; flowers relatively few (less than 30) per cyme, corolla tube slender (at least at base), longer than 1 cm... 2

2A. Corolla with a yellow eye; tube short (1.5–2 cm long), but free corolla segments long (2–3.5 cm); stamens inserted at or shortly above middle of locally widened tube; throat of corolla not closed by sparsely hairy connective ridges. Leaves apiculate, with a fine point, primary veins perpendicular to midrib, terminating in a fine marginal commissure........
....*C. odollam* Gaertn.
2B. Corolla with a red eye; tube long (2.5–4 cm), but free segments short (2–2.5 cm); stamens inserted in mouth of tube, which is closed by densely hairy connective ridges. Leaves bluntly pointed, primary veins arcuate, terminating in a well-developed commissure ....*C. manghas* L.

The status of *C. dilatata* S.T. Blake 1948 (*Proc. R. Soc. Queensl.* 59(8):161), is uncertain.

Cerbera manghas L. 1753
[*Sp. Pl.*: 208]

C. lactaria Hamilt. ex DC 1844 [*Prodromus* 8:353]
C. odallam Bl. 1826 [*Bidjr. Fl. Ned. Ind.*: 1032]

Tree to 15 m, with grayish smooth bark; twigs stout (pachycaulous), branching consistently below terminal inflorescences (Leeuwenberg's model) but with frequent reiteration. Leaves spirally arranged, often aggregated in terminal rosettes, ovate-oblong to elongate-obovate up to 30 cm long, 7 cm wide, apex bluntly pointed, base gradually narrowed to a short petiole 1 to 2 cm long. Inflorescence an irregular, often 1-sided cyme, with broad ephemeral bracteoles 1 to 2 cm long. Flowers perfect, pentamerous; calyx 5-lobed, greenish white, the lobes 1.5 to 2 cm long by 5 to 6 mm wide, narrowly lanceolate and often persistent. Corolla tube white, up to 4 cm long, scarcely 3 mm wide at the base, gradually widened upward to the expanded throat 5 to 6 mm wide; corolla lobes 5, shorter than the tube (to 2 cm), broadest about the middle and abruptly narrowed toward insertion. Stamens 5, sessile, inserted in the mouth of the tube with a hairy spurlike extension of each stamen more or less closing the throat, together with a glandular extension below each stamen. Ovary superior, with 2 free carpels, each with several ovules inserted on a thick ventral placenta. Style slender, expanded abruptly at level of stamens into a thick discoid, shortly bifid stigma. Fruit (often twinned) an ovoid or ellipsoid 1- or 2- seeded drupe 5 to 6 cm long and 3 to 4 cm wide with green flesh, ripening purple, the endocarp somewhat transitional to fibers of the inner mesocarp. Germination hypogeal; seedling leaves narrowly lanceolate.

Growth
Leaves are arranged in a dense 5/13 phyllotaxis. There is no development of bud scales, although inactive terminal buds are protected by a resinous varnish. Each leaf subtends an inconspicuous lateral bud also protected by the resin, which appears to originate from a series of glands adaxially on the petiole at its insertion. These glands shrivel early and may persist as a black fringe along the top of the leaf scar. The young leaves themselves may similarly be coated with the resin, which flakes off as they expand.

As the terminal inflorescence develops, from 1 to 5 renewal shoots expand by syllepsis, with the base of the branch and its subtending leaf commonly fused together for as much as 3 cm.

Floral biology

Inflorescence construction is irregularly cymose, the nodes developing a series of branches each with a single, somewhat petaloid and ephemeral bract that almost encloses the axis at its insertion. The inflorescence seems to develop with progressive overtopping of older by younger axes so that there is a rough acropetal sequence of flowering over an extended period.

The floral mechanism has been well described by Valeton (1895). The pollen is presented early, almost before the flower opens, with each half-anther depositing its pollen as a sticky mass on the stigma, which secretes mucilage over its upper surface. The pollen is further prevented from falling into the tube by a shallow rim at the base of the stigmatic disc. The mouth of the tube is partly occluded by the hairs of the 5 ridges, which protrude both above and below but leave 5 narrow channels with access to the lower, narrower part of the tube. Details of pollen transfer are not known, but clearly the flowers are attractive to animals; crossing is promoted by the flower visitor, which is likely to encounter the stigmatic disc before it touches the pollen mass.

The pollination mechanism of *C. odollam*, which is likely to be somewhat different in view of the different floral construction, has still to be clarified.

Nomenclature

Linnaeus, in his *Species plantarum* (1753), described 3 species. Confusion arose because the name he applied to his second species, *C. manghas*, was based on *Manghas lactescens* in Burmann's *Flora zeylanica* (150, t. 70, f. 1) and also on *Odollam* of Rheede's *Hortus malabaricus* (1, p. 79, t. 39). The other two Linnaean species refer to elements now included in *Thevetia*. *Cerbera odollam* was correctly circumscribed by Gaertner in his *Fructus et seminus plantarum* (2:193, t. 124) as a species distinct from the *C. manghas* of Linnaeus. Nevertheless, subsequent authors have used the names *manghas* and *odollam* for both entities, despite the clear distinction between them, so there is a confusing pattern of mutual synonymy.

Cerbera odollam Gaertner 1791 (Figs. B.4, B.5)
[*Fruct. Sem. Pl.* 2:193, t. 124]

This species differs from the previous one in the following important features: Flowers with lanceolate green or white sepals (15 by 2 to 5 mm), ephemeral, falling at or before anthesis; corolla trumpet shaped, equal to or scarcely exceeding calyx, inflated distally in region of attachment of stamens but not at the mouth, filaments short, scarcely hairy; mouth of corolla tube open with a yellow eye, corolla lobes asymmetrical and broadest above the middle, abruptly narrower

Figure B.4. *Cerbera odollam*
(Apocynaceae) foliage and fruits.
Cultivated at Singapore.

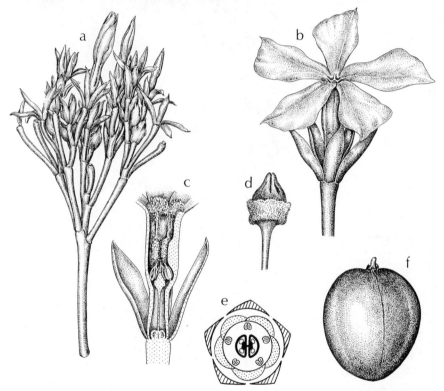

Figure B.5. *Cerbera odollam* (Apocynaceae) flower and fruit. (a) Inflorescence (× ½); (b) single
flower (× 1); (c) corolla tube in L.S. (× ³⁄₂); (d) detail of stigma (× 4); (e) floral diagram; (f) fruit
(× ³⁄₈). (Material cultivated at Forestry Department Headquarters, Kuching, Sarawak. P. B.
Tomlinson, 24.7.82)

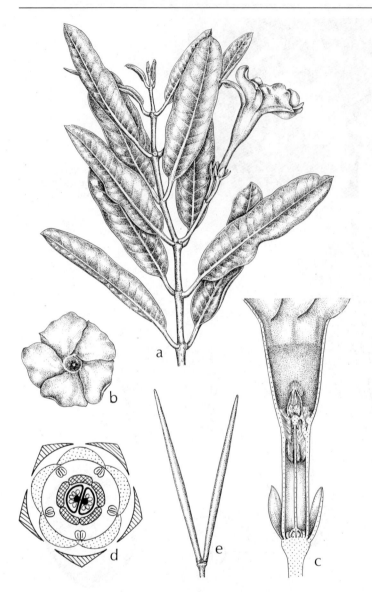

Figure B.6. *Rhabdadenia biflora* (Apocynaceae) flower and fruit. (a) Flowering shoot (× ½), flowering axis is terminal; (b) flower from above (× ½); (c) base of flower in L.S. (× ³⁄₂); (d) floral diagram; (e) fruits (× ½). (From Correll and Correll 1982)

basally with a narrow insertion. Fruit (always solitary) without clear distinction between mesocarp and endocarp.

According to Valeton (1895) there are also differences in leaf shape and venation as indicated in the key, but these have not been elaborated by subsequent authors. Corner (1939) says the species cannot be so distinguished.

Cerbera floribunda K. Schumann 1889
[Schumann-Hollrung, Fl. Kais.-Wilhelm Land:111]

C. micrantha Kanehira 1931 [*Bot. Mag. Tokyo* 45:343]

This species was originally described as endemic to New Guinea and is said to be distinguished by its numerous flowers (up to 100) in each compact cyme, the short corolla tube with the stamens inserted below the middle, and the elongated stigma. It is recorded from several coastal and estuarine stations, including an association with *Rhizophora* but otherwise chiefly inland (Markgraf 1960). Material from Palau identified by Markgraf as *C. floribunda* suggests an interesting range extension.

Rhabdadenia Muell.-Arg. 1860
[Mart. Fl. Bras. 6(1):173]

A viny genus of 3 species in subtropical and tropical America. The following species is widely distributed in the Caribbean region and occurs characteristically in fresh- or salt-water swamps. Typically it is found in back-mangrove communities.

Rhabdadenia biflora (Jacq.) Muell.-Arg. 1860   (Fig. B.6)
[Mart. Fl. Bras. 6(1):175]

Rhabdadenia paludosa (Vahl.) Miers 1878 *[Apoc. S. Am.*:119]

A slender vine with pliable stems several meters long. Leaves simple, opposite, often in distant pairs. Growth of flowering shoots is sympodial below-terminal cymes of flowers, with the renewal shoot displacing the inflorescence into an apparently lateral position. Flowers are conspicuous, white or pink, up to 7.5 cm long and 4 cm in diameter. Fruits are slender paired follicles up to 14 cm long, dehiscing to release the numerous hairy seeds. The plant is unusual in mangrove-associated communities in having primarily wind-dispersed seeds.

Family: Avicenniaceae

The genus *Avicennia* L. was initially included in the family Verbenaceae, from which it differs in important features such as its anomalous secondary thickening, leaf anatomy, placentation, incipient vivipary, and seedling morphology, which have been long recognized (van Tieghem, 1898); its segregation as a distinct family is now generally accepted. Its pollen also distinguishes it from Verbenaceae (Mukherjee and Chanda 1973). A thorough review and extended discusssion of *Avicennia* are provided by Moldenke (1960, 1967) based largely on a study of herbarium material. A modern revision of the genus based on field observations is much needed because it is one of the most consistently present constituents of mangrove vegetation. Some species and varieties have wide ranges, and there is additional geographical variation within individual taxa.

The family Verbenaceae itself is represented in back mangal by *Clerodendron inerme*, a diffuse shrub widely distributed in Australasia. Its floral biology has been described by Primack et al (1981).

Avicennia L.1753
[*Sp. Pl.*:110; *Gen. Pl. Ed.* 5:119, 1754]

> For a full synonymy, see Moldenke (1960, 1967) and Bakhuizen van den Brink (1921).

A pantropical genus of about 8 species, occupying diverse mangrove habitats, either within the normal tidal range or in back mangal and then with a high tolerance of hypersaline conditions. The genus is uniform in its gross morphology and anatomy and, although there are often useful diagnostic characters seen in the field, (e.g., bark color and texture and crown color), these are rarely transmitted on the labels of herbarium specimens. Consequently, available keys (notably that of Moldenke 1960) are not easy to use and identification of herbarium specimens may be difficult. The following description refers largely to the rather consistent and uniform generic characters; specific diagnoses are outlined in the later key and descriptions.

Generic description

Trees to 30 m, or when depauperate, low, much-branched, dense-crowned shrubs. Aerial stilt roots sometimes developed (e.g., *A. officinalis*). Underground roots extended, cablelike, and supporting as lateral branches erect, exposed pneumatophores and descending absorbing roots. Bark variously rough, black, hard, and fissured or checkered (e.g., *A. germinians*); gray and lenticellate (e.g., *A. marina*); to smooth, yellowish green, and flaking (e.g., *A. eucalyptifolia*, Fig. B.7). Secondary thickening anomalous; cylinders of xylem tissue alternating with cylinders including phloem, each cylinder produced by a new cambium. Shoots gla-

Figure B.7. *Avicennia* cf. *eucalyptifolia* (Avicenniaceae). Distinctive flaky bark; bark character may be helpful in understanding the ''*A. marina* complex.'' Townsville, Queensland.

brous or with a fine indumentum not readily recognized as made of separate trichomes; distinctly but finely woolly in *A. lanata*. Twigs somewhat jointed with swollen nodes, terete or more or less 4-angled; internode variable, long on vigorous shoots, for example, saplings, commonly short on distal laterals. Sequential branching by syllepsis, continuous or diffuse.

Leaves decussately arranged, the youngest leaf pair enclosing the apex within the petiolar groove; bud scales of terminal bud not developed. Leaves shortly petiolate (1 to 3 cm), the petiole with a deep basal groove, the groove with black marginal hairs, the hairs continuous in a line across the node. Blade leathery or somewhat fleshy with inconspicuous veins, the midrib prominent below; glabrous and light to dark green and shiny above with numerous close-set microscopic depressions; indumentum of a uniform dense palisade of microscopic club-shaped hairs below, imparting a characteristic gray, white, yellow, or olive green color. Blade 3 to 15 by 1 to 6 cm, varying from ovate to narrowly elliptic to lanceolate;

apex bluntly rounded and even slightly emarginate to acute or even attenuate; base acute. Margin entire, slightly thickened. Leaf shape generally useful in specific diagnosis (Figs. B.10, B.11).

Flowers protandrous, perfect, tetramerous to modified pentamerous, usually with 4 to 10 decussately arranged flower pairs in dense spicate to capitate units, the units themselves aggregated into terminal or axillary, somewhat paniculate assemblages on distal shoots, the lower units subtended by foliage leaves, the upper by reduced, often inrolled leaves. Each flower subtended by a convex, triangular bract and two lateral bracteoles, the three structures imbricate and enclosing the base of the flower. Sepals 5, free, ovate, and extensively imbricate, largely enveloped by bracteoles; the calyx and bracteoles persisting in fruit. Corolla tubular at the base, becoming equally or unequally lobed at the apex, the lobes either erect or reflexed; three main types of corolla may be distinguished, although this does not cover the total variation:

1. *Germinans* type (Fig. B.8e–h). Distinctly zygomorphic, white, 10 to 13 mm wide, corolla lobes at first slightly imbricate, later spreading, densely hairy on both surfaces, 2 lateral and abaxial (ventral) lobes rounded to at most slightly emarginate and somewhat narrowed toward insertion, adaxial (dorsal) lobe broad toward insertion and distinctly to conspicuously 2-lobed. *Avicennia bicolor* has smaller flowers (6 to 8 mm wide) with a distinct yellow throat.

2. *Officinalis* type (Fig. B.9f,g). Somewhat zygomorphic, yellow, 6 to 10 mm wide, corolla lobes at first slightly imbricate, later spreading and rather irregular, glabrous within, petal lobes as in type 1.

3. *Marina* type (Fig. B.9a–d). Strictly actinomorphic, yellow or orange, 4 to 8 mm wide, corolla lobes at first valvate, later spreading, glabrous within except for minute capitate glandular hairs toward base. Five-lobed flowers occasionally present in some populations.

Stamens 4, anthers about 0.5 mm, alternating with corolla lobes, dehiscing introrsely by longitudinal slits but retaining adherent pollen. Stamens in *germinans* and *officinalis* types inserted basally on corolla, slender, equal or slightly unequal filaments 2 to 3 mm long bent to orientate anthers adaxially; in *marina* type inserted regularly and equally on the corolla tube at its mouth, anthers sessile or at most with a short filament scarcely exceeding the anther. Anthers are markedly unequal in *A. bicolor*, with an inner pair enclosed by an outer pair.

Ovary completely hairy or hairy only at and above the middle, superior, unilocular, flask shaped, extending into a prominent longer or shorter style with a bilobed stigma, the lobes at first appressed, later diverging in dorsiventral plane. Placentation essentially axile, the 4 ovules pendulous from a central stalk that has a terminal umbo projecting into the base of the stylar canal without closing it.

Fruit a 1-seeded, leathery capsule, ellipsoidal to flattened ovoid, with a longer or shorter persistent stylar beak; pericarp thin, green or grayish green. Seed solitary, testa thin or fragmentary, ephemeral, endosperm absent; seed consisting mainly of ripe embryo with 2 fleshy cotyledons folded, one abaxially, the other adaxially around the short plumular axis, the radicle blunt and exposed, with a distal tuft of

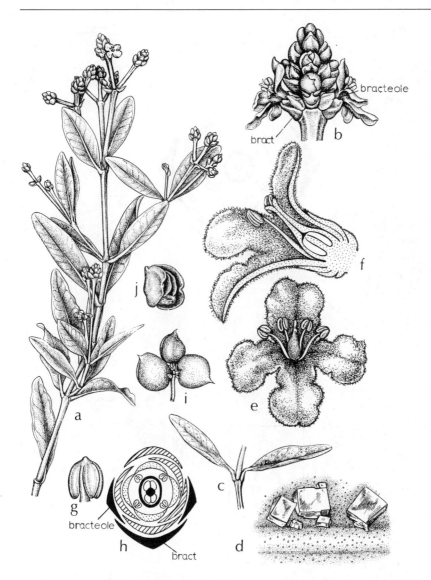

Figure B.8. *Avicennia germinans* (Avicenniaceae) shoot and flowers. (a) Distal shoot with flowers (× ½); (b) terminal flower cluster (× ⅔); (c) node (× ½); (d) leaf surface with salt crystals (× 10); (e) flower from front (× 3); (f) flower in L.S. (× 3); (g) isolated placenta with 4 pendulous ovules; (h) floral diagram; (i) fruit cluster (× ½); (j) seed (seedling) at time of release with folded cotyledons (× ½). (From Tomlinson 1980)

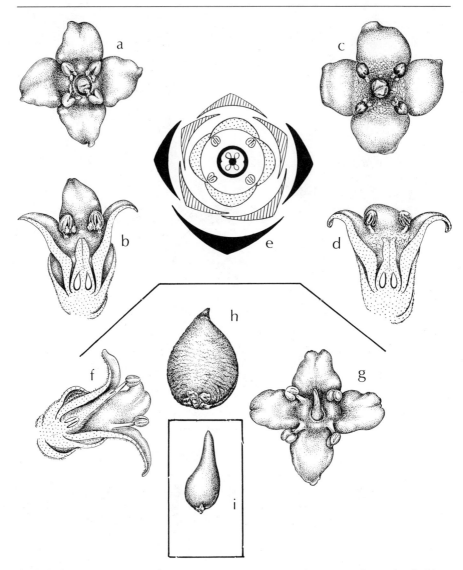

Figure B.9. *Avicennia* spp. (Avicenniaceae) flowers and fruit. (a–e) *A. marina* flowers (× 6): (a) flower from above; male stage, pollen extruded, stigmas appressed; (b) same flower in L.S.; (c) flower from above; female stage, pollen discharged, anthers turning black, stigmas diverged; (d) same flower in L.S.; (e) floral diagram. (f–g) *A. officinalis* flowers (× 3): (f) flower in L.S.; (g) flower from above, just prior to pollen release. (h–i) Fruits (× ¾): (h) fruit of *A. officinalis*; (i) fruit of *A. alba*. (Material a–d cultivated at Fairchild Tropical Garden from New Zealand collections; f–i from Sungei Santubong, Kuching, Sarawak. A. Juncosa 21.10.81A and H)

fine hairs. Cotyledons sometimes purple as in *A. officinalis* and populations of *A. marina*. Germination epigeal.

This general account may be amplified with a more detailed description of biological features; specific differences are described later.

Leaf morphology and systematics

The keys in Moldenke (1960) emphasize leaf morphology as a diagnostic feature. Examples from Indo-Malayan species (Fig. B.10, left) and the New World (Fig. B.11) give some indication of the range. Malaysian species are best distinguished by a combination of leaf shape (Fig. B.10, right) as well as color and density of the indumentum. Such characters are less useful for identification in the herbarium.

Architecture and growth

The tree conforms essentially to Attims's model but with irregular substitution of flowering axes; branching is usually discontinuous, but may be continuous in vigorous saplings. No resting terminal buds are developed, but the terminal meristem seems capable of extended inactivity (as in populations at high latitudes in both hemispheres or during drier seasons nearer the equator). In the inactive condition the bud is protected by the closely appressed margins of the petiolar grooves of the ultimate leaf pairs; axillary meristems are protected individually in the same way by older leaves. Axes vary much in the expression of their vigor; young saplings may have internodes 20 to 30 cm long, in depauperate shoots the internodes are scarcely developed. Shoot extension is often nonsynchronous in different parts of the plant.

Branching is at first always by syllepsis. On vigorous leaders or sapling shoots, all nodes may develop branches in a regular acropetal sequence, but on distal, less vigorous shoots, branching is diffuse with intervening nodes supporting unexpanded (reserve) meristems. Prolepsis characterizes all other branching, that is, the reiterative development of reserve buds or stump sprouts on badly damaged trees. A distinctive form of clonal development is by the rooting of reclining trunks in old open-grown trees, as in *A. marina* var. *australis*.

The architectural capabilities of trees are thus somewhat versatile; tall, single-trunked, broad-crowned trees occur in pure stands on good substrates; on adjacent open sites with poor soils, such as salt flats, the tree is represented by low, dense-crowned shrubs. This shrubby habit occurs exclusively in stands at the limit of the trees' range (e.g., in central Florida, southern New South Wales, and the North Island of New Zealand).

Inflorescence

The flower-bearing axes have a characteristic physiognomy that is not easily described using standard terminology, but it may be regarded as a panicle with leaves progressively reduced distally in size and number (Fig. B.8). In the flowering

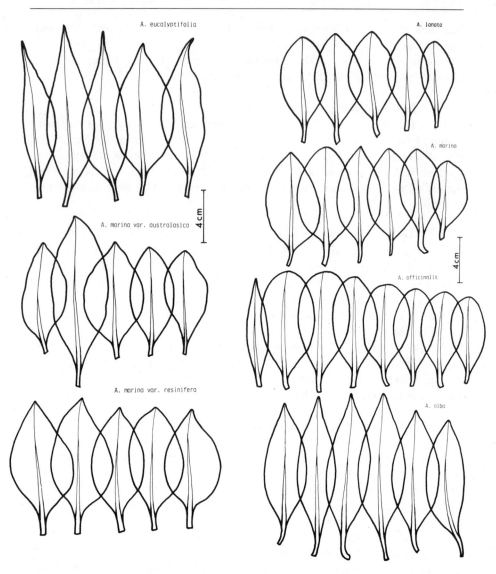

Figure B.10. *Avicennia*. Left: species of eastern Australia (*A. eucalyptifolia* and *A. marina* var. *australis*) and New Zealand (*A. marina* var. *resinifera*) showing relative constancy of leaf shape. Right: Malaysian species showing range of leaf shape and relative constancy of leaf shape for individual species.

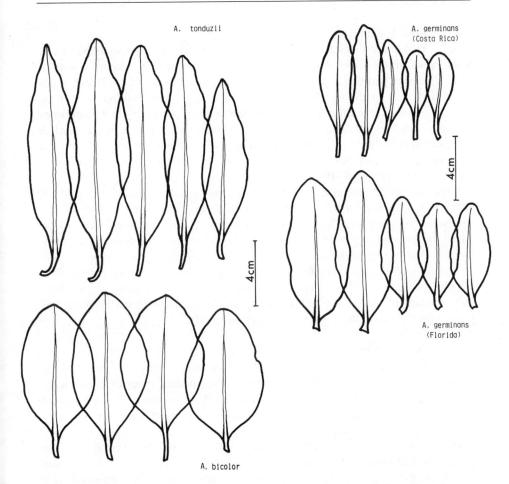

Figure B.11. *Avicennia* species from Caribbean and Central America. *A. tonduzii* is a form of *A. bicolor* with linear leaves. Size range for *A. germinans* represents variation due to soil conditions (scrub form in Costa Rica, tree form in Florida).

seasons branches toward the outside of the crown (but not the more vigorous leaders) produce determinate axes, which all end in the basic units of the inflorescence, each of which may be described as a flower cluster. Usually the lower inflorescence branches are subtended by foliage leaves, but distally the subtending structure may be a narrow leaf with characteristically inrolled margins. On vigorous shoots the lower axils may bear 2 flowering axes, one inserted above the other and apparently arising from separate axillary meristems. Because a whole branch complex is determinate, a renewal system must develop from a meristem at a much lower level, but this substitution growth is not precise in its expression. Inflorescence morphology can be diagnostic, for example, the spicate, extended inflorescence of *A. alba* contrasted with the globose inflorescence units of *A. officinalis*.

Root system

Two kinds of aerial roots develop. The least common are pendulous aerial roots, which in the saplings can lead to the development of a system of stiltlike roots. These are particularly characteristic of *A. officinalis* (see Fig. 5.2) and sometimes of *A. marina*. Similar aerial roots can develop on low branches that curve near the substrate or at other levels in response to injury. Chai (1982) demonstrates a support function for them in leaning trees. Aerial roots have been observed in American species (e.g., *A. germinans*) but are sufficiently infrequent to be treated as an anomaly. In one instance it is implied that they develop as a response to oil pollution (Snedaker et al. 1981).

More familiar are the extensive erect pencillike pneumatophores, up to 30 cm tall, which typically taper to a bluntly pointed tip, rarely branch, and have a somewhat corky texture. These pneumatophores arise from horizontal cable roots, which in turn originate at the base of the trunk and extend over distances of many meters. Descending lateral roots, so-called anchoring roots, arise more or less alternately with the pneumatophores from the cable roots. These are not long but usually become somewhat longer than pneumatophores. Both these second-order roots produce within the substrate a fine system of third-order (and higher-order) roots, which increase the absorptive area considerably.

Floral biology

The subdivision of the genus is based on three contrasted types of floral organization, which in turn seem correlated with contrasted types of floral mechanism. Protandry is universal, but the floral mechanism seems directed toward different-sized visitors in different species. The arrangement of bracteoles and calyx lobes is similar in all species.

In the *germinans* type, the flower is appreciably zygomorphic (Fig. B.8e). The stamens lie under the upper, somewhat enlarged corolla lobe; the lower lobe provides a platform for insect visitors so that pollen is most likely to be deposited on a visitor's back in the initial phase when the anthers dehisce. Subsequently the anthers shrivel somewhat and the two lobes of the stigma, which are at first appressed, diverge to expose their inner face, which appears to be the site of pollen reception. In any flower cluster there tends to be only one open flower at a time. Flower visitors are short-tongued insects, and visits by honeybees are sufficiently extensive for *Avicennia* to be a presumed source of honey ("mangrove honey") as on the Florida Keys.

In the *officinalis* type, the stamens are exserted and the style is toward the adaxial side of the corolla (Fig. B.9f). The floral mechanism is comparable to that in the *germinans* type.

In the *marina* type, the corolla is actinomorphic (Fig. B.9a–d) and the stamens either are sessile or have very short filaments. The pollen is displayed when the corolla lobes reflex; it is extruded introrsely through short longitudinal slits in each theca; these gape only slightly and the pollen remains adherent, presumably removed only by contact with an insect. In such shallow flowers with short stamens, the site of pollen deposition is likely to be the insect's abdomen. When pollen is finally dispersed, the stamens turn black and the stigma lobes diverge and become receptive. In this type protandry is more obvious, since the flower phases are more distinctly separate. Flower visitors are again recorded as bees, and the plant is considered to be a source of honey, as in *A. marina* var. *resinifera* in New Zealand.

In all three types, the floral reward is nectar, which accumulates in some quantity in the corolla tube. Its origin is almost certainly the minute glands on the corolla surface, since they visibly exude fluid; however, there are similar glands on the outside of the tube.

Seedling morphology and establishment

The fruit is incipiently viviparous, since the embryo will germinate promptly on release. The fruit is still shed as a unit, but the pericarp splits immediately or even at the time of detachment; because the testa is absent or lost with the fruit wall, the unit of dispersal is the embryo, which roots immediately on becoming stationary. The size and shape of the embryo are somewhat diagnostic for different species. The cotyledons are thick (to 3 mm), fleshy, and folded in opposite directions (Fig. B.8j). The cotyledons also may differ appreciably in shape; the outer is broadly emarginate or even bilobed, and the inner is rounded. The radicle and hypocotyl are well developed, with the radicle densely hairy and protruding slightly even in the fruit. The distinction between embryos with and without a prominent plumule, a diagnostic character used by some authors, seems obscure and has not been used in this account. The plumule is always short when the embryo is shed.

Germination is epigeal, the hypocotyl extends; the radicle may also contribute to extension, since the hairy region becomes longer, notably in *A. officinalis*. The shape of these hypocotylar hairs can be diagnostic; they are hooked in *A. alba*. Lateral root primordia become evident on the radicle apex; these develop and anchor the seedling. The cotyledons do not yet unfold so that there is a pronounced twist to the hypocotyl just below the cotyledonary node. The hypocotyl untwists and straightens so that the cotyledons are elevated and the plumule begins to elongate. The cotyledons expand during seedling extension but never become completely flattened. The first pair of plumular leaves have rather short blades and are borne at the end of a long internode.

Wood structure

A conspicuous feature of the wood of *Avicennia* is the uniform but anomalous growth rings, which consist of bands of xylem, including radial multiples of wide vessels, alternating with bands of conjunctive tissue, which include strands of phloem (Fig. B.12). Separating each band is a narrow band of sclerenchyma, 2 or 3 cells wide. The bands are quite regular, but there is radial connection between them at intervals where the conjunctive tissue of adjacent rings becomes continuous. These connections are most frequent in the region of branch insertions. Zamski

Figure B.12. *Avicennia* sp. T.S. young woody stem showing successive rings of xylem and phloem produced from successive cambia.

(1979) has shown that the phloem strands anastomose extensively in a tangential plane but that there is also continuity in a radial plane through the anastomosing connections. He suggests that phloem is active (conducting) in at least the 4 outermost rings.

It is now generally agreed (e.g., Studholme and Philipson 1966, Zamski 1979) that the rings are the products of division of successive cambia. The first cambium is formed in the usual way for dicotyledons with secondary thickening, and it cuts off xylem internally and phloem externally. It ceases to function after a limited period of activity, although it may remain locally as a nondividing layer on the inner side of phloem strands. The second cambium is formed from the innermost cortical layers; the third cambium partly from cortical cells and partly from outer parenchyma derivatives of the previous cambium; and all later cambia are derived entirely from the outermost parenchyma layer produced by each previous cambium.

In the development of each ring, the cambium produces a broad band of xylem to the inside and 6 to 10 layers of parenchyma cells to the outside. The xylem differentiates immediately, but the parenchyma differentiates centripetally, with differentiation being completed only after the cambium has ceased to function. The product of the first differentiation to the outside of the cambium is the layer of sclereids; the phloem strands appear later by cell divisions within the conjunctive parenchyma. Cambial initials remain inactive on the inside of each phloem strand, but the rest of the cambium differentiates as conjunctive parenchyma. The new cambium is then initiated by dedifferentiation of parenchyma cells immediately outside the sclerenchyma layer. Sclereids and the new xylem are always separated by 2 or 3 parenchyma layers.

Early studies suggested that the rings had a regular chronology in their appearance, either annually or semiannually, but the detailed investigation by Gill (1971) clearly shows that they are correlated directly with the stem diameter in a linear way and their formation is therefore under direct endogenous control. This can be demonstrated because the base of branches and the internode immediately above their insertion develop different numbers of growth rings, even though they are contemporaneous in development, because branches are sylleptic. The fact that the rings are nonannual can therefore be observed in a single specimen and provides an interesting exercise in deductive reasoning for students. In consequence, the number of rings cannot be used to determine age in *Avicennia*. Zamski (1981) observed that there is no correlation between the ring number at any level and the number of branches above that level, so that ring formation is not directly dependent on branch development. Most of this information relates to *A. germinans* and *A. marina* but is probably generally applicable, such is the evident uniformity of the wood structure in the whole genus.

The functional significance of this type of anomalous wood structure is not clear, although the vascular tissues of branches are highly integrated with that of stems through the higher frequency of anastomoses between rings in the region of their insertion (Zamski 1979). The anomalous structure of the wood of *Avicennia* renders it unsuitable for constructional timber, since the phloem decays relatively rapidly to leave extensive pores. The xylem, however, is persistent and quite hard.

Leaf and hair anatomy

The microscopic anatomy of the leaf of *Avicennia* seems very consistent and has been described by numerous authors (e.g., Areschoug 1902, Baylis 1940, Fahn and Shimony 1977, Mullan 1931a,b, Trochain and Dulau 1942). It is characteristic of the leaf of many mangroves in its markedly dorsiventral symmetry (see Fig. 4.15), with a well-developed colorless hypodermis below the adaxial epidermis, abundant terminal tracheids (Fig. 4.13), and few bundle sheath fibers (Fig. 4.14D). A feature is the glandular hairs on the upper and lower surfaces that secrete salt (Fig. 6.4G,H). Those on the lower surface are obscured by a dense layer of 3- or 4-celled uniseriate capitate hairs (Fig. 4.15). These two basic hair types occur in various guises on most parts of the plant. For example, glandular hairs in the flower apparently secrete nectar; capitate hairs on the calyx become enlarged and somewhat T-shaped; those on the seedling of *A. alba* are hooked. There is some suggestion in the literature that the Old World species (e.g., *A. marina*) differ from the New World species (e.g., *A. germinans*) in having 3-celled and not 4-celled stalks to the capitate hairs. I have not been able to confirm this suggestion. There is, however, appreciable variation in leaf anatomy among individuals of the same species in such features as epidermal wall thickness, density of glands, number of cells per gland, and density of uniseriate hairs, as reported for *A. germinans* by Schnetter (1978). Factors apparently responsible for this variation include soil salinity and degree of shading. Extremely shaded plants that grow in sites with low salinity had large, nearly hairless leaves with very thin external walls and a poorly developed mesophyll palisade. This kind of variation must be considered when comparative studies are done for systematic purposes.

Provisional key to species of Avicennia

1A. Corolla orange-yellow, glabrous within. Plants of the eastern mangroves, that is, in the area of East Africa, Indo-Malaya, and the Western Pacific . 2

1B. Corolla white, creamy white or with a yellow throat, rarely yellow, glabrous or hairy within. Plants of the western mangroves, that is, West Africa and tropical America 6

2A. Flowers 6–10 mm in diameter when expanded, appreciably zygomorphic, stamens exserted, 3–4 mm long, with the filament exceeding the anther. Ovary about 5 mm long, oblique, hairy throughout. Flower heads capitate, globose, the units 5–7 mm in diameter; bracts with the margin fimbriate and becoming black with age. Leaves usually rounded apically, never glaucous white below. Fruit 2–2.5 cm long (Fig. B.9h), abruptly narrowed to a short beak, russet brown . . . .*A. officinalis* L.

2B. Flowers 3–4 mm in diameter when expanded, actinomorphic, stamens included, 1–2 mm long, the filament about equaling the anther. Ovary about 3 mm long, erect, usually hairy only on the upper half. Flower heads never as wide as 5 mm, bracts without fimbriate or black margins. Leaves pointed or rounded apically, sometimes glacous white beneath. Fruit various, never russet, sometimes beaked 3

3A. Leaves usually rounded apically 4

3B. Leaves usually pointed apically 5

4A. Young twigs, petioles, midrib, leaf undersurface and upper surface of expanding leaves conspicuously but finely woolly hairy, giving a characteristic yellowish white color. Ovary hairy throughout, the apical hairs obscuring the stigma. Fruit rounded apically or with a short beak, covered with a dense woolly indumentum. . . . .*A. lanata* Ridl.

4B. Young twigs, etc., with a yellow but not woolly hairy indumentum. Ovary with a median ring of hairs not obscuring the stigma. Fruit rounded or at most shortly beaked, grayish green, never yellowish. Fruit abruptly narrowed to a short beak, silvery gray-green . . . . *A. marina* (Forsk.) Vierh.

5A. Leaves silvery gray or white beneath; inflorescence spicate. Fruit up to 4 cm long, grayish green, gradually narrowed to an extended beak (Fig. B.9i). Bark grayish; lenticellate smooth but often blackened by a sooty mould . . . .*A. alba*

5B. Leaves greenish yellow beneath, inflorescence capitate, never spicate. Fruit shorter than 3 cm, greenish yellow, without a conspicuous beak. Bark flaky, peeling in patches, mottled yellow brown or green. . . . .*A. eucalyptifolia* (Zipp. ex Miq.) Moldenke

6A. Corolla conspicuously hairy within, appreciably zygomorphic, stamens slightly to appreciably unequal. 7

6B. Corolla glabrous within, only slightly zygomorphic; stamens equal and
      included (restricted to Windward Islands, Trinidad, and northern South
      America to Uruguay)......*A. schaueriana* Stapf & Leechman ex Moldenke
7A. Flowers 10–15 mm long and almost as wide at anthesis, stamens exserted
      and only slightly unequal; style long, exserted from calyx when corolla is
      shed. Fruit distinctly beaked, glaucous, surface rough but not pitted. Leaf
      ovate to elliptic at most. Widely distributed from tropical America to West
      Africa ..... *A. germinans* (L.) Stearn (including *A. africana* P. de Beauv.)
7B. Flowers 5–6 mm long and about as wide at anthesis; stamens included and
      markedly unequal, with a short (filament 0.5 mm) inner pair and a longer
      (filament 1 mm) outer pair; style short, either deciduous or not exserted
      from calyx when corolla is shed. Fruit blunt, greenish yellow, surface
      irregularly pitted. Leaf blade oblong, elongate, often less than 3 times as
      long as wide. Restricted distribution in Costa Rica, Panama, and adjacent
      Colombia ....*A. bicolor* Standley (including *A. tonduzii* Moldenke)

Species of East Africa and the Indo-Pacific

Avicennia officinalis L. 1753   (Fig. B.9f,h)
[*Sp. Pl.* 1:110]

A. *tomentosa* Willd. 1800 [*Sp. Pl.* 3(1):395]

This species has a wide range from South India through Indo-Malaya to
New Guinea and eastern Australia. It is one of the more distinctive *Avicennia*
species of the eastern tropics and is recognized by its large orange-yellow flowers
that are 10 to 15 mm in diameter and therefore larger than any other of the eastern
group. The flowers have a rancid or fetid smell and are slightly zygomorphic, since
the adaxial (posterior) lobe of the corolla is the broadest and usually shallowly
bilobed, a feature that recalls the corolla of the white-flowered New World *A.
germinans*. In addition the ovary is slightly oblique, with the style below the adaxial
corolla lobe and not in the center of the flower. Unlike *A. germinans*, however,
the inner surface of the corolla is glabrous. The tips of the petals unfold in an
irregular way, blackening somewhat with age. The stamens are 3 to 4 mm long,
exserted, and prominently displayed with well-developed filaments that exceed the
anthers. The ovary is 0.5 mm long and densely hairy throughout, except the tip of
the well-developed style, which is 3 to 4 mm long and tapered to the unequal stigma
lobes. The flower heads are rather short (1 to 1.5 cm long and 0.5 cm wide) and
are always more or less globular (cf. *A. alba*), but they have up to 12 flowers per
unit. The bracts and bracteoles are black tipped with a fringed margin. The fruit
is densely hairy, about 3 cm long, somewhat longer than wide, and with a short
apical beak. The hypocotyl of the seedling often becomes hairy throughout. The
cotyledons can be either purple or green. The bark is smooth, lenticellate, light
colored, and not fissured.

Leaf shape is rather constant, being ovate or broadly elliptic to obovate, and the
apex is rounded, but this shape is not sufficient always to distinguish it from *A.
marina*. However, *A. marina* has a whitish, not greenish yellow, underleaf. Fre-

quently in *A. officinalis* the indumentum fails to develop in patches or even over the whole lower leaf surface. The seedling leaves are narrow with an acute apex, and this seedling character is illustrated in an exaggerated form by Biswas (1934, under "*A. marina*"); it also occurs in *A. alba*.

Aerial stilt roots seem more commonly developed in this species than in other *Avicennia* species.

Nomenclature

This species is the type of the genus (1 of the 2 described by Linnaeus), but as Moldenke (1960) points out, the name "*A. officinalis*" has been widely misapplied by herbarium and other workers and is a major contribution to the nomenclatural confusion with which species of this genus are invested. In the field, however, species can be distinguished with little difficulty. Examples of earlier confusion are the misapplication of the name "*A. officinalis*" to *A. marina* and the application of the name "*A. officinalis*" to one or more species of *Avicennia* in the New World even though the true *A. officinalis* does not occur there. The ultimate is the reduction of all species and varieties to synonymy under this one name. This confusion seems to have originated with Linnaeus himself; having described two species, the Asian *A. officinalis* and the New World *A. germinans*, in the first edition of the *Species plantarum*, he later extended his concept of the former to include both species, implying that *A. germinans* could be reduced to synonymy. His first assessment, however, was correct, because we now know that the two species are morphologically dissimilar as well as geographically separate.

Where good illustrations are available, it is sometimes possible to correct an author's nomenclature; thus Biswas (1934) describes *A. marina* as "*A. officinalis*" and the real *A. officinalis* as "*A. tomentosa*" (a common early mistake of Indian authors); his *A. alba* was correctly named. On the other hand, identification using illustrations of leaf shape alone may not be reliable.

As a consequence of this kind of confusion, descriptions by some authors are virtually worthless in the absence of voucher specimens, and few modern workers are likely to be in a position to verify identifications, even if specimens exist. The situation is particularly difficult for ecologists who may have no reliable basis for comparative study; geographical distribution still remains a valuable diagnostic character.

Avicennia marina (Forsk.) Vierh. 1907   (Fig. B.9a–e)
[*Denkschr. Akad. Wiss. Wien Math.-Nat.* 71:435]

A. *intermedia* Griff. 1846 [*Trans. Linn. Soc. Lond. Bot.* 20:pl. 1]
A. *mindanaense* Elmer 1915 [*Leafl. Philipp. Bot.* 8:2868]
A full bibliography is given by Moldenke (1960).

When the name is used in the widest sense, this species has the broadest distribution, both latitudinally and longitudinally, of the genus (indeed of any mangrove), with a range from East Africa and the Red Sea (the type locality) along tropical and subtropical coasts of the Indian Ocean to the South China Sea, throughout much of Australia into Polynesia as far as Fiji, and south to the North Island of New Zealand.

In this broad sense, this species may be recognized by its regular orange-yellow flowers not more than 6 mm in diameter with 4 equal lobes of the corolla that are glabrous within (except for minute glandular hairs). The stamens are short, scarcely 2 mm long, and not exserted, the filament about equaling the length of the anther. The ovary is rather short (2 to 3 mm), usually with a short indistinct glabrous style and two equal stigmas that diverge only in late anthesis. The ovary itself is glabrous below but hairy above as far as the style. The flowers are sweetly scented. The fruit is nearly spherical to slightly ovoid in outline but is always about as long as it is wide (2 cm); the short beak evident early in development is almost completely obscure in the mature fruit. The seedling hypocotyl is hairy only at its base, in the region of the short radicle. The leaves are rounded apically, more or less ovate, and whitish beneath. The bark is rather distinctive, being flaky, mottled greenish yellow, and peeling in patches. The flower heads are capitate and remain so, but with the units rather distant. The name *A. intermedia* Griff. sometimes used (incorrectly) is descriptively appropriate where *A. marina* co-occurs with *A. officinalis* and *A. alba*, since it has the leaf shape of the former and the indumentum of the latter.

Several varieties have been recognized, but morphological distinction is never clear cut and much of the segregation is geographic. Therefore, in the absence of precise morphometry on populations in widely separated localities and the recognition of other characters, it is impossible to provide keys with reliable diagnostic features.

The following represents the range of taxonomic variation within *Avicennia marina* in this wide sense, largely following Bakhuizen van den Brink (1921) and Moldenke (1960, 1967):

1. *Avicennia marina* (Forsk.) Vierh. var. *typica* Bakhuizen = *A. marina* (Forsk.) Vierh. var. *marina* of Moldenke.

In the opinion of Bakhuizen van den Brink, this has a limited range in East Africa and Arabia (Red Sea), which is the type locality (Yemen). However, Moldenke's view is much broader, and the variety in his interpretation ranges from East Africa through Indo-Malaya to Australia and New Caledonia.

2. *Avicennia marina* var. *resinifera* (Forst.) Bakhuizen used in much the same sense by both authors. This variety is distinguished by its compact inflorescence units with 2 to 12 flowers per unit, the elliptic oblong leaf blades with an acute or acuminate apex, and the ovary hairy only in the upper half. It is distributed in the Western Pacific, that is, Australia, the North Island of New Zealand, New Caledonia, New Guinea, and the Philippines. Again, Moldenke ascribes to it a wider distribution than does Bakhuizen and includes localities as far west as Sumatra in its range.

The uncertainty of these designations is suggested by the inclusion of New Caledonia in the range of both the typical form and var. *resinifera*. There is only one form in New Caledonia, and in my experience it clearly corresponds to the New Zealand taxon and is therefore presumably var. *resinifera*.

3. *Avicennia marina* var. *intermedia* (Griff.) Bakhuizen is included within *A. marina* var. *marina* by Moldenke. Bakhuizen records it throughout the Malay Archipelago, from the Malay Peninsula to New Guinea and the Philippines. It is said to be distinguished by its leaf shape (ovate-rotund, with an obtuse base and short petiole, less than ⅕ cm long) and the characteristically small fruits. Bakhuizen based his description on living material.

4. *Avicennia marina* var. *rumphiana* (Hall. f.) Bakhuizen is accepted by Moldenke. It is said to be distinguished by its long petioles (1.5 to 3 cm) and the rounded leaf apex. This has much the same range as the preceding variety.

In addition, Moldenke recognizes the following new varieties:

5. *Avicennia marina* var. *anomala* Moldenke 1940 (*Phytologia* 1:141); recognized by its "attenuated" inflorescences with scattered or opposite pairs of flowers and recorded for only Low Island, Port Douglas, and Queensland.

6. *Avicennia marina* var. *acutissima* Stapf & Moldenke, ex Moldenke 1940 (*Phytologia* 1:411); recognized by "its decidedly sharp-acute or acuminate leaf apex." It is restricted, according to its author, to the vicinity of Bombay.

7. *Avicennia marina* var. *australis* is a name often used by Australian field-workers, but it seems to have no taxonomic validity and is simply a geographic designation.

In view of the extreme uncertainty of the status of other varieties of *A. marina*, it seems inappropriate to recognize presumed forms of such limited occurrence as categories 5 to 7.

Avicennia alba Blume 1826   (Figs. B.9i, B.13, B.14)
[*Bijdr. Fl. Ned. Ind.* 14:821]

A. marina (Forsk.) Vierh. var. *alba* (Blume) Bakh. 1921 [*Bull. Jard. Bot. Buitenz.* Series 3(3):205, 207–19]

Avicennia alba has a wide distribution from India to Indochina, through the Malay Archipelago to the Philippines, New Guinea, New Britain, and northern Australia. It is distinguished within the *A. marina* complex by the long spicate distal flower units that are 1.5 to 3 cm long, with 10 to 30 flowers per unit and the lower flower pairs usually quite distant. The flowers are of the *marina* type, the ovary scarcely 2 mm long, and the style absent; the middle of the ovary is conspicuously hairy, but the lower part includes only glandular hairs. The leaves of *Avicennia alba* are characteristically lanceolate, pointed at the apex (in contrast to the rounded apex of *A. officinalis* and *A. marina*), and with a white undersurface.

The good field character is the conical shape of the fruit (and consequently seed), which is extended into a pronounced beak, especially in the early stages of development (Fig. B.13). The bark is slightly roughened, but not fissured, and is some-

Figure B.13. *Avicennia alba* (Avicenniaceae). The distinctive fruits were described by Watson (1928) as "resembling a gorged leech." Ponggol, Singapore.

Figure B.14. "As spent swimmers, that do cling together and choke their art": *Avicennia alba* (Avicenniaceae). Seedlings of this species have characteristic hooked hypocotyl hairs so that groups of entangled individuals are common. Ponggol, Singapore.

what brown. A frequent character, which may be diagnostic, is the development of a sooty mold on older stem parts.

Several observers (e.g., Chai 1982, Watson 1928) comment on the tendency of drifted seedlings in this species to become clumped, apparently because the hooked hairs of the hypocotylar axis of different seedlings become entangled (Fig. B.14). These hairs are also useful diagnostically.

Avicennia lanata Ridley 1920
[*J. Fed. Malay State Mus.* 10:151–2]

A. *officinalis* L. var. *spathulata* Kuntze 1891 [*Rev. Gen. Pl.* 2:502]

This species is readily distinguished in the field by the white woolly indumentum that covers leaves and younger stem parts so that even from a distance the crown is characteristically whitish. Flowers are of the *marina* type but with very hairy outer petals and calyx. The ovary is about 2.5 mm long and hairy; the hairs project distally and envelope the base of the short style, which has a bifid stigma.

According to Ridley, *A. lanata* is distinguished from *A. officinalis* by its recurved (not erect) calyx lobes (corolla lobes according to Moldenke, since the calyx lobes are always erect in all species), the filaments the same length as the anthers (not longer), the short (not long) style, and the agreeable (not disagreeable) floral fragrance. The flowers of the two species are possibly the same size, but this is nowhere reported unequivocably, and Moldenke speculates that this species is in fact close to *A. marina*.

The species has been recorded only for the states of Malacca and Pahang in the Malay Peninsula and more recently for extreme western Sarawak (Chai 1982). There are several records for Singapore, but most sites seem to have been destroyed by real estate development.

Avicennia eucalyptifolia (Zipp. ex Miq.) Moldenke 1960　(Fig. B.7)
[*Phytologia* 7(3):162–5; see Zippel ex Miquel, *Fl. Ned. Ind.* 2:912]

A. *officinalis* var. *eucalyptifolia* Valet. 1907 [*Bull. Dep. Agric. Ind. Neerl.* 10:53]

This taxon has been recognized by a number of authors entirely on the basis of its ovate-lanceolate to lanceolate leaves, which may be 3 to 5 times as long as they are wide or, if less, with the leaf apex gradually narrowed to a point. Moldenke regarded this feature as sufficient for recognizing a new species. *Avicennia marina* is largely separated by its rounded or at most bluntly or abruptly pointed leaves. However, the two seem to have identical flowers and fruits.

A distinguishing character may be the smooth bark, which is grayish, light brown, to greenish yellow and flaking in patches. *A. marina* seems to have similar bark characters in part of its range, however. Thus, Semeniuk et al. (1978) illustrate for the bark of *A. marina* what might in Queensland be thought of as that of *A. eucalyptifolia*. They suggest that *A. eucalyptifolia* is merely a northern variant of *A. marina* but do not describe its bark. Correlation of bark character with other features should be explored more fully as an aid to possible field identification.

Avicennia balanophora Stapf and Moldenke ex Moldenke 1940
[*Phytologia* 1:409–10]

The status of this species is dubious. It is based by Moldenke on material from Queensland (type F. Mueller *s.n.*), and its locality is said to be the mouth of the Brisbane River. It is also recorded from Keppels Isles (MacGillivray *s.n.*). It is said to be recognizable by its oblong, acornlike fruit and the corolla limb growing to 6 mm wide during anthesis. However, recent search of the mouth of the Brisbane River has not revealed any *Avicennia* except *A. marina*, with which this species may be equated in the absence of further evidence to the contrary. This type locality is now, in part, the site of an international airport.

Species of the New World and West Africa

Avicennia germinans (L.) Stearn 1958   (Fig. B.8)
[*Kew Bull*. 1958:34–6]

A. nitida Jacq. 1760 [*Enum. Syst. Pl. Carib.* 25]
A. tomentosa Jacq. 1760 [*Enum. Syst. Pl. Carib.* 25]

The full synonymy of this species is long and complex and has been provided in full by Moldenke (1960), but the resolution by Stearn is now generally accepted (see also Compere 1963).

Avicennia germinans is widespread from southern Florida and the Bahamas throughout the Caribbean to Mexico, Central America, and the Pacific coast of South America to Peru and the Galapagos; its distribution on the Pacific coast of the Americas seems discontinuous. It is recorded on the Atlantic coast of South America to northern Brazil, but some of this distribution may be in error because it has not always been clearly distinguished from *A. schaueriana*.

In its vegetative morphology *A. germinans* is characterized by its ovate leaves, usually with a blunt apex; the lower leaf surface and the twigs are conspicuously glaucous white in contrast to the dark green of the upper leaf surface. The leaves are 5 to 8 by 2 to 4 cm with a short petiole. The flowers are the largest in the genus, white and conspicuously zygomorphic, with the stamens long, equal, and exserted; the petals are conspicuously hairy inside; the fruits are somewhat variable in size (2 to 3 cm long) and always with a short lateral beak.

Nomenclature

This species was made known to European science by material collected by Patrick Browne in Jamaica that became incorporated into the Linnaean herbarium and used by Linnaeus, together with material from Venezuela, for the description of *Bontia germinans* in the *Systema naturae*. In the second edition of the *Species plantarum* (1763, 2:891) Linnaeus expanded his use of the name to include material from India that he had described as *Avicennia officinalis* in the first edition (1753, 1:110). This later was recognized to be a distinct species; because the confusion was soon evident, even to Linnaeus's contemporaries, the name *A. nitida* of Jacquin was adopted for the New World species and became well established until Stearn pointed out the need for clarification.

Avicennia africana Palisot de Beauvois 1805
[*Flore Oware Benin* 1:79–80, pl. 47]

This name is applied widely to material from West Africa, but in the absence of any clear-cut and consistent morphological distinction, it may be better to broaden the concept of *A. germinans* and include plants from both sides of the Atlantic. Schauer (in Moldenke 1960) suggests that there may be a difference in the color of the dried leaves and that the African species has narrower and more elongated leaves. No opinion is based on a comparison of living or fluid-preserved material from extensive natural populations, which should surely be the basis for a taxonomic resolution of the problem.

Avicennia bicolor Standley 1923
[*J. Wash. Acad. Sci.* 13:354]

A. tonduzii Moldenke 1938 [*Phytologia* 1:273–4]

Avicennia bicolor has a limited distribution and is recorded for Costa Rica and the Pacific coast of Colombia at the mouth of the Buenaventura River. When growing with *A. germinans*, this species can be recognized by the dense, dark green crown, which forms a discontinuous understorey to the more diffuse, grayish crowns of *A. germinans*, which overtops it. Otherwise it can form pure stands behind the *Rhizophora* zone, with trees of appreciable stature. Its flowers are distinctly zygomorphic and like those of *A. germinans* in having the petals hairy inside, but they are much smaller, scarcely 5 to 6 mm in diameter, and expanded. The white corolla sometimes has a yellow throat, hence the name, but it can be almost pure yellow in some populations. A distinctive feature is the inequality of the stamens; they are inserted at the same level in the flower, but the outer pair has filaments at least 1 mm long, compared with the inner pair where filaments are scarcely 0.5 mm long.

Although described by Moldenke as "very distinct," *A. tonduzii* seems to be only a variant of *A. bicolor* distinguished by the narrow leaves, (Fig. B.11) and the paniculate assemblages of floral axes with the individual flower pairs distant from each other. Material from Panama to southern Mexico (Chiapas) has been referred to it.

Avicennia schaueriana Stapf and Leechman ex Moldenke 1939
[*Lilloa* 4:336]

This species has a more or less continuous distribution from the lower Lesser Antilles and the Atlantic coast of northern South America from Guyana south to Uruguay. It is based originally on material sent by Leechman from Guyana (British Guiana) to Kew, where it was studied by Stapf, who wrote a lengthy unpublished account (quoted by Moldenke in full) of the contrasted sets of specimens at his disposal under the names "*A. nitida*" (*A. germinans*) and "*A. tomentosa*" (whose origins he discusses) and for which he suggests the name that was later formally applied by Moldenke.

The flowers are larger than those of *A. bicolor*, the corolla being 10 to 12 mm long and almost as wide, and so they approach those of *A. germinans* in size, but the inner face of the corolla is glabrous or at most slightly hairy. The lobes are narrow and are not reflexed and so enclose the equal stamens with filaments 1.5

to 2 mm long. The ovary is uniformly hairy and not beaked so that it is not exserted or only scarcely exserted when the corolla falls, unlike *A. germinans*. The fruit is a pale sap green, seldom with a purple tinge, and is flatter and more pointed than that of *A. germinans*.

Family: Batidaceae

A family of uncertain affinity containing the single genus *Batis*, this is a characteristic coastal plant throughout the tropics in salt flats, often extending into the mangal fringes.

Batis P. Br. 1756
[*Civ. Nat. Hist. Jamaica*:357]

A subshrubby genus of 1 or possibly 2 species with linear succulent leaves and spreading, prostrate, or scrambling woody stems. Plants are dioecious with reduced fleshy spicate inflorescences.

Batis maritima L. 1759 (Fig. B.15)
[*Syst. Nat. Ed.* 10:1289]

Leaves opposite, sessile, and subcylindrical to about 3 cm long. Spikes in leaf axils are up to 1 cm long with fleshy bracts (male), or in the female flower the carpels are fleshy and fused together. The flowers have no perianth; the male includes 4 stamens, the female a fleshy pistil with 4 apically inserted ovules. The fruit consists of the fleshy aggregate flowers with the numerous included seeds and is dispersed by water or possibly animals.

The plant is strictly a halophyte but is common in hypersaline areas adjacent to mangal and is commonly associated with more salt-tolerant species, for example, *Avicennia* in South Florida.

Family: Bignoniaceae

A tropical family of trees and lianes usually with opposite pinnately compound leaves; the flowers large and conspicuous with a tubular, trumpet-shaped corolla. A number of genera are recorded as having species that are mangrove associates;

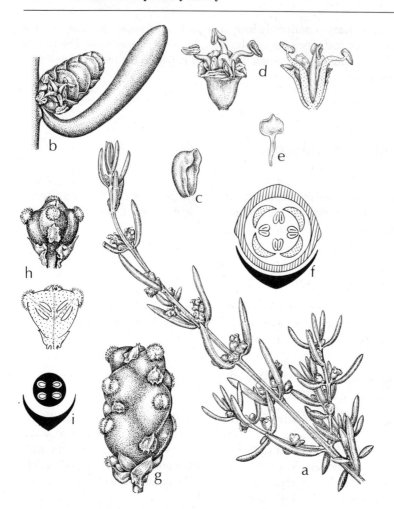

Figure B.15. *Batis maritima* (Batidaceae) shoots and flowers. (a) Flowering shoot ($\times$ ½); (b) male inflorescence ($\times$ 3); (c) single male flower in bud ($\times$ 6); (d) male flower at anthesis ($\times$ 6); left: side view; right: L.S.; (e) petal ($\times$ 6); (f) floral diagram of male flower; (g) female inflorescence ($\times$ 3); (h) aggregate of female flowers ($\times$ 3); upper: side view; lower: L.S.; (i) floral diagram of single female flower. (From Correll and Correll 1982)

Dolichandrone is an Old World genus of trees; 4 genera are New World: *Anemopaegma* and *Phryganocydia* are viny, and *Amphitecna* and *Tabebuia* are trees.

Dolichandrone Fenzl. ex Seeman 1862
[*Ann. Mag. Nat. Hist.* Series 3, 10:31]

A genus of about 9 species with a range from East Africa to New Caledonia. The Australian species are unusual in that their leaves have reduced, coriaceous, and even needlelike leaflets. *D. spathacea* has the widest range of any species in the genus and is a frequent constituent of back mangal, but only in areas flooded by the highest tides. Most typically it grows in swampy or beach communities such as dunes and dune slacks or on river banks.

Dolichandrone spathacea (L.f.) K. Schumann 1889 (Figs. B.16, B.17)
[*Fl. Kais.-Wilhelm Land*:123]

> See Chatterjee, D. (1948, *Bull. Bot. Soc. Bengal* 2:67) for a full synonymy. A modern account is provided by van Steenis (1977); see also van Steenis (1928).

This species is found from southern India throughout Malesia to New Caledonia. It is distinguished from *D. serrulata* of India and Ceylon, which has similar long-tubed flowers, by its entire (not serrate) leaflets.

Evergreen or briefly deciduous trees up to 20 m high with conspicuous perfect flowers; bark gray to dark brown, somewhat fissured in older trees. Twigs thick, conspicuously lenticellate with prominent and rather irregular pairs of leaf scars; bud scales not developed. Leaves opposite, decussate, imparipinnate, 20 to 30 cm long, petiole to 6 cm, with 2 to 4 (usually 3) pairs of leaflets and a somewhat large terminal leaflet, petioles and young leaves often reddish. Leaflets 5 to 10 by 3 to 5 cm, ovate to ovate-lanceolate, narrowed gradually to the pointed apex, base abruptly narrowed to a short (1 cm) petiolule; margin entire, often somewhat undulate, glabrous, or at most minutely hairy when young; lateral leaflets often asymmetric, with the basiscopic half narrowest. Domatia with a hairy fringe in abaxial angles between larger veins and midrib.

Flowers large, perfect, zygomorphic, solitary in the axil of minute scalelike bracts and forming terminal racemes of 2 to 6 flowers at the end of leafy twigs. Flowers develop in acropetal order; pedicel thick, about 2 cm long with bracteole inserted at about the midpoint. Calyx green, inflated tubular, 6 to 8 cm long, split adaxially, spathelike and recurved at anthesis, sometimes caducous; apex with a blunt mucro and a purple glandular patch on the abaxial side. Corolla at first green, maturing white, tubular, trumpet-shaped, 15 to 20 cm long, the tube only 7 to 8 mm wide, but about 10 cm long, abruptly enlarged to the 5-fringed lobes, spreading and about 12 cm in diameter at maturity, the 3 abaxial lobes 4 to 5 by 3 to 4 cm, the 2 adaxial lobes somewhat shorter, with a shorter median sinus between them. Stamens 4, with the fifth adaxial stamen represented by a filiform vestige; fertile stamens

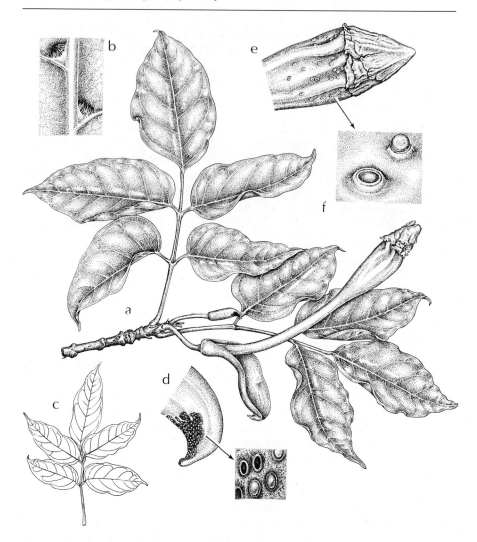

Figure B.16. *Dolichandrone spathacea* (Bignoniaceae) leaves and flowers. (a) Flowering shoot
(× ½); (b) domatia (× 6) in angle between midrib and main vein on lower leaf surface; (c) leaf
outline (× ⅙); (d) abaxial glands in tip of calyx (× ³⁄₂), inset: detail of glands (× 12); (e) flower
bud just before anthesis (× ³⁄₂); (f) surface glands on calyx (× 12). (Material cultivated at Fairchild
Tropical Garden, Miami, Florida)

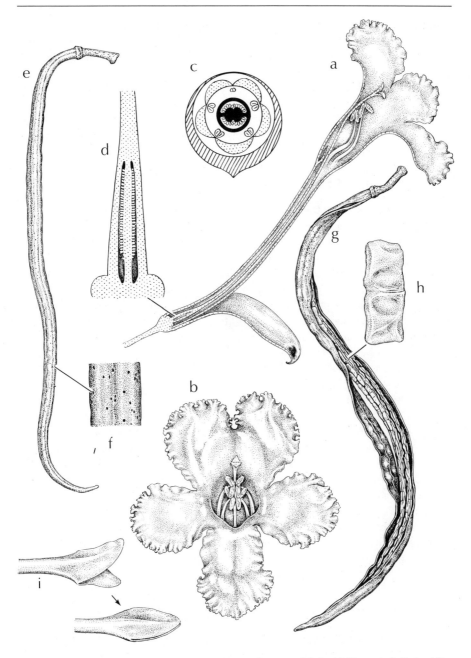

Figure B.17. *Dolichandrone spathacea* (Bignoniaceae) flowers and fruit. (a) Flower in L.S. (× ½); (b) flower from front (× ½); (c) floral diagram; (d) ovary in L.S. (× 3); (e) unripe fruit (× ¼); (f) surface of unripe fruit (× 1); (g) dehisced fruit (× ¼); (h) single seed (× ⅔); (i) stigma of receptive flower (× 3) in diverged (unstimulated) and closed (stimulated) position. (Material cultivated at Fairchild Tropical Garden, Miami, Florida)

inserted in the throat of the tube and enclosed by the corolla lobes, filaments about 3 cm long, anthers medifixed and horizontal. Ovary slender, cylindrical, on a short, narrow basal disc, tapered upward into the filiform style, stigma bilobed, peltate, extending beyond the stamens. Ovary bilocular, scarcely 1 cm long, with numerous ovules on the transversely axile placenta. Fruit a flattened, pendulous linear capsule, 50 cm or longer, drying white, with 2 valves splitting to reveal the central replum. Seeds numerous, packed in regular series, white, about 1.5 cm long, oblong-rectangular with a soft, corky winged testa; endosperm absent. Germination epigeal; cotyledons bilobed.

Growth and reproduction

Shoot construction. The species is deciduous in some climates; for ex-ample, cultivated specimens in South Florida lose their leaves in winter for several weeks. The tree is also recorded as deciduous within its natural range (e.g., Chai 1982), but the periodicity is not noted; van Steenis (1977) comments that it may "stand leafless in the dry season for many months." Shoot growth is somewhat articulate, with reduced leaves at intervals, but without the formation of organized, scaly, resting buds, despite the deciduous state. Distal shoots are sympodial through flowering, with the terminal raceme substituted by 1 or 2 lateral buds that develop by prolepsis. The architecture of the tree is not regular, but in open situations it becomes broad-crowned with a short trunk. Buttresses and aerial roots are not developed.

Floral biology

The flowers develop rapidly in acropetal order and expand in a few days from a stage where they are buds 2 to 3 cm long enclosed by the inflated calyx to the preflowering stage, after which the corolla reaches its maximum length of about 20 cm. Van Steenis (1977) emphasizes that the calyx is at first filled with water, a condition found in other Bignoniaceae and which he describes as one of the uniquely tropical features of plants. Flowers usually develop one at a time within a single inflorescence, but since there are many inflorescences at different stages on a single tree, the flowering period is extended. The corolla expands in the early evening and each flower usually lasts only 1 day, with the corolla falling on the morning of the following day. When the flower opens, a pervasive scent is produced, and nectar, presumably produced by the floral disc, accumulates copiously in the base of the corolla tube; the anthers dehisce. The stigmas protruding beyond the anthers diverge and become receptive. They are sensitive and close within 3 or 4 sec when the inner receptive face is touched (Fig. B.17i), as by the head of a suitable flower visitor. Because a visitor carrying pollen touches the stigma first, it is likely to receive foreign pollen, but the flower's own pollen, picked up as the visitor withdraws, is not likely to be deposited because the closed stigma lobes now conceal the receptive surface. Plants appear to be self-compatible so that self-pollination of different flowers on one tree is possible; isolated trees may set fruit. Pollination is presumably by long-tongued nocturnal animals, probably hawk moths, but no observations have been made.

Fruits develop rapidly, as is characteristic of many Bignoniaceae, maturing within 1 or 2 months. The fruits elongate and are at first green and beanlike, but they eventually dry and turn white and the valves separate by twisting to reveal the numerous oblong seeds. When released, these float readily and are dispersed by water but they germinate immediately on being stranded. The corky wing replaces the usually thin wing of the wind-dispersed species found otherwise in the family and provides a good example of specialization in relation to habitat.

Extrafloral nectaries

In addition to the nectar produced within the corolla tube, there are small discoid nectaries on exposed surfaces. The most obvious are those that occupy the tip of the calyx tube, but larger ones occur on the outside of the corolla lobe before it expands (Fig. B.16e,f). The function of these nectaries is not known, but van Steenis suggests that they are the source of a varnishlike material that covers the young flower bud.

Amphitecna Miers 1870
[*Trans. Linn. Soc.* 26:163]

Like the related calabash (*Crescentia cujete* L.) this genus is unusual in the family in its simple, spirally arranged leaves and nonwinged seeds.

Amphitecna latifolia (Mill.) A. Gentry 1976
[*Taxon* 25:108]

Crescentia cucurbitina L. 1771 [*Mant. Pl.* 2:250]
Enallagma latifolia (Mill.) Small 1913 [*Fl. Miami*:171]
A complete synonymy is given by A. H. Gentry [1980, *Fl. Neotrop. Monogr.* 25:66].

A small to medium-sized tree with simple, obovate, spirally arranged leaves, 5 to 10 by 15 to 20 cm, inserted on a short woody cushion. The flowers are solitary or in pairs in leaf axils or on the older wood, with a tubular, somewhat zygomorphic, yellowish green corolla. The fruit is gourdlike, 6 to 8 cm in diameter, and has a hard shell enclosing many seeds in a slimy pulp. Both the fruit and seeds are dispersed by water.

Gentry (1982) includes this as a mangrove associate and suggests that it is restricted to the Pacific coast but not the Atlantic coast of tropical America. Records of inland distributions may be due in part to its occasional cultivation, and it may not always have been distinguished from closely related species. Johnston (1949, p. 270) records it for the margins of mangrove swamps.

Anemopaegma Mart. ex DC. 1845
[*Prodromus* 9:182]

A genus of vines with about 30 species in the New World.

Anemopaegma chrysoleucum (HBK) Sandw. 1938
[*Lilloa* 3:459]

This and possibly other species of the genus have been recorded in association with mangroves in northern South America. The species is recognized by the showy, trumpet-shaped flowers and the opposite compound leaves with the rachis extended into a tendril. The capsules are rounded, 2 to 3 cm in diameter, and have numerous flattened papery seeds.

Other Bignoniaceae

Gentry (1982) records *Phryganocydia phellosperma* (Hemsl.) A. Gentry as restricted to the Pacific side of Colombia in the tropical American mangroves. This species is distinguished from the common and widely distributed tendrillous vine with large purple flowers [*Phryganocydia corymbosa* (Vent.) Bureau] by its simple tendril, short ovoid fruit with corky, almost wingless seeds, and floral details (Sandwith 1940). It is seemingly restricted to coastal, swampy habitats and might be termed a mangrove associate. Gentry also records *Tabebuia palustris* Hemsl. as a mangrove associate with a strictly Pacific distribution. It has a limited range into Panama. Johnston (1949, p. 274) records it as a shrub or small tree up to 4 m tall in the drier marginal parts of mangrove swamps, an "unobtrusive plant and not very common." It may be recognized by the trifoliate leaves and conspicuous white flowers with a yellow throat. It is deciduous.

Family: Bombacaceae

A family of soft-wooded trees with simple or digitately lobed leaves. Flowers are often large with numerous stamens. It includes a number of species of minor economic importance, including the durian fruit (*Durio zibethinus* Murr.).

Camptostemon Masters 1872
[*Hook Icon. Pl.* 12:18, Pl. 1119]

> *Cumingia* Vidal 1885 [*Cuming. Philip.*:211]
> See also Masters [1875, *J. Linn. Soc. Bot.* 14:505].

A genus with 2 species of evergreen trees in back mangal from Borneo, the Moluccas and northern Australia, to New Guinea and the Philippines. Within the Bombacaceae the genus is included in the tribe Durioneae because of its scaly pubescence, cupular epicalyx, and closed calyx, and it is distinguished from other members of the tribe by the 5 or more stamens united into an androecial tube and particularly the 2 (rarely more) locular ovary with 1 or 2 ovules in each loculus, the capsular, 2-valved fruits, and 1 or 2 hairy seeds without endosperm. The surface roots may bear knobby pneumatophores. Masters suggested that the genus formed a link between Bombacaceae and Malvaceae because of its stamen tube, which is not divided into separate phalanges. The inclusion of *Cumingia*, which broadens the concept of the genus, further emphasizes this transitional status.

The 2 species, which may not overlap in distribution, are distinguished as follows:
1. Leaves glabrous above, stamens numerous, dithecate .... *C. schultzii* Masters
(northern Australia and New Guinea)
1A. Leaves sparsely but uniformly lepidote above, stamens 5, polythecate ......
.... *C. philippinensis* (Vidal) Becc. (Borneo, Philippines)

Camptostemon schultzii Masters 1872   (Fig. B.18)
[*Hook. Icon. Pl.* 12:18, Pl. 1119]

Evergreen hermaphroditic trees to 30 m tall, trunk flanged at the base, bark dark, scaly, unfissured; exposed roots with blunt rounded pneumatophores to 30 cm tall. Younger exposed surfaces (except upper surface of leaf) with a continuous indumentum of minute peltate scales. Twigs with prominent round leaf scars. Leaves spirally arranged, petiolate, the petiole 2 to 3 cm long, terete but finely grooved above, blade 8 to 10(-15) by 2.5 to 3(-4.5) cm oblong-ovate to broadly lanceolate, margin entire, apex acute to rounded or even slightly emarginate, base cuneate but always with an abrupt notch at the petiole insertion. Flowers in axillary 3- to 6-flowered cymes, as either regular dichasia or umbellate; peduncle 4 to 8 mm long, pedicels 3 to 4 mm long. Epicalyx cupulate, 2 to 3 mm long, with an

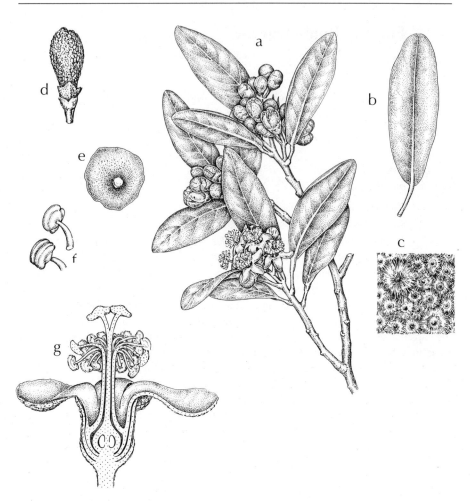

Figure B.18. *Camptostemon schultzii* (Bombacaceae) leaves and flowers. (a) Flowering shoot
(× ½); (b) leaf (× ½); (c) scales on lower surface of leaf (× 35); (d) flower bud (× 4) with
bracteoles, calyx, and scaly petals; (e) single peltate petal scale from below (× 70); (f) stamens
(× 8); (g) flower in L.S. (× 3). (Material from Missionary Bay, Hinchinbrook Island, Queensland;
part a after Masters 1872)

irregular margin; calyx campanulate, about twice as long as epicalyx, more or less
distinctly 3-lobed but margin becoming irregular with age. Petals 5, imbricate,
bluntly pointed to 6 mm long, white but densely scaly without. Stamens numerous,
united into a narrow tube about the same length as the petals, free portion of filament
1 to 2 mm long, anthers bithecate. Ovary 2 mm long, superior, globular, scaly,
included within inflated base of stamen tube, with 2 locules and one campylotropous
ovule in each locule on an axile placenta. Style slender, 8 mm long, with 2 exserted

flattened peltate stigmas, each obscurely 3-lobed. Fruit scaly, a somewhat laterally flattened ellipsoidal capsule 15 by 8 mm with persistent basal epicalyx, dehiscing loculicidally into 2 valves, each with a narrow persistent placental replum. Seeds 2, conspicuously hairy; embryo green.

Ecological and geographic distribution

The species is more characteristic of open rocky shores than estuarine mangroves and is commonly described as occurring on sandy beaches, but within the tidal range. On the coast between the Oriomo and Fly rivers in Papua New Guinea it is described (*Brass 6465*) as a "dominant species of extensive swampy rain forests; 30 m or more." It is recorded for southern New Guinea and northern Australia, but also into Indonesia as far as Borneo.

Growth

Shoot construction. The shoots are not morphologically articulate, but extension is periodic, with the exposed but unexpanded leaf primordia of resting terminal buds protected by the scaly indumentum.

Scales. The peltate scales that form the indumentum are about 200 μm in diameter, but there may be some dimorphism, since there are a few uniformly scattered, larger scales. These may represent a preliminary population that covers the leaf in early stages of development; after a period of rest, the majority of narrower scales may be produced. Scales on calyx and petals may correspond to the larger size class.

Root system. This was described by Troll (1933a). The above-ground roots are knobby, woody structures up to 30 cm tall. They originate as slight irregularities of the extending horizontal roots, much as in *Bruguiera* but less prominent. They are augmented by secondary growth, the wood being very thin walled and with few tracheids. They are not the site of feeding roots, as in *Bruguiera*. Since the cortex of the pneumatophore is not aerenchymatous, Troll says that its role as an aerating organ has still to be demonstrated.

Camptostemon philippinense (Vidal) Becc. 1898
[*Malesia* 3:273]

See also K. Schumann (1890) in Engler and Prantl [*Die natürlichen Pflanzenfamilien*, ed. 1, t. 3, ab. 6:67].

This species differs from *C. schultzii* in the presence of lepidote scales on the upper as well as the lower surface of the leaves, and the 5 stamens, which are described as polythecate. The ovary may more often have 3 or 4 instead of 2 locules; the stigma may then have more than 2 lobes, the number corresponding to the number of locules, with the capsule separating into a corresponding number of valves. The 2 species are said by Bakhuizen van den Brink (1924) also to differ in leaf shape (elliptic in *C. schultzii* as distinct from obovate-oblong to lanceolate in *C. philippinense*), but they both have a similar range of leaf size.

Nomenclature and other notes

The genus is based on *Camptostemon schultzii* (the specific name referring to the collection of *Schultz 511* Port Darwin, North Australia; the specimen had no fruit). The genus *Cumingia* was erected by Vidal (1885, *Phan. Cuming. Philip.*:212, Fig. 1) in his description of plants collected by Cuming in the vicinity of Luzon in the Philippines. He did comment on its proximity to the *Camptostemon* of Masters, but he considered it sufficiently distinct in its "unilocular" anthers. He also noted the prior use of *Cumingia* Don (in Benth. and Hook. *Gen. Pl.* 3:69) in the synonymy of *Conanthera* Ruiz et Pavon, which he dismissed in view of the remoteness of this taxon (contrary to today's rules of botanical nomenclature) in his desire to commemorate the collector's name. Schumann retained the 2 genera in his treatment of the Bombacaceae but still without a description of the fruit of *C. schultzii*.

Beccari recognized the 2 species as cogeneric, in part from having better material to study, and made the necessary transfer of Vidal's species.

There is some confusion about the structure of the androecium, even though this is a primary factor in the distinction of species or even genera in the Bombacaceae. Masters described the anthers of *Camptostemon* as bilocular, and his illustration shows a typical tetrasporangiate anther that is medifixed. Vidal described the anther as "unilocular." In Schumann's account, the distinction is between polythecate (*Cumingia*) and dithecate (*Camptostemon*) anthers, although it is not clear from the illustration what is the precise difference. The illustration in Bakhuizen van den Brink (1924) suggests that *C. philippinensis* has a tetrasporangiate anther. The difference may relate to the number of microsporangia but needs clarifying.

The geographical distribution of the 2 taxa is not clear. *Camptostemon philippinensis* is recorded from the Philippines, Borneo, and the Celebes (Sulawesi); *C. schultzii* was described first from Darwin in northern Australia and subsequently southern New Guinea, but its further distribution needs precise analysis. It is not certain whether the 2 taxa co-occur.

Family: Celastraceae

A large, mainly woody but commercially unimportant family with over 50 genera and 1000 species, having a wide distribution, particularly in the tropics. The following species has been recorded as a mangrove associate.

Cassine L. 1737
[*Gen. Pl.*:338; *Sp. Pl.* 1753:268]

> *Elaeodendron* Jacq.f. 1782 ex Jacq. [*Icon. Pl. Rar.* 1:48]

A genus of about 80 species, mainly African. The status of the genus is discussed by Ding Hou (1963).

Cassine viburnifolia (Juss.) Hou (1963)
[*Fl. Males.* 1, 6:286]

A small tree to 8 m tall with smooth gray bark. Leaves opposite, simple, shortly petiolate; blade obovate, about 7 by 4 cm, undersurface light green to glaucous; margin obscurely crenate, initially with glands in each sinus. Petiole short (1 to 1.5 cm). Flowers small (2 to 3 mm in diameter), in axillary long-stalked cymes toward the ends of the shoots, 4 or more, white; stamens alternate with the petals. Disc inside the stamens green; ovary conical, scarcely 1 mm long, with a sessile bilobed stigma. Fruit a 1-seeded drupe, ripening yellow, about 1 cm long, narrowed at the base, with corky rather than fleshy mesocarp; endocarp somewhat bony.

This species occurs in wet coastal communities and has been recorded in mangrove swamps and along tidal rivers. It has been noted (e.g., Ridley 1930) that the fruit structure is unique in the genus and clearly adapted to water dispersal. It has a range from the Andaman Islands through the Malay Peninsula to northern Sumatra, the island of Borneo, and into Sulawesi.

Family: Combretaceae

A moderately large tropical woody family of some 20 genera and 500 species, the 2 large genera *Combretum* and *Terminalia* each accounting for 200 species. The family is characterized by flowers with an inferior, unilocular ovary with usually 2 pendulous ovules, a well-developed floral disc, and a 1-seeded drupelike fruit (pseudocarp) without endosperm.

Two genera are typical constituents of mangrove communities: *Lumnitzera*, with 2 species, is an Asian genus most characteristic of back mangal; and *Laguncularia* is monotypic and common and even locally dominant in West Africa and Caribbean–tropical American mangal. Also described here are 2 other species: *Conocarpus erectus* (''buttonwood'') is an associate of the New World and West African mangal; and *Terminalia catappa* (''sea almond'') is an Asian strand species that is introduced and widely naturalized in other parts of the tropics.

Within the family Combretaceae, *Languncularia* and *Lumnitzera*, the 2 strict mangrove genera, are quite closely related and have, in the Australian genus *Macropteranthes*, a terrestrial relative closer to them than to other genera in the family (Exell and Stace 1966). On this basis, the 2 mangrove Combretaceae could have evolved from a common ancestor, though not necessarily the extant *Macropteranthes*. The leaf anatomy of *Lumnitzera* and *Laguncularia* is very similar (see Stace 1965a,b).

Artificial key to genera treated in this account

1A. Leaves opposite; plants dioecious, flowers in loose terminal panicles, petals present............................*Laguncularia* Gaertn.f. (mangrove)
1B. Leaves alternate; other characters various 2
2A. Flowers in globose clusters aggregated in loose panicles, fruit a cluster of laterally compressed nutlets...........*Conocarpus* L. (mangrove associate)
2B. Flowers in axillary or terminal spikes, not in globose clusters, fruits not clustered 3
3A. Flowers without petals, in axillary panicles, leaves large (up to 30 cm long) mostly in terminal erect rosettes; fruit a slightly winged drupe 5–7 cm long................. *Terminalia* L. (*T. catappa*) (mangrove associate)
3B. Flowers with petals in either terminal or axillary spikes; leaves 5–8 cm long, not in erect rosettes; fruit an unwinged drupe 2–3 cm long...........
.... *Lumnitzera* Willd. (mangrove)

Laguncularia Gaertn.f. 1805
[*Fruct. Sem. Pl.* 3:209, pl. 217]

A monotypic mangrove genus with a distribution restricted to America and West Africa.

Laguncularia racemosa (L.) Gaertn.f.   (Fig. B.19)
[*Fruct. Sem. Pl.* 3:209, pl. 217]

This species is typically restricted to the landward fringe of the mangrove community but also pioneers readily into disturbed sites where it can form pure stands. Pneumatophores are only facultatively developed; in some situations they are abundantly developed, in others they may be absent, but the precise stimulus for their development is not known. Good field characters are the opposite, bluntly

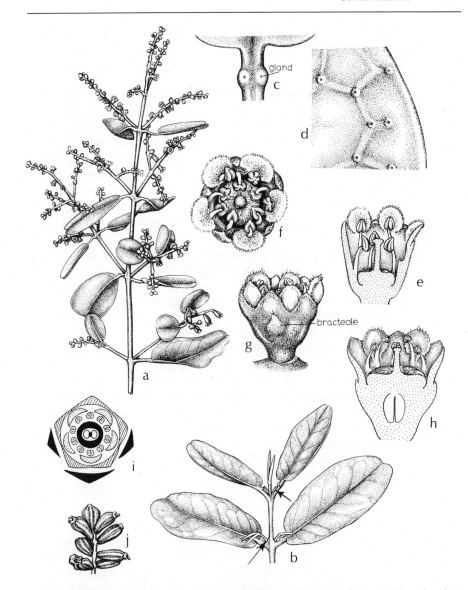

Figure B.19. *Laguncularia racemosa* (Combretaceae) leaves, flowers, and fruits. (a) Flowering shoot (× ½); (b) vegetative shoot (× ½) with developing axillary branches (arrows); (c) detail of petiole with paired glands (× 3); (d) detail of periphery of leaf blade from below (× 3), with minute peripheral glands; (e) male flower in L.S. (× 6); (f) male flower from above (× 6); (g) perfect flower from side (× 6), subtending bract represented by its scar; (h) perfect flower in L.S. (× 6); (i) floral diagram; (j) cluster of ripe fruits (× ½). (From Tomlinson 1980)

ovate, somewhat fleshy leaves and the conspicuous pair of glands on the distal adaxial part of the petiole.

Description

Dioecious or hermaphroditic trees, trunk solitary or clustered. Bark rough, fissured, gray. Shoot growth continuous or irregular, with diffuse branching (as in Attims's model but with terminal inflorescences), lacking distinct resting terminal buds or articulations. Phyllotaxis decussate, stems terete, twigs with numerous slightly prominent lenticels. Leaves with a short (1 to 2 cm) petiole, slightly grooved above and supporting 2 circular glands on the adaxial surface toward the blade. Leaf blade revolute in bud, somewhat fleshy, elliptic to oblong, 5 to 8 by 3 to 5 cm; margin entire; rounded to slightly emarginate apically, base truncate; glabrous but with several minute pits distally and submarginally on the abaxial surface at the junctions of minor veins.

Inflorescence spicate, aggregated to form loose terminal panicles. Flowers unisexual or perfect, slightly to appreciably dimorphic. Each flower 4 to 5 mm in diameter, subtended by an inconspicuous caducous bract and with a pair of lateral scalelike bracteoles on the hypanthium. Calyx cup shallow to deep, hairy without, supporting 5 shallow, pointed calyx lobes. Petals 5, rounded, greenish yellow, and hairy. Stamens 10, in 2 series of 5 inserted at different levels. Disc yellow, conspicuous. Style simple, 1 to 2 mm long. Male flowers with functional stamens, 2 mm long but included in the calyx cup; ovary absent, the base of the flower narrow; pistillode represented by style. Female flowers with nonfunctional stamens, the ovary well developed with 2 pendulous functional ovules, the base of the flower correspondingly broad. Perfect flowers recorded, with the dimensions of female flowers and apparently both functional stamens and ovules.

Fruit produced abundantly on functionally ovulate plants; a ribbed, 1-seeded nutlet 2 cm long, with the persistent remains of the calyx; seed coat spongy. Endosperm absent; germination epigeal.

Growth and reproduction

In South Florida this species is distinctly seasonal in its development, with shoot extension and branching occurring in summer months but with an extended period of inactivity in winter. Branching is mostly by syllepsis and vigorous shoots can produce a branch at every node, but branch abortion is correspondingly frequent so that trunks are characterized by numerous dead or dying branches, with only a few of the more robust ones surviving. Nodes that do not produce a sylleptic branch support a suppressed bud that functions as a reserve bud.

In the strongly seasonal climate of South Florida and probably elsewhere in its range, this species has a distinct flowering season in the warmest months, resulting in a heavy fruit set within 2 to 3 months. There seems to be a clear distinction between seeding and nonseeding trees, and the species is predominantly dioecious. Sexual segregation is not exclusive, however, because perfect flowers are recorded, possibly exclusively on trees with otherwise female flowers. The presence of self-fertile individuals would be consistent with this plant as a weedy, pioneering species, but the point needs to be verified unequivocally. Flowers are visited by bees, and insect pollination seems to be usual.

Fruiting is usually abundant (at least in South Florida) and can lead to the development of extensive carpets of seedlings from stranded fruits, but mortality in the first year is almost 100 percent.

Secretory structures

Biebl and Kinzel (1965) describe the three kinds of secretory structures ("Drüse" of German authors) on leaves of *Laguncularia racemosa*. Most conspicuous is the pair of glands in the petiole of each leaf (Fig. B.19c). These function as extrafloral nectaries (they secrete a sweet solution) and are irrigated indirectly via a supplementary petiolar vascular strand, with a branch that terminates blindly at the base of the gland. In addition there are large submarginal glands along the leaf in the marginal loops of lateral veins (Fig. B.19d). These may function temporarily as hydathodes in young leaves, secreting water (or mucilage), but the loose cells that occupy the cavity soon disorganize and the surface becomes protected by an incipient periderm. These structures become conspicuous in older, succulent leaves because the depression is exaggerated by an increase in leaf thickness. The smallest glands are microscopic and sparse to abundant but can be seen as translucent dots. They are represented by a prominent group of densely cytoplasmic cells at the base of a deep, irregular, epidermal depression (Fig. 4.16C,D). They function as salt glands; the salt solution may crystallize so rapidly that crystals are extruded in chains from the mouth of the cavity. A fourth secretory structure may be represented by glandular trichomes on the midrib and young leaves.

Lumnitzera Willd. 1803
[*Neue Schr. Ges. Naturf. Fr. Berl.* 4:186]

> *Pyrrhanthus* Jack 1822 [*Mal. Misc.* 2, no. 7:57]

A genus of 2 ecologically associated species ranging from East Africa to the Western Pacific (Fiji and Tonga), tropical Australia, and Indochina. Other taxa have been named, but their status is uncertain in the absence of further field study, and the present situation is conveniently accepted.

The species are similar in vegetative appearance and may be described collectively, although *L. littorea* tends to have a more diffuse sprawling habit, with the lower branches frequently taking root. The following key indicates the features in which they consistently differ.

> 1A. Inflorescences terminal, flowers red, shortly pedicellate and slightly zygomorphic, 16–18 mm long; stamens twice as long as the more or less erect petals. Leaf blade glabrous . . .*L. littorea* (Jack) Voigt. (not ranging to East Africa)

Figure B.20. *Lumnitzera racemosa* (Combretaceae). Fibrous fissured bark. Cape Ferguson, Queensland.

1B. Inflorescences or flowers axillary; flowers white, sessile, actinomorphic, 7–8 mm long; stamens equaling or only slightly exceeding the petals. Leaf blade sometimes hairy when young.....*L. racemosa* Willd. (ranges to East Africa)

The species may also differ in fruit structure, with the sclerenchyma strands of the pericarp either dispersed (*L. littorea*) or in a single series (*L. racemosa*).

Evergreen hermaphroditic trees up to 30 m high, trunk to 65 cm diameter or small, much-branched shrubs, often with red twigs. Bark of large-trunked specimens dark and deeply fissured (Fig. B.20). Pneumatophores commonly developed as looped above-ground laterals from the main horizontal roots, especially in *L. littorea*. Leaves spirally arranged, more or less sessile or with a short, rounded petiole 3 to 5 mm long, the leaf base pouched to enclose the axillary bud. Blade revolute in bud, 4 to 6 by 2 to 4.5 cm, succulent, veins obscure; obovate, gradually tapered at the base, rounded or slightly emarginate at the apex, with a minute abaxial apical pore and obscure marginal glands; glabrous at maturity. Flowers in short terminal or axillary spikes, perfect, actinomorphic or slightly zygomorphic, sessile or shortly pedicellate, a pair of short bracteoles inserted on the calyx tube. Calyx lobes 5, short, rounded, sometimes minutely hairy on the margins or gland-tipped; petals 5, free, red or white (rarely yellow or pink). Stamens usually 10 but often fewer, inserted on the inner rim of the calyx cup; anthers versatile, sometimes with a minute apical appendage. Ovary inferior, the calyx tube forming a deep nectar cup. Style simple, filiform, persistent, and arising from the center of the calyx cup (*L. racemosa*) or partially adnate to it on one side (*L. littorea*). Ovules 2 to 5 (perhaps 7), pendulous from the upper part of the ovary cavity. Fruit a flattened 1-seeded pseudocarp (i.e., a drupelike structure) with the sclerotic portion developed from the inner tissues of the calyx tube. Petals caducous, style and calyx lobes persistent, outer parts of dispersed (floating) fruits decaying to leave fibrous tissues of fruit wall. Germination hypogeal.

Lumnitzera littorea (Jack) Voigt. 1845   (Fig. B.21f–i)
[*Hort. Suburb. Calc.* 39]

L. coccinea Wight and Arnold 1834 [*Prodromus* 316]

Flowers in terminal spikes 2 to 3 cm long; petals red, erect; flowers slightly zygomorphic, the calyx tube curved and the style slightly adnate to the inner adaxial side of the tube. Stamens up to 10 mm long and prominently exserted at anthesis.

Pneumatophores usually well developed and consisting of looped laterals that protrude up to 10 cm above the substrate, the loops remaining spongy and with little secondary thickening (see *Bruguiera* and *Ceriops*). This species is said to be glabrous at all times. Since it has terminal inflorescences, their scars between branch forks are useful diagnostic features.

Lumnitzera racemosa Willd. 1803   (Fig. 21a–e)
[*Neue Schr. Ges. Naturf. Fr. Berl.* 4:187]

Flowers in short axillary spikes 1 to 2 cm long, petals white, spreading (or yellow in var. *lutea* Presl.); actinomorphic, the stamens somewhat exserted. Pneumatophores as looping lateral roots sometimes developed, but much less commonly than in *L. littorea*.

Lumnitzera rosea (Gaud.) Presl. 1834
[*Rep. Bot. Syst.* 1:156]

Laguncularia rosea Gaud. 1827 [*Voy. Uranie* 1:155]
Freycinet [*Voy. Bot.* 481, t. 105, f. 2]

A form with pink flowers, which is interpreted by Tomlinson et al. (1978) as a hybrid *L. racemosa* × *L. littorea*, has been recorded on Hinchinbrook Island, Queensland. This form is intermediate in morphology between its putative parents and sterile. These authors equate this with *L. rosea* (based on *Laguncularia rosea*

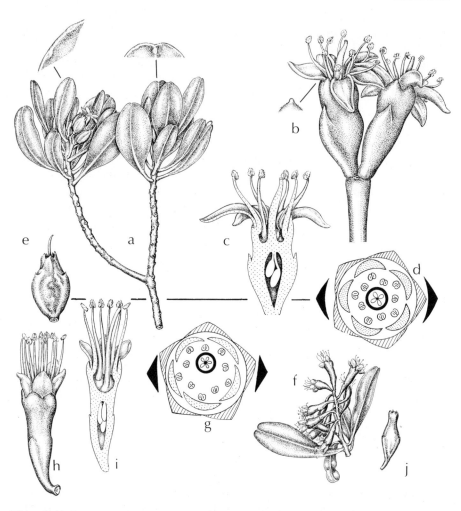

Figure B.21. *Lumnitzera* (Combretaceae) leaves, flowers, and fruit. (a–e) *Lumnitzera racemosa*: (a) habit with axillary flowers (× ½), left inset: detail of leaf margin, right: detail of leaf tip; (b) two-flowered inflorescence (× 3); inset: detail of tip of sepal (× 4); (c) flower in L.S. (× 3); (d) floral diagram; (e) immature fruit (× ³⁄₂). (f–i) *Lumnitzera littorea*: (f) terminal axis (× ¼); (g) floral diagram; (h) flower in side view (× ³⁄₂); (i) flower in L.S. (× 3). (j) Fruit (× ½). (Material a–e cultivated at Fairchild Tropical Garden, Miami, Florida; f–j from Labu Lagoon, Lae, Papua New Guinea. P. B. Tomlinson 23.10.74C)

Gaud.) on the basis of the description and illustrations by Gaudichaud-Beaupré. Unfortunately the type is not identifiable because all flowers have been lost, although the specimen is clearly the one from which the illustration (t. 105) in Freycinet's *Voyage* is drawn. This specimen is said to come from the Philippines in a locality where presently only one species (*L. racemosa*) occurs. Furthermore, Gaudichaud-Beaupré described the flowers as axillary, but they may have been terminal, as in the Queensland form.

Growth

Lumnitzera racemosa is a good example of Attims's model (see Fig. 4.1A). The sapling shows continuous or diffuse branching, with no evidence of articulations. The twigs are glabrous, reddish brown, with close-set regular leaf scars. Each leaf subtends a single but obscure axillary bud; branching is always by syllepsis and is continuous on vigorous shoots; flowers are axillary. Branches are orthotropic and repeat the construction of the parent axis, but with fewer laterals and these mostly toward the outside of the tree. Consequently at an early stage of development the plant is much branched. Further development as a single-trunked tree involves the loss of branches. This is somewhat selective, and at a later stage the tree shows a confused intermixture of dead and persistent branches. In open situations the tree may retain the rounded, low-crowned shape, especially since lower branches can root basally when in contact with the substrate. Otherwise, where the tree suffers crown competition, the lower branches die and a single-trunked form with a narrow, conical crown appears.

The two species differ in the position of the inflorescences, and by definition *L. littorea* conforms to Scarrone's model. This is seen in the forking of distal axes. In younger stages the terminal inflorescences are closely substituted by lateral branches and the resultant sympodial growth does not influence the crown form strongly.

Floral biology

The differences in floral structure and inflorescence position reflect contrasted methods of pollination. *Lumnitzera littorea* (Fig. B.22) seems to be predominantly pollinated by birds, specifically sunbirds and honey eaters, although bees and wasps are additional flower visitors. Bird pollination is suggested by the exposed position of the flowers, which are slightly zygomorphic to accommodate a probing beak; the petals are red and remain erect, effectively lengthening the tube somewhat and providing some protection from short-tongued insects. The more prominent anthers thus brush the bird's head beyond the beak. The calyx tube is relatively deep and nectar is abundant. *Lumnitzera racemosa* (Fig. B.23), on the other hand, is visited by a variety of day-active wasps, bees, butterflies, and moths, for which the white spreading petals, actinomorphic flower, and somewhat shallower calyx cup are suited. Pollen is presented on the first day, with the stigma not becoming receptive until the second and subsequent days, so protandry exists.

Figure B.22. *Lumnitzera littorea* (Combretaceae). Terminal inflorescence; no obliging bird visitors. (From a color transparency by A. M. Juncosa)

Figure B.23. *Lumnitzera racemosa*. Lateral inflorescence and obliging butterfly visitor. (From a color transparency by A. M. Juncosa)

Plants, however, may be self-compatible, since single trees in cultivation set viable seed, and seed set in wild populations is often high, with all the flowers in a head setting fruit. Cross-pollination can be promoted by visits of more discriminate pollinators.

Fruits and seedlings

Although the fruit set seems high, there is a high percentage of fruits with aborted embryos so that seemingly mature fruits are commonly empty, sometimes also because the embryo has been eaten by a small grub that originates from eggs laid by the parent insect early in fruit development.

In normally formed fruits, the embryo is well protected by the layer of sclerenchyma within the outer corky or fleshy layers of the fruit wall. There is no evidence of vivipary. Fruits are dispersed by water and lose the softer outer layers, which exposes the sclerenchymatous fibers. In nature most floating fruits lose their viability, but those taken directly from trees germinate fairly readily. The seedling is hypogeal, but without other obvious adaptive devices, which aids its establishment.

Ecological and specific isolation

Van Slooten (1924) and Exell (1954) have both produced maps showing a similar distribution of the two species in Malesia. Van Slooten suggested that the two species occupy different ecological sites, and this is confirmed by Tomlinson et al. (1978), who comment that stands in north Queensland almost without exception contain one species to the exclusion of the other and that the two species never intermix. None of these authors gives a precise explanation for this differential distribution, although Tomlinson et al. suggest that *Lumnitzera littorea* is better suited to less saline, well-drained sites and is most vigorous on highly organic substrates, whereas *L. racemosa* is more resistant to saline conditions, as at the margin of bare salt pans, in association with *Ceriops* and *Avicennia*. In Queensland *L. littorea* has a more extensive southerly range than *L. racemosa* (Byrnes 1977).

It is interesting to consider why the two species remain distinct, especially if we accept the interpretation of Tomlinson et al. (1978) that *L. rosea* is a hybrid, since this suggests that the reproductive barriers between them can be broken. The ecological isolation assists in the continued separation, which could be further strengthened if the pollinators were quite specific or had limited ranges. The evidence for this interpretation is circumstantial and presently slight.

Systematics and nomenclature

The genus is based on Willdenow's recognition of it as distinct from *Laguncularia* in his description of *Lumnitzera racemosa*. Presl made the necessary combinations for a number of names included in *Laguncularia* by Gaudichaud-Beaupré, but of these, *Laguncularia lutea* (as *Lumnitzera lutea*) from Timor is now considered to be a variety of *L. racemosa* (van Steenis-Kruseman 1950, p. 187). *Laguncularia coccinea* is a synonym of *L. littorea*. *Laguncularia purpurea* has been interpreted as a name used in error by Gaudichaud-Beaupré in the plate and legend of the entity he called *L. coccinea*. *Laguncularia rosea*, as we have seen, has been resurrected by Tomlinson et al. simply to draw attention to the fact that Gaudichaud-Beaupré, in creating this name, could have been referring to a hybrid between the two now generally accepted species.

Gaudichaud-Beaupré used the presence or absence of calyx glands in his diagnoses, but in agreement with Exell (1954) this seems an inconstant character.

Terminalia L. 1767
[*Syst. Nat. Ed.* 12(2):674]

A pantropical genus of about 200 species, including a few commercially important timber species. Many species have 2-winged fruits, the wings formed from a lateral extension of the fruit wall. In other species (including the following) the wing is represented by a narrow ridge, and the fruits are dispersed by water, creating a degree of reduction that parallels that in *Heritiera*. No species are mangrove constituents, but the following is a characteristic strand plant of tropical Asia that is widely naturalized in other parts of the tropics.

Terminalia catappa L. 1767   (Fig. B.24)
[*Syst. Nat. Ed.* 12, 2:674]

For a complete citation, see Exell (1954, p. 566).

An andromonoecious deciduous tree with broadly obovate leaves 10 to 20 (perhaps 30) cm long with a rounded or bluntly pointed apex. Flowers of 2 kinds, male and perfect, are borne in axillary spikes, the male distally and the perfect proximal. Each flower has a calyx of 5 lobes and 10 stamens with an inconspicuous hairy disc. Male flowers lack any conspicuous pistillode; perfect flowers include a well-developed ovary with 2 functional ovules and a slender style. The flowers are rendered relatively conspicuous by the yellowish calyx and massed stamens. Fruits develop in clusters at the base of the spikes. Each is almond shaped, 5 to 7 cm long, with a shallow lateral ridge. The mesocarp is fleshy and fibrous, the fruit floats, and the fibrous part rots. Germination is hypogeal.

Of interest is the architecture of this species (Aubréville's model), which is a "type" for the characteristic shoot physiognomy known as "Terminalia" branching in tropical trees. Appreciation of the form is ancient; its architecture was illustrated by Rheede (1678)! The trunk axis is orthotropic and bears spirally arranged leaves. After one or more flushes of growth, a tier of sylleptically produced branches establishes a series of lateral axes with obligate plagiotropy and apposition growth. This results in a sympodial system in which each unit extends horizontally for a limited distance and then turns erect, but remains as a functional short shoot with a rosette of closely set, spirally arranged leaves. The system is extended by 1 or 2 renewal shoots that repeat the process. The total result is a trunk with widely spaced

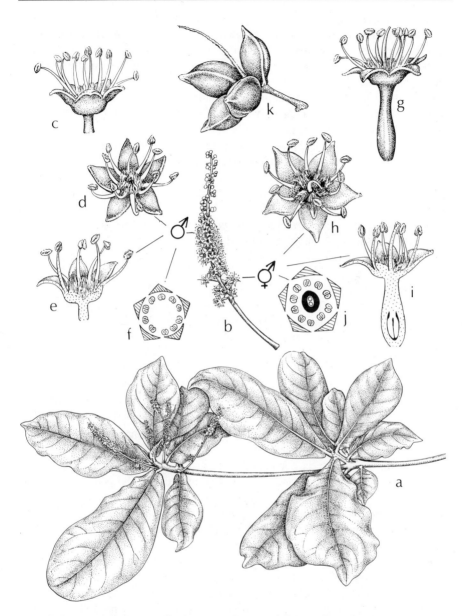

Figure B.24. *Terminalia catappa* (Combretaceae) flowers and fruit. (a) Flowering shoot (× ¼) representing two sympodial units produced by apposition growth. (b) Flower spike (× ½), perfect flowers (♀) at base, male flowers (♂) distally. (c–e) Male flower (× 3): (c) from side, (d) from above, (e) L.S. (f) Floral diagram of male flower. (g–i) Perfect flower (× 3): (g) from side, (h) from above, (i) L.S. (j) Floral diagram of perfect flower. (k) Fruit (× ⅜). (From Tomlinson 1980)

rosettes of leaves in a regular hexagonal pattern of horizontal axes. Trees are deciduous, and in some parts of the range lose their leaves regularly twice a year, with the leaves turning yellow and then red as they senesce.

The shoot system has been investigated extensively both directly and by computer simulation (Fisher 1978, Fisher and Honda 1977, 1979a,b). These authors have shown that this branch system minimizes mutual shading between rosettes arranged in an axis system with minimal path length. Fisher's first careful description celebrates the 300th anniversary of Rheede's first illustration! The same physiognomy is achieved by *Bruguiera* and *Rhizophora*, with the former also representing Aubréville's model and the latter Attims's model, as described elsewhere. In all these examples the regular symmetry of the basic architecture is confounded by reiterative responses in damaged trees. It remains reasonable to accept the original architecture as a highly adaptive mode of branch expression. This is probably equally effective in a tree in open sites, like *Terminalia catappa*, as in a tree in the forest understorey, like *Bruguiera* saplings that grow in the shade of parents.

Conocarpus L. 1753
[*Gen. Pl. Ed.* 5:81]

A genus of 2 species: one with a very restricted distribution in East Africa and the other widely distributed in coastal communities in tropical America and West Africa.

Conocarpus erectus L. 1753   (Figs. B.25, B.26)
[*Sp. Pl.* 1:176]

The species is frequently considered a "true" mangrove, but it is better regarded as a mangrove associate because it lacks any of the morphological and biological features (such as pneumatophores and vivipary) that characterize true mangroves; furthermore, it occurs in inland communities. The point is somewhat pedantic, however; the tree's somewhat weedy tendencies in part account for its mangrove association. In the mangrove community, *Conocarpus* is a back-mangal constituent, but only within the limits of the highest tides. It is tolerant of high salinities and rather dry soils, but it also grows in or near fresh water. In South Florida, for instance, it can form almost pure stands in saline marly soils, but it also occupies low hammocks inland, commonly as a fringe to open ponds or artificial canals in the company of *Salix, Chrysobalanus,* and *Myrica.*

A form of this species that has silvery leaves because of a dense indumentum of long hairs is sometimes given a varietal name (var. *sericeus* Grisebach – *C. pubescens* Schumach.), and it is commonly cultivated in South Florida as "silver-leaved buttonwood" in beach plantings. However, the silver-leaved condition is not genetically fixed (Semple 1970).

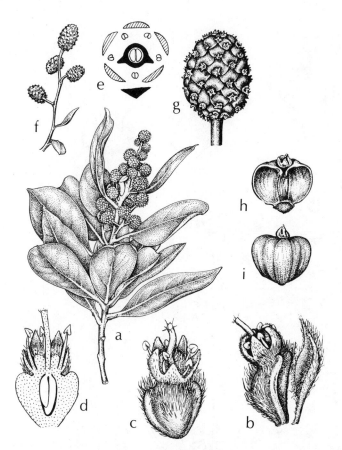

Figure B.25. *Conocarpus erectus* (Combretaceae) female flowers and fruit. (a) Flowering shoot (× ½). (b–e) Female flowers (× 10): (b) side view with bract (× 10), (c) face view (× 10), (d) L.S. (× 10), (e) floral diagram. (f–i) Fruits: (f) fruiting heads (× ½), (g) single fruiting head (× 2), (h) detached fruit from front (× 5), (i) detached fruit from back (× 5). (From Tomlinson 1980)

Description

Dioecious trees, usually with several spreading trunks, or a low shrub. Bark rough and fissured. Branching frequent, but diffuse and irregular (Petitt's model). Shoots orthotropic, without articulations; leaves spirally arranged, scattered, or somewhat clustered distally. Twigs angled with a ridge beneath each leaf. Leaves either glabrous at maturity or persistently slightly to densely silver hairy (var. *sericeus*). Petiole short (to 1 cm) with a pair of circular glands, one on each side on the decurrent extension of the blade. Blade folded in bud; ovate-lanceolate, 4 to 9 by 2 to 3 cm, apex acute to acuminate, margin entire. Lower leaf surface with several domatia (minute shallow pits) in the angles between the midrib and

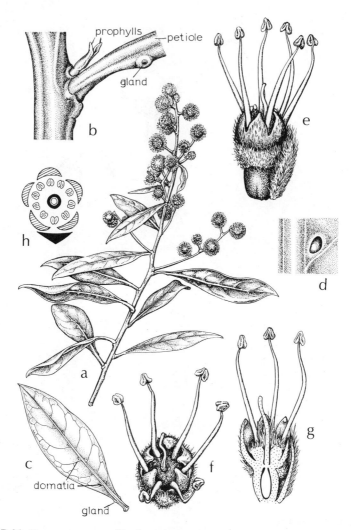

Figure B.26. *Conocarpus erectus* (Combretaceae) male leaves, flower, and fruit. (a) Flowering shoot
(× ½). (b) Detail of node (× 3), dormant axillary bud enclosed by prophylls, one of a pair of basal
petiolar glands (see part c). (c) Leaf from below (× ½) showing glands and domatia.
(d) Single domatium (× 4). (e–h) Male flowers: (e) side view (× 10); (f) from above (× 10); (g)
L.S. (× 10); (h) floral diagram. (From Tomlinson 1980)

the major veins. Inflorescences terminal, compact or rather diffuse little-branched
panicles; the flowers clustered in terminal globose heads 5 mm in diameter, the
branches of the inflorescence subtended by either foliage leaves or obscure bracts;
female inflorescences usually somewhat more compact than male. Male heads with

prominent stamens at anthesis. Each head with about 25 flowers, each flower subtended by a short (1 to 2 mm) hairy bracteole. Flowers somewhat dimorphic, each with an inferior ovary hairy above and 5 acute calyx lobes with sparse hairs; petals absent; disc well developed and hairy. Male flowers distinguished by 5 to 10 functional stamens, at first recurved but erect and about 3 mm long at anthesis; style simple, ovary narrow and commonly including 1 or 2 nonfunctional ovules. Female flowers with a well-developed laterally compressed ovary; staminodes usually 5, 1 to 2 mm long and inconspicuous; style with a tuft of short sparse hairs below the stigma. Fruiting head of short laterally compressed nutlets about 5 mm long, released by shattering of the head. Seeds small, angular; germination hypogeal.

Growth

Plants are ever-growing in the sense that they produce no protected terminal buds, but, at least in South Florida, they undergo an extended dormancy in the winter. Branching is by both prolepsis and syllepsis. Little is known about the reproductive biology. Seeds are dispersed by water because they float. The seed set by female trees is abundant, but either many seeds are aborted or germination is difficult, since seedlings are not common compared with the dense swards that are produced in most years by *Laguncularia*.

Family: Compositae (Asteraceae)

The largest family of dicotyledons, cosmopolitan and with more than 20,000 species, mainly herbs and often weedy. The following genus has been recorded as a mangrove associate (Gentry 1982). Species of *Pluchea* that occupy coastal areas may be encountered, like *Pluchea indica* Less. in the Asian tropics.

Tuberostylis Steetz 1953
[Seeman, *Bot. Voy. Herald* 142, t. 29]

The 2 species of this genus have been recorded as mangrove associates. The spathulate leaves are opposite, about 2 to 4 mm long, with a flattened petiole and an expanded, ovate blade.

Tuberostylis axillaris Blake 1943
[*J. Wash. Acad. Sci.* 33:265]

Recorded as a semiscandent shrub in dense tidal forest in Colombia.

Tuberostylis rhizophorae Steetz 1853
[Seeman, *Bot. Voy. Herald* 142, t. 29]

The original description refers to this species as growing "epiphytically on the roots of Mangrove Trees, Southern Darien" (i.e., Colombia). The "plant affixes itself to the aerial roots of *Rhizophora* by means of adventitious roots which arise in pairs immediately below the node." The species has a wider distribution in Central America and has been recorded on the base of trunks of *Mora oleifera*.

Family: Ebenaceae

A sizable family of 3 genera and about 500 species, mostly in *Diospyros*. A few species occur in warm temperate regions (e.g., *Diospyros virginiana* L., persimmon), but the family is primarily tropical. Ebony is primarily the wood of *D. ebenum* Koen.

Diospyros L. 1753
[*Sp. Pl.*:1057]

Maba J. R. & G. Forst. 1775 [*Char. Gen. Pl. Ed.* 2:61]

A pantropical and subtropical genus of mainly dioecious trees with over 400 species that are often difficult to distinguish, since apomixis is suspected in some forms. The genus *Maba* now is usually included in *Diospyros*. The following elements may include populations in back mangal or beach communities and have been described as mangroves by some recent authors.

Diospyros ferrea (Willd.) Bakhuizen 1932
[*Gard. Bull. Straits Settl.* 7:162; see also 1937, *Bull. Jard. Bot. Buitenz.* 3, 15(3):1–515 and 1941, ibid. 3, 15(4):428–44]

A moderate-sized evergreen dioecious tree, up to 25 m tall, with dull gray, finely fissured bark; buttresses scarcely developed. Trunk axis with spirally arranged leaves; branches distributed continuously to discontinuously on the trunk (either Massart's or Roux's model), strongly plagiotropic with distichously arranged leaves. Leaves on branches ovate-oblong, 6 to 12 by 2.5 to 4 cm, with a short (to 5 mm) petiole; apex acute to rounded or sometimes emarginate, base rounded cuneate. Flowers hypogynous, unisexual in few-flowered condensed axillary cymes, the female inflorescences with fewest flowers. Male flowers subsessile, calyx campanulate, 3-lobed, corolla white or pale yellow, 3-lobed, stamens 6 to 12 (rarely more), pistillode slender, densely hairy. Female flowers similar, the calyx and corolla 3– (− 5); staminodes absent; ovary ovoid, 3-locular, each locule with 1 or 2 ovules; style simple, short. Fruit an ellipsoid or globose 1- to 6-seeded berry, about 1 cm wide, with the calyx persistent and cupuliform. Seeds oblong, with a thin testa, endosperm present, the embryo straight.

The species is exceedingly polymorphic and occupies diverse habitats, from back mangroves to lowland montane rain forests, with a geographic range from East Africa, throughout the Asian tropics to Polynesia and northern Australia. The modern tendency is to segregate local species on the basis of careful study (e.g., Smith 1981, p. 730), but identification of specimens in areas where there has been no field study is still exceedingly difficult. For example, Fosberg (1939) recognizes 12 nameable taxa in his treatment of the Hawaiian forms.

Elsewhere the mangrove forms have been referred to var. *littorea* (R. Br.) Bakh. and var. *geminata* (R. Br.) Bakh.

Growth rings

The species has been the subject of a detailed study correlating growth rings with climate in North Queensland (Duke et al. 1981). A specimen about 12 m high and 10 cm in diameter at a height of 1.3 m was shown to have distinct growth rings differentiated by differences in fiber wall thickness. The width of these was shown to have unusual correlations with rainfall. First, there is a general correlation with extended changes in average annual rainfall, so that a period of decreasing rainfall is matched by a period of decreasing ring width. Second, the rings are not annual but the numbers come close to an average production of 7 rings every 4 years (i.e., 1.77 rings per year). The authors account for this in terms of the erratic rainfall in this part of Australia and emphasize that, although this complicates the problem of determining the age of mangrove trees, it indicates that it is not always insurmountable.

This study remains the only one in which (1) growth rings of a standard anatomical kind have been shown clearly in mangroves and (2) the growth rings have some correlation with climate (in this instance rainfall).

Family: Euphorbiaceae

One of the largest plant families with over 7000 species, most commonly tropical. The family is very diverse; the tribes delimit the most natural units (Pax and Hoffmann 1931). Three unrelated genera are considered here: two Old World and the other New World:

1A. Plants without milky latex, fruit a capsule 1.5–2 cm in diameter with 12–
15 locules . . . . *Glochidion*
1B. Plants with a milky latex; fruit fleshy or 3-lobed. 2
2A. Old World, dioecious; fruit a 3-lobed schizocarp . . . . *Excoecaria*
2B. New World, monoecious; fruit indehiscent, resembling a small apple . . . . . .
. . . . *Hippomane*

Excoecaria L. 1759
[*Syst. Nat. Ed.* 10, 1288, and *Sp. Pl.* 2:1451, 1763]

A genus of 35 to 40 species in tropical Africa and Asia to the Western Pacific. The following species, which is the type of the genus, occurs in mangroves but also occasionally in inland stations; one other species may be considered a mangrove associate.

Excoecaria agallocha L. 1759   (Fig. B.27)
[*Syst. Nat. Ed.* 10, 1288]

(*Stillingia* (sect. *Excaeria*) *agallocha* (L.) Baill. 1858 [*Et. Gen. Euphorb.* 518.]

From East Africa, India, and Ceylon to Hainan and the Ryu-Kyu Islands, through Malesia and Papuasia, including tropical Australia and into the Pacific as far as Niue and Samoa.

A dioecious tree to 15 m high with abundant white latex, sometimes branched from the base and somewhat shrubby. Bark gray, becoming fissured; lenticels prominent on younger twigs. Leaves spirally arranged, sometimes somewhat clustered toward the ends of erect shoots. Shoots orthotropic, with infrequent diffuse branching (Attims's model); articulations indistinct, terminal buds inconspicuous with, at most, an envelope of persistent stipules. Stipules minute, ephemeral as lateral triangular scales on each side of petiole. Leaves simple, coriaceous or somewhat fleshy, with a terete petiole 1 to 2 cm long, blade ovate-elliptic to obovate, up to 9 cm long and 6 cm wide. Apex rounded, slightly emarginate, or at most bluntly acuminate. Margin inconspicuously notched, with a minute gland in the notch in young leaves. Basal glands 2 (− 4), usually 1 on each side of blade at its insertion on the petiole; the glands usually circular and often associated with additional outgrowths like the marginal glands. Each leaf subtending 1 (− 2) axillary bud with a series of minute, decussately arranged, woody bud scales.

Inflorescences axillary, pale green, 3 to 7 cm long, initiated as catkinlike structures within the leaf-bearing portion of the shoot, but sometimes persistent and not expanding until the subtending leaf falls. Male inflorescence up to 7 cm long at maturity, with a series of spirally arranged, often glandular bracts, each bract subtending a male flower. Male flower almost sessile with 3 narrow laciniate tepals (calyx) below the 3 yellow stamens, each anther bilocular, basifixed to almost versatile and longitudinally dehiscent. Pistillode absent. Filament initially obscure, but rapidly extending to 5 mm at maturity. Female inflorescence usually shorter

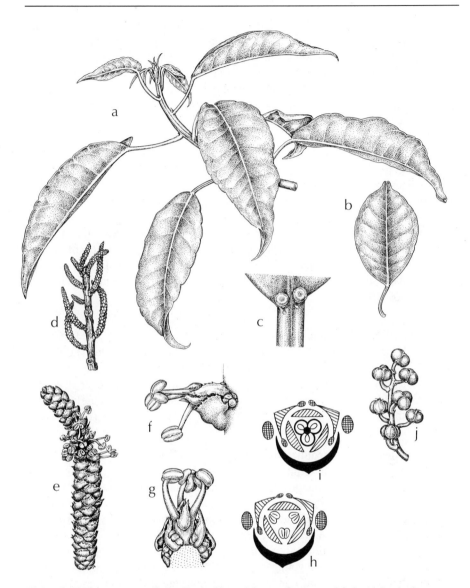

Figure B.27. *Excoecaria agallocha* (Euphorbiaceae) leaves, flowers, and fruit. (a) Leafy shoot
(× ½); (b) single leaf (× ½); (c) detail of pair of glands at leaf base (× 4); (d) deciduous axis with
immature axillary male spikes (× ¾); (e) male spike at early anthesis (× 4); (f) single male flower
from the side (× 8); (g) male flower from above, including axis in T.S. (× 8); (h) floral diagram of
male flower; (i) floral diagram of female flower; (j) immature fruiting head (× ¾). (Vegetative shoot
from Bako, Sarawak, P. B. Tomlinson, 28.7.82B; female flowers from Bootless Bay, Papua New
Guinea, P. B. Tomlinson, 31.10.74F; male flowers from Kinoya, Fiji, P. B. Tomlinson, 18.11.74C)

than male, bracts glandular; flowers initially sessile and with a pair of basal bracteoles, pedicel extending to 5 mm in fruit. Tepals 3, initially somewhat cupulate, wider than those in male flowers; staminodes absent. Ovary trilocular with 3 short, spreading or recurved simple styles, each locule with a single basal ovule. Fruit 3-lobed about 7 mm in diameter, dehiscing into 3 cocci (schizocarp) to release the solitary seeds; the pericarp somewhat leathery but not fleshy. Seeds black about 3 mm in diameter. Endosperm absent. Germination epigeal; cotyledons somewhat cuneiform.

Geographic and ecological distribution

This species is recorded in the Asian tropics throughout much of the eastern range of the genus *Excoecaria*, but is not known for certain to occur in Africa. It is a characteristic associate of the mangrove community but only in open sites in the back mangal. It occurs more commonly on exposed beaches and in sandy estuaries and is also recorded in disturbed or open sites to an elevation of 400 m, so it cannot be regarded as an exclusively mangrove component. In keeping with this, the plant shows no obvious morphological adaptation to mangal; the root system is not obviously specialized.

Field identification

The tree is easily recognized by the catkinlike inflorescences and the copious white latex that exudes from wounds. This latex is irritating, and contact with it, especially around the mouth and eyes, should be avoided. However, the latex is water soluble (as in *Hippomane*). The simple petiolate leaves, obscurely notched margins of the ovate blade, and especially the basal pair of glands, are also distinctive.

Growth and reproductive biology

The tree grows intermittently and irregularly. There are obvious articulations associated with the inconspicuous resting terminal buds. The plant may be briefly deciduous in dry seasons, with the senescent leaves becoming orange or even scarlet. Branching is diffuse, irregular, and by prolepsis. Axillary buds are present, either solitary or sometimes in vertical paris, but most remain inconspicuous and undeveloped. Reiteration can occur, leading to somewhat shrubby forms. Inflorescences are developed on younger shoots, usually from the upper of the axillary bud pair (but sometimes both) and seemingly independent of the time of shoot extension. The inflorescences may remain unextended for a long time, but the final extension at anthesis is rapid.

Flowers are pollinated by insects because the pollen is sticky; bees are particularly common flower visitors and may be the chief pollinators. They are attracted by the yellow nectar-secreting glands, which characteristically occur at the margin of the catkin bracts. These glands may be homologous with the glands on the foliage leaf blades, whose function is not known. Leaf glands are larger and more or less circular; inflorescence glands are irregular and almost peltate. The distribution of glands in association with the inflorescence scales is somewhat irregular, and additional glands within the bract axil may represent those belonging to bracteoles, if the suggested homology is correct.

Plants are typically unisexual, but there are records of male catkins with a few basal female flowers, a condition that recalls the bisexual catkins of related monoecious genera (e.g., *Ateramnus* and *Hippomane*).

Taxonomy and nomenclature

Excoecaria is a typical member of the subfamily Euphorbioideae in the simple leaves, absence of petals, trilocular ovary with 1 ovule per loculus, inarticulate laticifers with white latex, and biglandular bracts on the inflorescence. It is included by Webster (1975) within the tribe Hippomaneae, which includes a number of genera with similar inflorescences, including *Hippomane* itself, which has a coastal distribution.

Excoecaria is distinguished from closely related members of this group by a combination of characters, including the dioecious condition, axillary inflorescences, male flowers with only (2-) 3 stamens, and the absence of a caruncle from the seed. The persistent columella of the schizocarp fruit lacks any horizontal projection at its base as is found in the related genus *Stillingia*.

As with many common plants of the Asian tropics, the species *E. agallocha* was first described by Rumphius in *Herbarium amboinensis* in 1741 (as *Arbor excoecans*); the spelling *Excaecaria* is used by a number of authors. The remaining species of the genus occupy inland sites and are not necessarily restricted to the coast. There is no recent treatment of the genus that discusses the evolutionary relationships of *E. agallocha* with other members of the genus. The most similar inland species seems to be *E. philippinensis* with much larger leaves, up to 20 cm long, and longer (to 10 cm) lax inflorescences. However, Airy-Shaw (1975) comments that generic limits in the tribe Hippomaneae are not clear, and the distinction between *Excoecaria, Stillingia,* and *Sapium* has not been maintained by all authors (e.g., Baillon). *Excoecaria* itself is the oldest generic name of the three.

Excoecaria dallachyana (Baill.) Benth. 1873
[*Benth. Fl. Austr.* 6:153]

E. agallocha var. dallachyana Baill. 1866 [*Adansonia* 6:324]

This taxon is described from south Queensland and New South Wales, but seems chiefly distinguished by its inland habitat in scrubby vegetation, for example, Sandiland Ranges (Pax and Hoffman 1912, p. 168). The relation between *E. agallocha* and *E. dallachyana* would merit comparative study if, as seems likely, the latter is simply an inland form of the former. Within the species a number of varieties have been named based on leaf dimensions and the size of the reproductive parts (Pax and Hoffmann 1912).

Excoecaria indica (Willd.) Muell.-Arg. 1863
[*Linnaea* 32:123]

Airy-Shaw (1975, p. 114) indicates that this species occupies "primary *Nypa* forest in salt water, in seasonal swamps, on tidal river banks and seashores, on black soil or yellow clay alluvium from sea-level up to 10 m alt." It ranges from south and east India, Southeast Asia, Malesia, and the Solomon Islands, but not the Philippines. It is distinguished from *E. agallocha* by its thorny trunk, regularly crenulate, almost lanceolate leaves, and globose smooth (not lobed) capsular fruit, up to 3 cm in diameter, which dries black.

Hippomane L. 1753
[*Sp. Pl.* 2:1191 and 1754 *Gen. Pl. Ed.* 5:499]

A genus of perhaps 3 species of the New World tropics.

Hippomane mancinella L. 1753   (Fig. B.28)
[*Sp. Pl.* 2:1191]

This species ("manchineel") is a characteristic constituent of seashore communities throughout the Caribbean (including the Florida Keys), Mexico, and northern South America, extending to the Galapagos and Revillagigedo islands (Webster 1967). It is not a mangrove but is mentioned here because it may be encountered, and its white but water-soluble latex is poisonous (Howard 1981). Its fruits resemble small apples and are very dangerous. The Latin name means literally "the little apple that makes horses mad." This species is a common but scattered constituent of drier, sandy, coastal communities within its range, and it may be encountered in back mangal or adjacent beach communites in the New World.

Hippomane mancinella forms a low, much-branched, usually wide-spreading tree or sprawling shrub. It is recognized by its terminal inflorescences and the small, several-seeded, applelike fruits that are borne later in the crotch of the branch forks. These fruits develop from a multilocular ovary, each locule with a single ovule; there are 3 or more recurved stigmas. The leaves are stipulate, long petiolate, and have a single conspicuous gland at the base of the ovate blade that has a shallowly toothed margin (Fig. B.28b). The inflorescence bears unisexual flowers, with male flowers in axillary aggregates above, and 1 or more female flowers below, often also associated with further male flowers. The structure and arrangement of the flowers are somewhat variable, and the conspicuous yellow glands associated with each flower cluster are presumably attractive to pollinators. Perhaps because of its sinister reputation the plant has not been studied carefully; details of the pollination mechanism are not known and the method of seed dispersal is unknown. Germination has apparently not been studied.

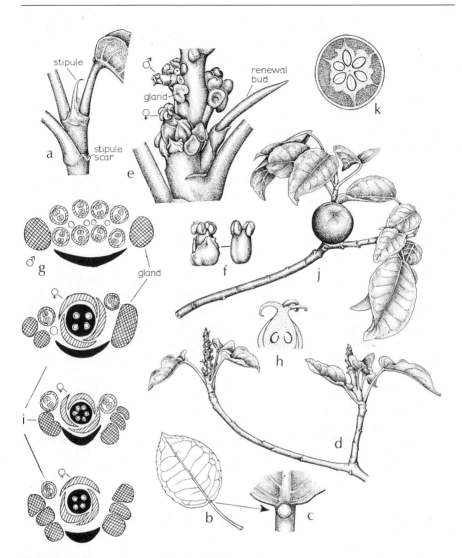

Figure B.28. *Hippomane mancinella* (Euphorbiaceae) shoots, flowers, and fruits. (a) Terminal bud (× 3) with stipule and stipule scar; (b) outline of foliage leaf (× ½); (c) detail of gland at base of leaf (× 4); (d) shoot with terminal inflorescences (× ½); (e) base of flowering shoot with male and female flowers (× 3), renewal bud in axil of leaf to right; (f) male flower (× 6) with (left) and without (right) perianth; (g) diagram of male flowering unit (glands crosshatched); (h) female flower in L.S. (× 5); (i) diagrams of lower flowering units with 3 examples of flower arrangement (glands crosshatched); (j) fruiting branch (× ½); (k) fruit in T.S. (× ½). (From Tomlinson 1980)

The genus is in the same tribe as the eastern *Excoecaria* but is distinguished by the terminal inflorescence, monoecious condition, multilocular ovary, and several-seeded indehiscent fruit.

Glochidion J. R. & G. Forst 1776
[*Char. Gen.* 113, t. 57]

A large genus of about 300 species mainly in tropical Asia, distinguished by its broadly ovate, glabrous leaves with rounded apex, flowers with a multilocular ovary, producing a multilocellate fruit. The following coastal species is recorded in back-mangal communities, but it also has a wide distribution inland, including "elfin woodland" (var. *culminicola*, Airy-Shaw 1975).

Glochidion littorale Blume 1825
[*Bidjr.* 585]

A shrub or small tree growing to 6 m with brown or gray, somewhat flaky or fissured bark. Leaves spirally arranged, glabrous, obovate with a rounded apex, about 5 to 7 by 2.5 to 4 cm, narrowed basally to the short (2 to 3 mm), somewhat fleshy petiole, margin entire or slightly crenulate. Fruit a globose capsule about 1.5 to 2 cm in diameter with numerous (12 to 15) locules, becoming lobed before dehiscence; green, flushed crimson and splitting between the lobes to reveal the two red seeds per locule.

A species with a wide distribution in India, Ceylon, Indochina, and West Malesia. Airy-Shaw (1975) distinguishes a number of varieties, of which the following are coastal in their distribution: var. *littorale*, the usual form, with the leaf apex rounded and the pedicels of the female flowers short; and var. *caudatum*, with the leaf apex pointed and the female pedicels long.

Family: Flacourtiaceae

A diverse tropical family of woody plants with more than 1000 species in 90 genera. The family has many disparate elements and is seemingly unnatural.

Scolopia Schreber 1789
[*Gen.*:335 nom. cons.]

A genus of about 40 species with a distribution from West Africa eastward to the Solomon Islands and eastern Australia.

Scolopia macrophylla (Wight and Arnold) Clos 1857
[*Ann. Sci. Nat.* Series 4, 8:253]

For a detailed account, including full synonymy, see Sleumer (1972).

A small tree growing to 10 m with smooth grayish pink bark, the trunk described as having spirally arranged simple spines. Leaves spirally arranged but appearing 2-ranked on the branches. Petiole slender, 5 to 10 mm long, scarcely 1 mm wide. Blade simple, ovate, and about 8 to 11 by 4 to 6 cm, either tapered gradually to a point or rounded apically; margin minutely and regularly toothed; base acuminate with 2 orange glands at the insertion of the blade. Flowers in axillary spikes up to 6 cm long, sometimes aggregated distally to make a loose terminal panicle; flowers white, about 4 mm in diameter, at first green and then becoming red or orange, style persistent and up to 4 mm long; base enclosed by persistent perianth.

This tree is typically a species of river sides or marshes to an altitude of almost 1000 m but is recorded in "inland mangroves with consolidated soil and numerous crab mounds" (Anderson and Chai, Sarawak Forestry Department #S29344). It has a distribution from Indochina and the Malay Peninsula through Sumatra and Borneo to Java.

Family: Goodeniaceae

A family of 12 genera and about 300 species mostly in Australia. *Scaevola*, a genus of some 90 species, is also concentrated in Australasia but has a pantropical distribution by virtue of 2 species; one Old World and the other New World, which are coastal beach plants.

Scaevola L. 1771
[nom. cons. *Mant. Pl.* 2:145]

The 2 species considered here are erect, shrubby plants with succulent leaves, pithy stems, and axillary cymes of conspicuous flowers with undulate petal margins. They are both typical of beach communites, especially sand dunes, where they can form extensive colonies apparently by subterranean branching of the stems. They sometimes occur in mangrove communities, but only in sandy, well-drained areas. They are recognized by their white zygomorphic flowers, with the corolla split completely down one side to expose the curved style. Fruits are fleshy, with 1 or 2 seeds.

1A. Leaves usually larger than 10 cm; calyx segments obvious, pointed; petals often with violet stripes on the inside; fruit white at maturity. Asian tropics. . . . . *S. taccada* (Gaertn.) Roxb.
1B. Leaves usually shorter than 10 cm; calyx segments not obvious or at most short and rounded; petals without internal markings; fruit black at maturity. American tropics. . . . . *S. plumieri* (L.) Vahl.

Scaevola plumieri (L.) Vahl. 1796   (Fig. B.29)
[*Symb. Bot.* 2:36]

This species ranges from Florida throughout the West Indies to Central and South America.

Scaevola taccada (Gaertn.) Roxb. 1824
[*Fl. Ind. Ed.* 1, 2:146]

This species has a wide range from the Malay Peninsula to the South Pacific, including Hawaii. It has been introduced to the New World and occasionally escapes from cultivation.

Family: Guttiferae

A large tropical family characterized by simple opposite leaves with finely parallel lateral veins and colored (usually yellow) latex. The flowers have numerous stamens. One species may be included here as a commonly encountered mangrove associate.

Calophyllum L. 1753
[*Sp. Pl.* 1:513]

A pantropical genus of about 200 species. The following, which is the type of the genus, has wide coastal distribution, typically in beach communities in the eastern tropics, with a range from East Africa to Polynesia. Its range further extended artificially into the Pacific. It is a commonly planted coastal ornamental in the tropics, including the New World.

Calophyllum inophyllum L. 1753   (Fig. B.30)
[*Sp. Pl.*1:513]

A full synonymy and extensive citations are provided by Stevens (1980).
Trees grow to 30 m without buttresses or root modifications; bark dark,

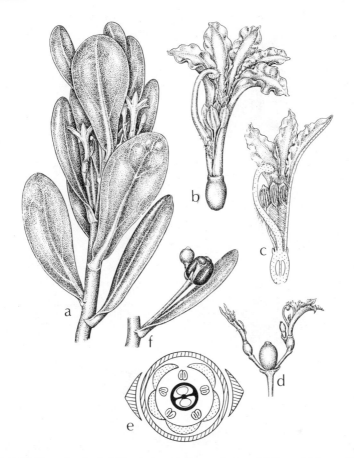

Figure B.29. *Scaevola plumieri* (Goodeniaceae) flowers, fruit, and foliage. (a) Apex of flowering shoot (× ½); (b) flower from side (× ⅔); (c) flower in L.S. (× ⅔); (d) part of flowering shoot with young fruit (× ½); (e) floral diagram; (f) mature fruit on axillary branch (× ½). (From Correll and Correll 1982)

fissured. Latex yellow to whitish, sticky, from all cut surfaces. Shoots with naked terminal buds except for obscure indumentum; stems angular. Leaves glabrous opposite, simple, with a short petiole (1 to 2 cm); blade elliptic to oblong, about 15 by 8 cm, margin entire; apex always rounded, base cuneate or rounded. Lateral veins uniformly parallel. Inflorescence up to 15 cm long, axillary, usually a simple raceme with 5 to 15 flowers. Flowers about 1 cm in diameter, apparently bisexual, tetramerous with 8 (-13) tepals; 2 outer pairs decussately arranged, 4 inner in an imbricate whorl, little differentiated from outer. Stamens numerous, the filaments shortly connate below. Ovary globose with a simple style 4 to 8 mm long; stigma peltate; ovule solitary, basal, anatropous. Fruit a 1-seeded globose drupe, 2 to 4

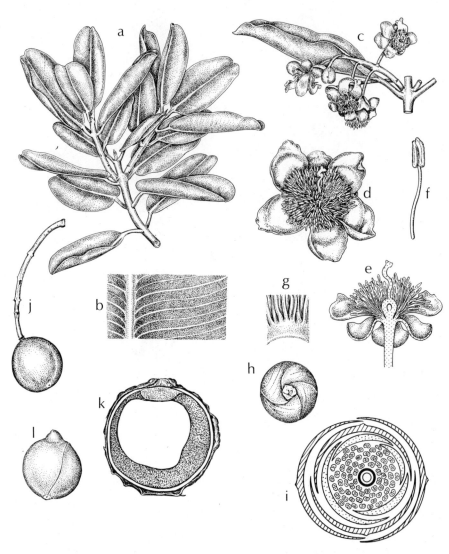

Figure B.30. *Calophyllum inophyllum* (Guttiferae) leaves, flowers, and fruit. (a) Vegetative shoot
(× ¼); (b) detail of leaf, midrib region to show venation (× ⅔); (c) node with single inflorescence
(× ½); (d) flower from above (× ⅔); (e) flower in L.S. (× ⅔); (f) single stamen (× 4); (g) fused
bases of stamens, detail (× 6); (h) flower bud from above (× 3), petal tips folded under the stigma;
(i) floral diagram, overlapping of petals varies; (j) fruit (× ¼); (k) fruit in L.S. (× 1);
(l) seed (× 1). (From specimens cultivated at Fairchild Tropical Garden, Miami, Florida)

cm in diameter; the single embryo enclosed by a spongy layer and a thin stony layer derived from the testa (Corner 1976). Germination hypogeal, the radicle both breaking the stony layer and pushing aside a basal plug.

The tree is most characteristic on sandy soils and is a good example of a beach tree but has been recorded to an altitude of 200 m. It therefore occurs only occasionally in association with true mangroves and then always in transitional habitats. Stevens (1980) provides many biological details. The tree characteristically has a leaning or even prostrate trunk; flowering is probably continuous, although there may be one or more peaks per year; the flowers smell sweet and are almost certainly pollinated by insects; the fruit is dispersed by sea currents as it floats for extended periods, as does the seed itself, but there are also records of it being dispersed by bats, which eat the fleshy outer layers. The tree has many local uses as a source of dye, oil, timber, and medicine.

Family: Lecythidaceae

A small tropical family of 15 genera and 325 species with large, showy flowers. The most familiar product is Brazil "nuts," the seed of *Bertholletia excelsa* H.B.K.

Barringtonia J. R. & G. Forst. 1776
[*Char. Gen.* 75, t. 28]

A genus of about 40 species distinguished by its large simple leaves clustered at the ends of the branches and the large angled fruit. It has a distribution from East Africa to Polynesia. Two species are characteristic of coastal communities in the Asian tropics, and others may be encountered in back-mangal communities. A full discussion is provided by Payens (1967). The following species are the ones most likely to be encountered. The flowers open at night and are pollinated by night-flying animals. The bark and crushed fruit of both of them contain saponins and are used as a fish poison (Tattersfield et al. 1940). The fruit of *B. edulis* Seem. is edible.

1A. Leaf margin always entire. Racemes usually terminal, erect, relatively short (to 8 cm) with few (− 10) flowers; pedicels 5–9 cm long. Flowers white, petals 6–8 cm long, filaments white, red-tipped. Fruit cubical, 10–15 cm in diameter . . . . *B. asiatica* (L.) Kurz.

1B. Leaf margin usually toothed. Racemes terminal or from axils of fallen leaves on older wood, pendulous, long (more than 10 cm), many-flowered; pedicels at most 5 cm long. Flowers pink, petals at most 3 cm long, filaments red, white, or yellow-tipped. Fruit narrow, longer than wide, (5–7 by 3–4 cm). . . . . *B. racemosa* (L.) Bl.

Figure B.31. *Barringtonia asiatica* (Lecythidaceae). Germinated seeds. Bako National Park, Sarawak.

Barringtonia asiatica (L.) Kurz 1876   (Fig. B.31)
[*J. Asiat. Soc. Bengal* 45:131, see also 46:70]

Tree grows to perhaps 20 m with a short, buttressed bole and dense crown. Bark rough and thick in older trees. Leaves obovate, somewhat fleshy, 30 to 40 by 12 to 30 cm, apex rounded, base slightly auriculate, petiole short, stout. Flowers in short, erect terminal racemes, showy, up to 10 cm in diameter with 4 white petals. Fruit massive with 2 persistent calyx lobes, floating by virtue of the fibrous pericarp.

This is a characteristic strand plant of the Indo-Malayan and Polynesian region. Brass (field notes of *Brass 2610*) describes it as "the largest of the beach trees."

Barringtonia racemosa (L.) Spreng. 1826
[*Syst. Veg.* 3:127]

The synonymy of this species is long and complex, and Payens (1967), who has monographed the genus, provides an extensive list but says that it is impossible to interpret pre-Linnean names with certainty so that the extensive early synonymy serves no useful purpose in citation.

A usually small tree, but up to 10 m; leaves obovate-lanceolate, 20 to 30 by 5 to 8 cm, tapered below to the short, somewhat fleshy petiole scarcely 1 cm long.

Flowers on slender pendulous spikes 40 to 50 cm long; the flowers about 2 cm long with a slender filiform style. Fruit narrow, pointed at each end and 4-angled or grooved.

This tree has a distribution in Indo-Malaya to Polynesia; it is variously recorded as a mangrove and is abundant along tidal rivers and in areas subject to tide and salinity. It also occurs in and on the edge of peat swamp forests and on hillsides to altitudes of 200 m.

Barringtonia conoidea Griff. 1854
[*Notul.* 4:656]

A species with a more limited distribution than the previous 2, recorded as growing along river banks but only seaward at the limit of saline influence. It is distinguished from *B. racemosa* by its few-flowered but pendulous raceme that is 5 to 10 cm long, but most characteristically by the top-shaped fruit about 7 cm long with 8 projecting basal flanges. A suggested field character (Chai 1982) is that the leaves of *B. conoidea* wither yellow, whereas those of *B. racemosa* wither red.

Seedling and embryo in *Barringtonia*

The embryo is unusual, has a distinctive method of germination, and bears some comparison with viviparous seedlings. The embryo is solid and undifferentiated and the mature seed is usually interpreted as lacking endosperm. The shoot apex is enclosed by a series of spirally arranged minute scales, but there is no recognizable pair of cotyledons. The radicular end is blunt and undifferentiated. The bulk of the axis can be interpreted as a hypocotyl. On germination, the plumule elongates and the scales expand, with a transition to the normal foliage leaves. Scales apparently subtend buds, which can grow out and form a replacement shoot if the apex is damaged. The "radicular" end produces roots that may be entirely adventitious, although one seems dominant.

Although comparison with Rhizophoraceae has been made, suggesting a close relationship (e.g., Miers 1880), Payens does not agree. The embryology and development in Lecythidaceae seem quite diverse but provide no obvious clues to the morphology of this seedling.

Family: Leguminosae (Fabaceae)

One of the largest tropical families, it is divided into 3 main subfamilies, Mimosoideae, Caesalpinioideae, and Papilionoideae (sometimes recognized as separate families), but a characteristic fruit, the legume, occurs in all 3. Although no member of the family grows in the front-mangal communities, the following genera are characteristic mangrove associates with a wide distribution. Four vines are mentioned here: they are the more frequent species with this habit in mangrove associate communities.

Artificial key to legumes treated in this account

1A. Flowers more or less regular, not like those of sweet pea.................
    .... 2 (subfamily Caesalpinioideae)
1B. Flowers zygomorphic, like those of either sweet pea or, if regular, plants
    with simple leaves. .... 4 (subfamily Papilionoideae)

[*N.B.*: The most consistent feature distinguishing these 2 subfamilies is somewhat technical and contrasts the petals in bud that overlap either from below upward (imbricate-ascending–Caesalpinioideae) or from above downward (imbricate-descending–Papilionoideae).]

2A. Trees, unarmed.. 3
2B. Climbers with numerous recurved prickles ....*Caesalpinia*
3A. Leaves uni- or bijugate with 2 or 4 leaflets; fruit about 3–4 cm long, Old
    World ....*Cynometra*
3B. Leaves pinnate, usually with 2 pairs of leaflets; fruit large, up to 25 cm
    long; New World ....*Mora*
4A. Vines with twining stems... 5
4B. Trees or shrubs; if climbing, without twining stems.................... 6
5A. Inflorescences obviously paniculate, flowers borne on extended second-
    order axes, corolla green and white...........................*Aganope*
5B. Inflorescences in contracted spikelike panicles, the flowers on short second-
    order (often dischasial) axes, corolla white or pink.................*Derris*
6A. Shrubs or semiscandent plants with grapnel-like short shoots. ... *Dalbergia*
6B. Trees, plants always self-supporting. 7
7A. Leaves simple .... *Inocarpus*
7B. Leaves pinnately compound... 8
8A. Leaf equal pinnate with 2–4 leaflets. .... *Intsia*
8B. Leaf unequal pinnate with 5–7 leaflets. .... *Pongamia*

Subfamily Caesalpinioideae: *Cynometra* L. 1753
[*Sp. Pl.*:382; *Gen. Pl. Ed.* 5, 1754:179]

A pantropical genus of about 70 species of trees or shrubs, mainly restricted to moist lowland forest. The genus is included within the subfamily in the very large tribe Detarieae DC 1825 (Cynometreae Benth.) in recent accounts (e.g., Cowan and Polhill 1981) and is considered to be a somewhat "basal" (i.e., unspecialized) type from which more specialized genera may be derived. It is distinguished by the uni- or bijugate leaves, small bud scales, short inflorescence axis, 4 calyx segments, usually 10 stamens, usually 5 short petals, and thick, 1-seeded fruit. The following account is based on the monograph by Knapp-van Meeuwen (1970), where a complete synonymy is given.

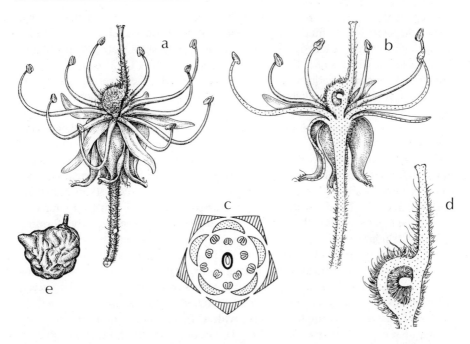

Figure B.32. *Cynometra iripa* (Leguminosae-Caesalpinioideae) flower and fruit. (a) Single flower (× 4); (b) flower in L.S. (× 4); (c) floral diagram; (d) ovary in L.S. (× 9); (e) mature fruit (× 1). (From Sunday Creek River, Deluge Inlet, and Hinchinbrook Island, Queensland. N. C. Duke and J. S. Bunt, 3.1.76 and 3.8.76)

The 2 species distinguished in the following key occur somewhat infrequently in back mangal but are also recorded inland to an altitude of 400 m. A related species, *C. cauliflora* L. (known only in cultivation but possibly originating in East Malesia), which is the type of the genus, is grown for its edible fruit ("nam-nam").

1A. Style straight, in line with the dorsal part of the ovary, fruit without a prominent lateral beak. Ovary wall glabrous inside. Sepals not curved distally when reflexed at anthesis... *C. ramiflora* L. (from India throughout Southeast Asia and Malesia to the Pacific but not Australia)

1B. Style bent, oblique to the dorsal line of the ovary, becoming a prominent lateral beak in fruit. Ovary wall hairy inside. Sepals curved distally when reflexed at anthesis. ...... *C. iripa* Kostel (Fig. B.32) (with much the same distribution but much more localized, occurring in North Queensland)

Cynometra ramiflora L. 1753
[*Sp. Pl.*:382]

Trees to 25 m high with smooth, thin, gray bark; the trunk fluted somewhat at the base but not to more than 1 m. Leaves distichous; shoots somewhat zigzag, developing rapidly from lateral buds with numerous overlapping distichous bud scales, the shoot at first limp and pendulous with white or reddish leaves. Stipules intrapetiolar, narrow, caducous, and leaving no scar. Leaves glabrous, essentially paripinnate but with a short (1 to 2 cm or longer) axis and 1 (− 2) pair of opposite leaflets and so usually unijugate, less commonly bijugate, the lower pair of leaflets then often very small. Leaflets 10 to 13 by 4 to 6 cm, but commonly smaller, asymmetrical, ovate, lanceolate, the apex usually acute or bluntly rounded, margin entire, the acroscopic side almost straight, the basiscopic side curved. Petiole and leaflets each with a short (1 to 2 mm) corky or fleshy pulvinus.

Flowers perfect, slightly zygomorphic, in short, axillary, hairy racemes, sometimes with 2 inflorescences per node, racemes up to 1 cm long but usually almost capitate. Flowers about 15 per axis, each subtended by a short (to 10 mm), narrow scarious bract, the lower bracts often empty. Pedicels 6 to 15 mm long with 1 or more basal bracteoles. Floral envelope white, but turning brown, the calyx forming a shallow, slightly asymmetric hypanthium about 1 mm deep; calyx deeply 4-lobed above, the lobes narrow, 3 to 4 mm long, at first imbricate but becoming reflexed. Petals 5, free, to 5 to 8 mm long, lanceolate, narrower than the calyx lobes. Stamens 10, rarely fewer (− 8) or more (− 13), free with small anthers on slender curved filaments up to 7 mm long. Ovary inserted slightly eccentrically with a short (1 mm) stalk, asymmetric-elliptical, hairy, with a slender style 3 to 4 mm long and a somewhat capitate stigma. Ovule 1 (− 2), inserted dorsally. Fruits 1 or more on each raceme, each 1-seeded, elliptic to semiorbicular, about 3 to 4 by 2 to 3 cm, brown with a roughened, wrinkled, often hairy surface; indehiscent, distributed by water currents. Seed without endosperm. Germination epigeal.

The species is somewhat variable and much of the synonymy refers to local forms. A number of varietal or subspecific names have also been proposed, but only one, var. *bifoliolata*, like *C. ramiflora* var. *bifoliolata* (Merr.) van Meeuwen (*C. bifoliolata* Merr. 1917, *Philipp. J. Sci. Bot.*12:272), is accepted by Knapp-van Meeuwen. It is distinguished by its petiolules, 5 to 8 mm long. The most clearly related species seems to be *C. glomerulata* Gagnep., which is distinguished by the consistently bijugate leaves, with a short (less than 2 cm) axis, the acuminate-cuspidate leaflets, and the rhombic outline of the pod. It is not likely to be encountered, since it has a limited distribution in Laos and is known only from 3 collections.

Cynometra ramiflora has been most frequently confused with *C. iripa*, which apart from being similar morphologically, can also occur in back mangal. According to Knapp-van Meeuwen (1970), all material from Australia named as *C. ramiflora* should be identified as *C. iripa*, so that Australia can no longer be included in the range of *C. ramiflora*. The 2 are distinguished primarily by the characters used in the key.

Subfamily Caesalpinioideae: *Caesalpinia* L. 1753
[*Sp. Pl.*:380]

A number of species in this genus of about 40 species (which can include segregate genera like *Guilandina, Mezoneuron,* and *Poinciana*) are climbers and often spiny. Two of these with twice-compound leaves extend into beach vegetation and are frequent mangrove associates; one is pantropical, the other Asian. Although they have been confused and so given rise to a complex synonymy (Dandy and Exell, 1938; Hattink 1974), they are readily distinguished as follows:

1A. Stipules conspicuous, pinnate; leaflets opposite or subopposite, 16–24 per pinna, base unequal. Inflorescence always axillary; flowers unisexual. Fruit armed with rigid spines, seeds gray.........*Caesalpinia bonduc* (L.) Roxb.
(pantropical)

1B. Stipules obscure; leaflets always opposite, 2–4 pairs per pinna, base equal. Inflorescences often terminal; flowers bisexual. Fruit unarmed; seeds black.
.... *Caesalpinia crista* L. (Indo-Malaya)

Caesalpinia bonduc (L.) Roxb. 1832   (Figs. B.33, B.34)
[*Fl. Ind.* 2:362, emend Dandy and Exell (1938) in *J. Bot. Lond.* 179]

A full synonymy is given in Hattink (1974).

A coarse scrambling vine. Leaves twice pinnate, up to 1 m long, with 6 to 11 pairs of pinnae, leaflets 16 to 24 per pinna, 2 to 4 cm long, asymmetric. Rachis and stem with hooked prickles; stems with additional numerous stout to soft prickles. Stipules conspicuous, pinnate. Inflorescences lateral, supra-axillary, often several per axil and serial, branched, axes up to 50 cm long. Flowers unisexual by abortion, petals yellow with reddish streaks; stamens 10, anthers hairy. Fruit 1- to 2-seeded, ellipsoid, 6 to 9 by 3 to 4 cm, with numerous stout spines, ultimately dehiscent. Seeds ovoid, smooth, gray.

This species is essentially pantropical; it is so widely distributed partly because seeds can float and retain their viability in water for extended times. It is familiar in a diversity of coastal communities, including back mangal, especially in disturbed sites, but also occurs inland chiefly in secondary forests to an altitude of about 850 m.

A similar and also widely distributed species, which may be confused with it and can also occur in beach vegetation, is *C. major* (Medik) Dandy & Exell, distinguished by its simple stipules, if they are present at all, fewer (6 to 14) symmetrical leaflets per pinna, longer pedicels, and yellow seeds; the ovary has 4 ovules (not 2 like *C. bonduc*).

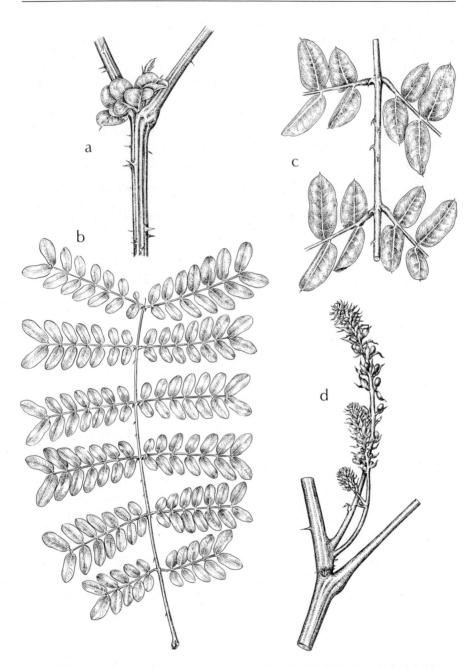

Figure B.33. *Caesalpinia bonduc* (Leguminosae-Caesalpinioideae) leaf and habit. (a) Node with pinnate stipules ($\times$ 1); (b) leaf ($\times$ ¼); (c) part of leaf with recurved spines ($\times$ ½); (d) node with axillary complex of young inflorescences ($\times$ ½). (Material from mangal at Fairchild Tropical Garden, Miami, Florida)

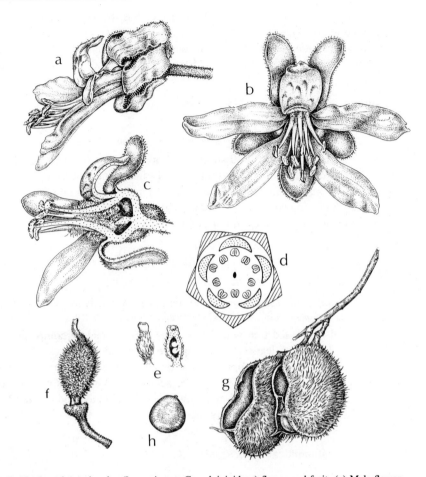

Figure B.34. *Caesalpinia bonduc* (Leguminosae-Caesalpinioideae) flower and fruit. (a) Male flower from side (× 3); (b) male flower from front (× 3); (c) male flower in L.S. (× 3), ovary aborted; (d) floral diagram of male flower; (e) functional ovary of female flower (× 3); (f) young fruit (× ½); (g) mature fruit (× ⅓); (h) seed (× ⅓). (Material from mangal at Fairchild Tropical Garden, Miami, Florida)

Caesalpinia crista L. 1753
[*Sp. Pl.*:380]

Caesalpinia nuga (L.) Ait. 1811 [*Hort. Kew*, ed. 2, 3:32]

Hattink (1974) provides a full synonymy; see also Dandy and Exell (1938). A climber to 15 m. Leaves bipinnate, to 30 cm long, with 2 to 4 pairs of pinna, leaflets per pinna about 4 to 6, opposite, Stipules obscure or absent. Rachis and stem armed with recurved prickles. Racemes either axillary or aggregated into terminal inflorescences. Flowers bisexual, petals yellow; stamens 10, with woolly filaments. Fruit ellipsoidal, 4 to 7 by 2 to 3 cm, flat, beaked with 1 ($-$2) seed. Seeds ovoid and black.

Commonly recorded as a back-mangal constituent from India and Ceylon through most of Southeast Asia to the Ryu-kyu Islands, Queensland, and New Caledonia.

Subfamily Caesalpinioideae: *Mora* Schomb. ex Benth. 1839
[*Trans. Linn. Soc.* 18:210, t. 16]

> *Dimorphandra* Schaft [in Sprengel 1827 *Syst.* 4; App. 404]

A genus of trees with 20 species in tropical South America and the West Indies. *Mora excelsa* Benth. is a common and often dominant swamp species in Trinidad and northern South America, where it may extend into the mangrove fringes. The following species is recorded as growing with *Pelliciera, Rhizophora,* and *Acrostichum* on the Pacific coast of Colombia.

Mora oleifera (Triana) Ducke 1925
[*Arch. Jard. Bot. Rio de Janeiro* 4:45]

> *Dimorphandra oleifera* Triana ex Hemsl. 1885 [*Bot. Voy. Challenger* 3:301]
> *M. megistosperma* Britt. & Rose 1930 [*North Am. Flora* 23(4):218]

A tall buttressed tree with large, alternately pinnately compound leaves usually with 2 pairs of leaflets, and long dense pendulous spikes of mimosoid flowers. There are 5 stamens opposite the petals alternating with 5 staminodes. Leaflets are 10 to 25 by 5 to 10 cm ovate, acuminate. The fruits are massive 1-seeded, woody structures up to 25 cm long, the seed flattened and about 12 cm in diameter said to be the largest of any dicotyledon.

Subfamily Papilionoideae: *Inocarpus* J. R. & G. Forst. 1776
[*Char. Gen.* 66, t. 33]

> *Inocarpus* is a name conserved against *Aniotum* Parkinson 1773 [*J. Voy. Endeavour* 39]

A genus of perhaps 3 species in Malesia and the Pacific Islands, distinguished from related genera (e.g., the American *Etaballia*) by its fruit, the almost equal petals, and the stamens free or only basally joined. Verdcourt (1979) comments that the systematic position of these two genera within the Papilionoideae is uncertain. One species of *Inocarpus*, readily recognized in the field by its red sap, is recorded on river banks subjected to tidal influence, brackish swamps, and sandy foreshores, but is essentially a lowland swamp forest species. The seeds are roasted or boiled and eaten, and the young leaves are said to be edible. The tree is sometimes planted, presumably for its edible seeds. It is one of the very few examples of a plant in the mangrove associations that provides human food, as its older name *I. edulis* indicates. In English-speaking parts of the Pacific it is known as "Tahitian chestnut."

Inocarpus fagifer (Parkinson) Fosberg* 1941
[*J. Wash. Acad. Sci.* 31:95]

I. edulis J.R. & G. Forst. 1776 [*Char. Gen.* 66, t. 33]

Tree grows to 30 m tall with a fluted trunk, exuding a red sap from cut stem surfaces. Leaves alternate, simple, oblong to oblong-lanceolate, about 20 by 7 cm, but up to 30 cm or longer, apex usually rounded, base cordate; petiole short (1 cm). Flowers scented, in axillary spikes to 12 cm long, the spikes either single or in clusters on a common stalk, often on older wood. Calyx pinkish white; petals white or yellow, 1 to 1.5 cm long, recurved at the tips. Petals almost equal and without characteristic differentiation usual in papilionoid flower. Stamens 10, in 2 series, alternately long and short. Fruit 1-seeded, indehiscent, irregularly globose, flattened, 5 to 10 cm in diameter, and variously keeled, ribbed, or smooth.

The precise distribution of the plant is uncertain, since its range may have been artificially extended through cultivation. Corner (1939) concluded from its abundance in seemingly natural situations that it was native at least to Johore in the Malay Peninsula and that part of its eastward distribution was the result of human migration and exploration. Presently it occupies the full range of the genus.

A closely related species (or possibly simply a variety), *Inocarpus papuanus* Kosterm. has a limited distribution in New Guinea. It is distinguished by its smaller, inedible fruit and the usually pubescent undersurface of the leaf, and occurs exclusively in rain forest.

Subfamily Papilionoideae: *Intsia* Du Petit-Thouars 1806
[*Gen. Nova Madag.*:22]

A genus of perhaps 3 species along coasts and on islands of the Indian and Pacific oceans. It is distinguished by its 3 fertile stamens and exarillate seeds, 3 to 4 in the flat, oblong, woody fruit. The following widely distributed species is recorded as a mangrove associate.

Intsia bijuga (Colebr.) O. Kuntze 1891
[*Rev. Gen. Pl.* 1:192]

A tree up to 40 m tall with a long bole, but little buttressed, crown spreading, foliage deciduous. Leaves paripinnate with (2 −) 4 leaflets; leaflets ovate, about 10 to 12 by 4 to 5 cm, apex obtuse, glabrous except for hairs on midrib beneath. Flowers numerous in dense terminal, finely hairy spikes or panicles. Sepals 4, unequal, up to 10 cm long. Petal solitary, clawed, 2 to 3 cm long, at first white but turning red or orange. Stamens red, 3 plus 7 staminodes. Fruits flat, oblong,

---

* In the original account the spelling *fagiferus* was used, but *fagifer* is orthographically correct (B. Verdcourt, personal communication).

to 20 by 7 cm, seeds 1 or more, black, flattened with a white fleshy funicle that turns brown.

Subfamily Papilionoideae: *Pongamia* Vent. 1803
[*Jard Malm.* 1, t. 28]

A genus of probably 2 species sometimes included as a section *Pongamia* (Adans.) J. Bennet in the larger genus *Derris*. It is recognized by its panicles of white or pink flowers and round, 1-seeded fruits.

The following species is widespread in tropical Asia in coastal environments, including back mangal, but it is also recorded inland to altitudes of over 500 m. It is commonly planted elsewhere in the tropics in coastal areas because it is resistant to salt and exposure and may be found as a street tree.

Pongamia pinnata (L.) Pierre 1899
[*Fl. For. Cochinch.* sub. t. 385]

P. glabra Vent. 1803 [*Jard. Malm.* 1, t. 28]

A tree growing to 25 m without buttresses, the crown in open-grown trees typically spreading with dense foliage and producing deep shade; briefly deciduous in seasonal climates. Leaves with 5 to 7 ovate-elliptic glabrous leaflets on the order of 10 to 15 by 5 cm but variable in size. Inflorescences 20 to 25 cm long as lax panicles from the axils of distal leaves on each shoot. Fruit oblong-ellipsoid, beaked, 4 to 6 by 2 to 3 cm, seed single, compressed, with a smooth coat.

Pongamia velutina (C. T. White) Verdcourt 1977
[*Kew Bull.* 32:250; see also Verdcourt (1979, pp. 311–14)]

This species, known from a very limited distribution in Papua New Guinea (Madang District and Central District), is distinguished by the velvety lower surface of the leaflets and the dense gray spreading hairs on the calyx. The fruit is said to be dehiscent, whereas in *P. pinnata* the fruit is always indehiscent, somewhat larger, and perhaps never so elongated.

Subfamily Papilionoideae: *Dalbergia* L.f. 1782
[nom. cons. *Suppl.* 52:316]

A genus of about 300 species, well developed in Africa. At least 2 species are commonly recorded in mangrove communities.

Dalbergia ecastophyllum (L.) Taub. 1894
[Engler and Prantl, *Nat. Pflanzenfam.* 3(3):335]

A shrub or small tree with reclining or scandent branches. Leaves with a short (to 1 cm) petiole, the flowers in axillary clusters. Legume 2 to 3 cm long and

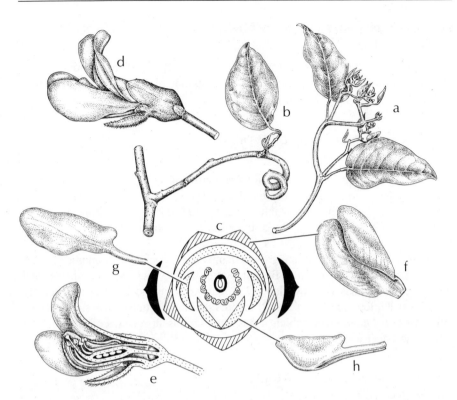

Figure B.35. *Dalbergia amerimnion* (Leguminosae) shoot and flowers. (a) Apical region of major axis (× ½) to show sylleptic branching and characteristic zigzag configuration of extending internodes; (b) determinate lateral branch (× ½) modified as a woody tendril; (c) floral diagram; (d) flower from side (× 4); (e) flower in L.S. (× 4); (f) standard petal (× 4); (g) wing petal (× 4); (h) keel petals (× 4). (Material cultivated at Fairchild Tropical Garden, Miami, Florida)

usually 1-seeded. The species has a wide distribution in the Caribbean and South America, with an extension to West Africa.

> ***Dalbergia amerimnion*** Benth. 1860   (Fig. B.35)
> [*J. Linn. Soc.* 4 Suppl:36]

Resembles the previous species closely but is distinguished by its longer petioles (to 2 cm), more numerous ovules, and the frequently 2-seeded fruits. It is more characteristically scandent and supported by short, lateral determinate shoots that function as grapnels (Fig. B.35b). The growing ends of the main axes are characteristically zigzag. This species is restricted to the New World.

Subfamily Papilionoideae: *Aganope* Miq. 1885
[*Fl. Ind. Bat.* 1(1):151]

    A genus of 6 species ranging from West and Central Africa to India, China, the Philippines, and New Guinea. For a discussion of the distinction between *Aganope* and *Derris*, see Polhill (1971).

Aganope heptaphylla (L.) Polhill 1971
[*Kew Bull.* 25:268]

    Woody climbers or scrambling shrub growing to 15 m; shoots hairy when young. Leaflets 5 to 7, elliptic or oblong elliptic, 5 to 10 by 3 to 6 cm. Flowers about 1.5 cm long in long narrow inflorescences up to 30 cm long. Standard green, wings white. Fruit elongate, up to 20 by 3 cm, flat, often constricted between the 2 to 6 flat seeds.

    This species is recorded in mangrove swamps and associated communities, with a range from Sri Lanka through Bengal, South China, Malaysia, Indonesia, and the Philippine Islands to New Guinea. It may grow inland but has a decidedly coastal distribution. Verdcourt (1979) records that it sometimes grows intertwined with *Derris trifoliata*, with which it has been confused; mixed collections have been made under a single number.

Subfamily Papilionoideae: *Derris* Lour. 1790
[*Fl. Cochinch.*:433]

    A genus of about 50 species in the tropics and subtropics, mainly Asian.

Derris trifoliata Lour. 1790
[*Fl. Cochinch.*:433]

Dalbergia heterophylla Willd. 1800 [*Sp. Pl.* 3:901]
Derris uliginosa Benth. in Miq. [*Pl. Jungh*:252]

    Erect shrub or a rambling climber to 15 m or more, spreading by root suckers. Leaflets 3 (−5 or −7), ovate to elliptic-lanceolate, 10 to 12 by 4 to 5 cm but variable. Inflorescence slender, about 20 cm long; flowers about 1 cm long, corolla white or pale pink. Fruit elliptic, 3 to 4 cm by 2 cm, seeds 2 or 3, wrinkled.

    This species has a coastal distribution from East Africa, Madagascar, and throughout tropical and subtropical Asia to tropical Australia. It is recorded in coastal communites such as beaches, strand vegetation, and coastal swamps and is a frequent constituent of back mangal throughout its range. It is distinguished from the frequently associated *Aganope heptaphylla*, with which it is sometimes confused, by its fewer glabrous leaflets and smaller flowers and fruits. A good field character is the dark red, strongly ridged younger stems with prominent lenticels.

Other legumes

Gentry (1982) includes *Muellera moniliformis* L.f. as a mangrove associate common to the Atlantic and Pacific mangal of the New World. It is a small tree, most characteristic of river banks, with imparipinnate leaves 15 to 25 cm long and 5 to 11 rather delicate pulvinate leaflets. The fruit is a lomentum; that is, the pod is constricted between the seeds. It is most commonly recorded well away from the sea.

Family: Lythraceae

A cosmopolitan, mainly herbaceous family of 25 genera and 500 species. Two genera unrelated in the family, one Old World and the other New World, are represented in mangrove communities. They are distinguished as follows:

1A. Shrubby or subarborescent, Old World; widely distributed........*Pemphis*
1B. Herbaceous, New World; localized. ....*Crenea*

Pemphis Forst. 1776
[*Char. Gen. Pl.*:67, t. 34]

See Koehne 1880–5 [*Engl. Bot. Jahrb.*1–7:132] and 1903 [*Engl. Das Pflanzenreich* 4, 216:185]

A tropical genus of coastal shrubs or at most small trees, ranging from East Africa through southeastern Asia to northern Australia, Polynesia, and northward to Hong Kong but with an apparent disjunction in East Malaysia. The wide range of *P. acidula* seems related to its ecological status, somewhat intermediate between a strand plant and a mangrove. The genus is distinguished by its perfect hexamerous distylous flowers, a campanulate calyx persistent in fruit, the stamens united basally into a short tube, and the globose, many-seeded capsule with circumcissile dehiscence.

Two species are recognized:

1A. Coastal, evergreen. Plant without black punctate dots. Flowers dimorphic (heterostylous). Petals white, stamens 12. Style long, slender, capitate; ovules inserted only on base of placenta ....*P. acidula* Forst.
1B. Inland, deciduous. Leaves and calyx with black punctate dots. Flowers monomorphic (homostylous). Petals rose colored, stamens 18. Style short or very short, with a conspicuous capitate stigma; ovules inserted throughout the placenta..............*P. madagascariensis* (Bak.) H. Perr.

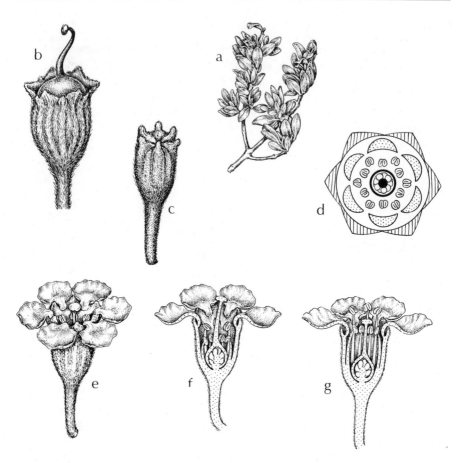

Figure B.36. *Pemphis acidula* (Lythraceae) shoot, flowers, and fruit. (a) Flowering shoot (× ¾); (b) fruit (× ½); (c) flower bud (× 3); (d) floral diagram; (e) flower from side (× 3); (f) flower, "pin" form in L.S. (× 3); (g) flower, "thrum" form in L.S. (× 3). (Material from Idler's Bay, Papua New Guinea. P. B. Tomlinson 1.11.74C)

Pemphis acidula Forst. 1776   (Fig. B.36)
[*Char. Gen. Pl.* 68, t. 34]

Small tree 7 to 10 m high, or more usually a low shrub 1 to 2 m high of strands and rocky foreshores and more exposed mangrove associations. Bark light gray to brown, becoming deeply fissured with age and shredding into long curling strips. Young twigs angular, densely hairy as on the leaves, irregular branched by syllepsis. Each node supporting an axillary bud, the bud with a lateral series of minute, ephemeral linear colleters. Leaves decussate, sometimes fleshy (to 2 mm thick), very shortly petiolate (1 to 2 mm), 10 to 30 by 3 to 10 mm, narrowly ovate to obvate, apex bluntly pointed to rounded, margin entire, base shortly cuneate; both surfaces densely covered with stiff white hairs. Flowers axillary, solitary, less commonly in few-flowered bracteolate sessile cymes. Flowers hexamerous, perfect, perigynous, distylous; pedicel 5 to 10 mm long. Calyx 5 to 8 mm long, becoming somewhat larger and persisting in fruit, greenish red campanulate 12-angled or ribbed with 6 pointed triangular lobes, alternating with 6 thickened accessory lobes, one in each sinus, the calyx lobes valvate and at first forming a flat lid to the calyx cup. Petals 6, inserted on the mouth of the calyx cup below each sinus, shortly clawed, 5 to 8 mm long, crumpled in bud, ephemeral. Stamens 12 (-18) in 2 series inserted on lower third of calyx cup, the antesepalous stamens longer and inserted higher, filament 2 to 4 mm long, anthers dorsifixed, dehiscence introrse. Ovary superior, globular, 2 mm in diameter, unilocular but with 3 (-4) obscure basal dissepiments, placentation essentially free central. Ovules numerous, crowded, orthotropous. Style simple, stigma capitate. Short-styled (about 1 mm) flowers with stamens enclosing stigma, long-styled (about 4 mm) flowers with stigma exceeding stamens. Fruit enclosed in somewhat enlarged calyx, a spherical capsule (4 to 5 mm in diameter), with persistent style, dehiscence circumcissile; seeds numerous (20 to 30), flattened, angular, with a corky margin or wing and narrowed below. Embryo minute, straight; endosperm present.

Growth and reproduction

Architecture. This appears to correspond to Attims's model because the shoots are not articulate, although there may be some periodicity of extension correlated with the frequent but diffuse or irregular branching. Though often recorded as a small tree, the characteristic habit of this plant is a diffuse low-growing shrub. Every node seems capable of supporting a vegetative branch, since even flowering nodes have an additional (supernumerary) vegetative bud.

Floral biology. Floral dimorphism has been recognized at least since the time of Koehne's monograph [1880–5; see also Koehne (1903, cited earlier)], but a more modern understanding of the significance of this in relation to the genetics of incompatability is only recent.

Lewis and Rao (1971) indicated distyly is not a local phenomenon because they recorded it in both Malaysia and East Africa. Gill and Kyauka (1977) made a quantitative study of populations in Tanzania. They describe the two forms in the traditional terminology of pin (long styled) and thrum (short styled), although the two sets of stamens (inner and outer whorl) and the stigma occupy 3 levels. The stigma then stands at a level either above the 2 sets of stamens (pin, Fig. B.36f) or between them (thrum, Fig. B.36g). The relative levels are such that the style of one flower form corresponds approximately to the level of at least one set of stamens of the other form. Styles are either 1 mm or 4 to 5 mm long, stamen length is either 3 mm or 4 to 5 mm. These authors conclude that the proportion of pin to thrum in natural populations does not differ significantly from a 1:1 ratio.

Pollen grains from the 2 forms differ appreciably in size and rate of germination on artificial media, with thrum pollen growing faster. Stigmatic papillae are larger in thrum (16 to 18 m) than in pin (12 to 14 m) flowers. Artificial crosses between pin and thrum result in normal seed sets (20 to 30 seeds per capsule), whereas selfed flowers set no seed. This efficient outbreeding mechanism is said to be maintained by a differential or disruptive selection mechanism, which has eliminated intermediate forms in favor of the two extremes.

Pemphis madagascariensis (Baker) Koehne 1903
[Engl. Bot. Jahrb. 29:164]

P. madagascariensis (Baker) H. Perrier 1954 [*Fl. Madagas. Fam.* 147–22]

This endemic Madagascan species differs from *P. acidula* in the characters cited in the key. According to Perrier, this species occurs only inland in an inter-montane region of Madagascar. This contrast between species, one of which is a mangrove, is unusual.

Crenea Aubl. 1775
[Pl. Guia. 1:523, t. 209]

A genus of 3 species restricted to Trinidad and northern South America. All seem to have an association with salt water and the following species is more or less restricted to the muddy tidal flats of the Pacific coast of Colombia (Gentry 1982). It could well be described as a mangrove but is exclusively herbaceous and has no morphological adaptations to the mangal environment.

Crenea patentinervis (Koehne) Standley 1947
[Publ. Field Mus. Nat. Hist., Chicago Bot. Ser. 23:218]

C. surinamensis (L.f.) Koehne ssp. *patentinervis* [Koehne, 1882, *Bot. Jahrb.* 3:320]

A herb growing to about 50 cm with basal stolons. Leaves opposite, simple obovate-spathulate (3 to 4 by 0.5 to 1 cm), with a blunt, rounded apex. Flowers white, axillary. Capsules about 5 mm in diameter.

Family: Malvaceae

A large, mainly tropical family of trees, shrubs, or, especially in temperate regions, woody herbs; including about 80 genera and over 1500 species. Characterized by simple alternate stipulate leaves and often stellate hairs or a scaly indumentum. The flowers are often large and showy and commonly have an epicalyx of free or fused scales below the calyx; the numerous stamens are fused to form a central staminal column and have spiny pollen; the ovary is superior with 5 or more locules; the fruit is usually dehiscent as a capsule or schizocarp.

Species in 2 genera are characteristic of tropical seashores and are often associated with mangroves but never occur on strict mangal substrates. They both have very similar conspicuous flowers that are yellow on the evening they open but turn purple by the end of the following day. A third genus is added from the New World.

These taxa may be distinguished as follows:

1A. Shrubs, tropical America; flowers scarcely 2 cm in diameter ...... *Pavonia*
        spicata Cav. and *P. rhizophorae* Killip.
1B. Trees, pantropical (at least in cultivation); flowers 5–7 cm in diameter ... 2
2A. Calyx conspicuously 5-toothed, persistent in fruit; flowers usually with a continuous maroon eye; receptacle without yellow latex. Capsule regularly loculicidally dehiscent and gaping widely. Seeds smooth. Leaves usually with 9–11 main veins and always 1–3 glands on the abaxial surface near the petiole insertion; blade with an indumentum of stellate hairs ...*Hibiscus*
        tiliaceus L.
2B. Calyx entire or at most with minute teeth; flowers with an eye of 5 separate dark spots; receptacle with yellow latex. Capsule shattering irregularly or if dehiscent not gaping widely. Seeds finely hairy. Leaves usually with 7 main veins, often lacking glands on the lower surface; blade with an indumentum of scales. ..*Thespesia populnea* (L.) Soland ex Correa [and *T.*
        populneoides (Roxb.) Kostel.]

Hibiscus L. 1753
[*Sp. Pl.*:693]

The largest genus in the family with over 200 species in the tropics and subtropics; most of them are shrubs or woody herbs. The following species is a tree and unusual in the genus because of its large, broad stipules and maritime habitat.

Hibiscus tiliaceus L. 1753   (Fig. B.37)
[*Sp. Pl.*:694]

For a detailed treatment, see Borassum Waalkes (1966); see also A.C. Smith (1981).

This is a spreading tree growing to a height of 10 to 15 m in coastal communities, especially adjacent to beaches. It is recorded inland up to altitudes of 800 m but is widely planted, and its precise natural ecological and geographic distribution is uncertain. A closely related species, *H. elatus* Sw., is distinguished by the deciduous (not persistent) calyx and epicalyx.

The tree may be described as ever-growing, and flowers are present throughout much of the year, singly in the axils of foliage leaves, with the base of the long peduncle enclosed by the stipule pair. Stipules fall later and leave a conspicuous annular nodal scar. The flowers are typically malvaceous with the surfaces of the petals, stamen tube, and style covered with glandular hairs. In addition to the glands on the foliage leaves, there are similar glands on the back of the epicalyx lobes,

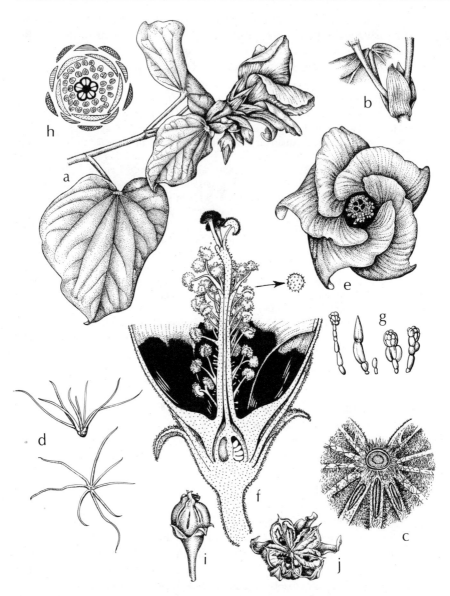

Figure B.37. *Hibiscus tiliaceus* (Malvaceae) shoot, flowers, and fruit. (a) Flowering shoot (× ⅜);
(b) leaf insertion with stipules (× ½); (c) leaf blade insertion from below (× ³⁄₂), glands on major
veins; (d) stellate hairs from leaves (× 15); (e) flower from above (× ½); (f) base of flower in L.S.
(× ³⁄₂), inset: pollen grain (× 35); (g) glandular hairs from the style (× 30); (h) floral diagram,
epicalyx crosshatched; (i) unripe fruit (× ½); (j) dehisced fruit (× ½). (From Tomlinson 1980)

but they seem to have no function in pollination. The seeds are exposed when the capsule splits; they float and will withstand extended immersion in sea water.

Borassum Waalkes (1966) recognizes 5 subspecies; these are largely distinguished by leaf shape and degree of marginal lobing.

Thespesia Solander ex Correa 1807
[*Ann. Mus. Hist. Nat. Paris* 9:290]

> For a full synonymy, see Fosberg and Sachet (1972).

A genus of 15 species, restricted to the tropics, with the following 2 widely distributed species associated with mangroves but always in dry, unexposed situations. *Thespesia populnea* is widely planted as an ornamental, particularly as a street tree.

Two closely related species occur in marine environments, one of them pantropical and the other with a very limited distribution. Their distinction, distribution, and nomenclatural history have been discussed in detail by Fosberg and Sachet (1972), who separate them as follows:

1A. Leaves green, deeply cordate; pedicels erect, 1–5 cm long with a bracteate joint very near the base; fruit indehiscent; seeds with long soft hair, especially on angles............ *Thespesia populnea* (L.)Soland. ex Correa (pantropical on sea coasts, widely planted and occurring inland)

1B. Leaves somewhat bronzed or coppery, very shallowly cordate to subtruncate; pedicels tending to droop, 5–12 cm long without a bracteate joint; outer layer of fruit dehiscent, seeds with short clavate or bulbous hairs..... *Thespesia populneoides* (Roxb.) Kostel. (coasts of Indian Ocean, extending to Indochina and Hainan, uncommon, sparingly introduced elsewhere)

Thespesia populnea (L.) Solander ex Correa 1807   (Fig. B.38)
[*Ann. Mus. Hist. Nat. Paris* 9:290]

The Latin generic name refers to the change in color of the flower from yellow to purple, usually within 24 hours. The flowers are borne singly in the axils of distal foliage leaves. The calyx is cupular and either entire or at most with a series (5) of short teeth. The "nectar guides" of the flower are usually represented by basal spots on the petals. The stigma is elongated and longitudinally grooved. The fruit becomes dry and normally remains undehisced on the tree, or at most with the wall rupturing irregularly. Consequently the fruit itself, which floats, is the primary organ of dispersal, although the seeds themselves also float when released by shattering or decay of the capsule. A feature not usually recorded in descriptions of this species is the yellow latexlike fluid that exudes from the receptacle when it is cut. This exists in a series of canals in the peripheral tissues, and chemically the fluid is largely gossypol. In some locations this species is further distinguished from *Hibiscus tiliaceus* by the absence of the glands on the lower surface of the leaf (e.g., Tomlinson 1980). However, Fosberg and Sachet dem-

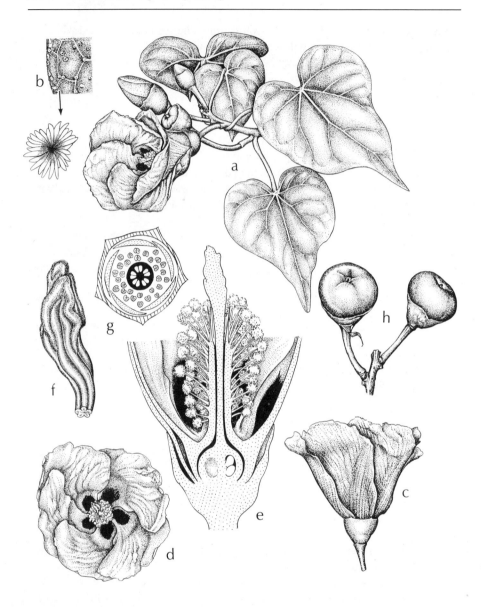

Figure B.38. *Thespesia populnea* (Malvaceae) shoot, flowers, and fruit. (a) Flowering shoot ($\times$ ½);
(b) details of leaf surface ($\times$ 6) and enlarged single scale ($\times$ 90); (c) flower from side ($\times$ ½); (d)
flower from above ($\times$ ½); (e) base of flower in L.S. ($\times$ 3⁄2); (f) stigma ($\times$ 3); (g) floral diagram;
(h) fruits ($\times$ ½), one epicalyx scale persistent to left. (From Tomlinson 1980)

onstrate that they may occur in some collections, so that the distinction is not absolute.

Thespesia populneoides (Roxb.) Kostel. 1836
[*Allg. Med. Pharm. Fl.* 5:1861]

For a discussion of the history of this taxon and a full synonymy, see Fosberg and Sachet (1972). According to these authors this species is distinguishable relatively easily from *T. populnea* in the field where the 2 occur together by the above mentioned characters. Specimens with a range of intermediate characters have been collected, however, and it is suggested that these intermediates represent hybrids between them. The hybrids may persist in areas where they are propagated vegetatively, for example, as living fence posts.

The mature fruit of this species has the pericarp differentiated into 3 distinct layers, a smooth exocarp separated from a hard fluted endocarp by a spongy-fibrous mesocarp. The exocarp dehisces into usually 5 valves but never gapes widely as in *Hibiscus tiliaceus*.

Pavonia Cav. 1786
[*Diss.* 2: App. 2, and 3, 132, t. 45, 1787]

A pantropical genus of about 200 species with a dry, 5-carpellate, usually spiny fruit. A number of species have a coastal distribution; 2 closely related species (Kearney 1954), distinguished as follows, have been recorded as mangrove associates:

1A. Leaf blades broad and shallowly cordate, 5–7 nerved from the base. Each carpel with 3 apical cusps. Involucel long relative to the calyx .. *P. spicata* Cav.

1B. Leaf blades narrow, contracted, and subcuneate, 3-nerved from the base. Each carpel with a single apical cusp. Involucel short relative to the calyx ....*P. rhizophorae* Killip.

Pavonia spicata Cav. 1787
[*Monad. Cl. Diss. Dec.* 3:136, t. 46]

Collections of this species, which has a wide distribution in tropical America, seem always to be made in mangal, salt marshes, and coastal communities.

Pavonia rhizophorae Killip. in Kearney 1954
[*Leafl. West. Bot.* 7:118]

This species is recorded only for Colombia, but more or less exclusively in mangal (Gentry 1982).

Family: Melastomataceae

A large pantropical family with over 3000 species in about 200 genera, mainly in South America. Members of the family are usually sparsely branched shrubs with opposite (decussate) leaves with 3 to 5 prominent veins diverging from the base of the leaf blade.

Ochthocharis Bl. 1831
[*Flora* 14:523]

A genus of 5 species in the Malay Archipelago, distinguished by its spurred stamens and the 3-sided seeds, which are irregularly keeled. The following species has been recorded in back-mangal communities (Chai 1982).

Ochthocharis bornensis Bl. 1849
[*Mus. Bot. Lugd. Bat.* 1:40]

A shrub growing to a height of about 6 m, the leaves opposite and narrowly ovate, 5 to 10 by 2 to 4 cm, the petiole slender and from as short as 1 cm to as long as 5 cm. Flowers axillary, pentamerous, small (about 3 mm in diameter), the petals white with pink tips. Fruits light green, 3 to 4 mm in diameter, spherical, calyx persisting as a rim. The plant is recognized as a member of the family by the characteristic 5 conspicuous principal veins connected by numerous transverse, parallel secondary veins.

Within mangrove communities this species is easily recognized by its stature and leaves; in adjacent communities there may be other members of the family that are often weedy (notably *Melastoma malabathrica* L. with much larger flowers).

Family: Meliaceae

A large tropical family of 50 genera and over 1000 species, commercially important for the large number of species with high-quality timber (Pennington and Styles 1975). The family is recognized by the unisexual flowers with the stamens (or staminodes) united by their expanded filaments to form a staminal tube. It is divided into a number of tribes; the Xylocarpeae consists of *Xylocarpus* and *Carapa*. Two of the three species of *Xylocarpus* occur in the landward mangrove communities of the tropics from East Africa to Polynesia. The other genus, *Carapa*, is entirely fresh water and New World. In addition *Aglaia cucullata* (Roxb.) Pellegrin (*Amoora cucullata* Roxb.) (tribe Aglaieae), a swamp forest species with pneumatophores, may occur in tidally influenced areas of the Sundarbans.

Tribe: Xylocarpeae Roemer 1846
[*Synopses Monogr. Fasc.* 1 : *Hesperides* 78]

Carapeae Harms 1896 Meliaceae [in Engler and Prantl, *Nat. Pflanzenfam.* 3(4):276]

Flowers without a gynophore. Petals free. Stamens 8 to 10, filaments completely united to form an urceolate or cup-shaped tube with appendages on the margin. Capsule subwoody or leathery with a rudimentary columella; seeds unwinged with a corky or woody sarcotesta; cotyledons large, fused together; endosperm absent; germination hypogeal.

Key to genera of Xylocarpeae

1A. Sepals 4 or 5, imbricate, free; seed with a woody sarcotesta; inflorescence unit (panicle) more than 25 cm long (to 90 cm); rachis long (to 80 cm); leaflets in (4-)6 to 12(-18) pairs. Tropical rain forests from west and central Africa to tropical America; $2n = 58.*$ .... *Carapa* Aublet (7 species)

1B. Sepals 4, valvate, remaining connate in the lower half; seed with a corky sarcotesta; inflorescence unit (panicle) less than 25 cm long; rachis shorter (to 20 cm); leaflets in (1-)2 to 3(-4) pairs. Coastal habitats and mangroves, from East Africa to Polynesia; $2n = 52.$ .... *Xylocarpus* Koen. (3 species)

Noamesi (1958) included the 2 genera in a separate tribe Xylocarpeae within the subfamily Melioideae; Pennington and Styles (1975, p. 444) follow this, but they consider that the tribe should be transferred to the subfamily Swietenioideae (together with the tribes Cedreleae and Swienteniae) producing, in their opinion, a more natural arrangement.

There has been a confusing synonymy between these 2 genera, which are distinct both morphologically and in their distribution. Noamesi (1958) seems to have resolved the essential difficulties and his treatment is followed here. It must be emphasized that his account is based entirely on herbarium material together with accounts provided in the extensive literature. Furthermore, all descriptions up to the publication of the monograph by Pennington and Styles (1975) have proceeded on the assumption that the flowers are perfect, rather than unisexual. Sexual floral dimorphism is slight but should be taken into consideration where species have been segregated partly on the basis of floral features, as in Ridley (1938).

---

* From Pennington and Styles (1975); 2 species in each genus examined. There is a record of $n = 21$ for *Xylocarpus granatum* from Tanzania (*Taxon* 27:229, 1978).

Xylocarpus König 1784
[*Naturforscher* 20:2]

Three species with a coastal distribution in the East African and Indo-Malayan tropics. They are very similar except for bark and root characters and may be described collectively.

Monoecious (perhaps sometimes dioecious), evergreen or deciduous trees, less commonly shrubs; trunk base with or without buttresses; root system often developing either pneumatophores or ribbonlike surface roots; bark smooth or rough and fissured, inner bark red or pink. All parts glabrous. Leaves alternate, without stipules, paripinnately compound, the rachis about 10 to 20 cm long. Leaflets (1-) 2 to 3(-4) pairs each with a short corky petiolule, sometimes slightly asymmetric. Inflorescence a lateral panicle at the base of the current increment, but often elsewhere; ultimate lateral units more or less regular 3-flowered cymes, bracts and bracteoles minute and usually ephemeral; cymes sometimes 2-flowered by abortion. Flowers functionally unisexual, pedicellate, tetramerous, 3 to 5 mm in diameter. Calyx lobes 4, valvate, shortly united below, rounded, green and much shorter than the petals at anthesis; petals 4 (3 to 6 mm long), creamy white, valvate, more or less free, rounded. Stamen tube white, spherical, supporting 8 sessile orange stamens each situated opposite a sinus in the mouth of the tube, the intervening lobes acute, rounded, or emarginate. Disc well developed, orange-red, its outer surface warted, pitted, or roughened; ovary globose, inserted on the disc, 4-locular, with 2 to 4 ovules on an axile placenta in each loculus; short style ending in a capitate stigma with a flattened apex, more or less occluding the mouth of the stamen tube. Male flower with a nonfunctional, rather slender ovary; female flower with nonfunctional stamens either never dehiscent or with sterile pollen. Fruit a woody capsule, globose, with 4 indistinct furrows indicating the future limits of thick valves, up to 25 cm in diameter, with a short woody stalk. Seeds several, up to 16 or more, more or less tetrahedral with outer surface against fruit wall rounded, up to 6 cm long, brown, with a corky testa, and future radicle evident. Seedling hypogeal, radicle developing as a taproot, seedling axis initially with a long series of scale leaves. Seedling foliage leaves simple.

Distribution and relationships
In coastal sandy, rocky, or typically mangrove environments in the Old World tropics, from East Africa and northern Madagascar as far north as the Bay of Bengal, eastward to tropical Australia and Polynesia.

Key to species

1A. Mangrove plants. Leaflets usually 4, more or less elliptic, the apex rounded or at most narrowed to a blunt point. Inflorescence less than 8 cm long. Root systems elaborated above ground. Bark fissured or scaly. Fruit about the size of an orange or much larger 2

1B. Plants of sandy or rocky beaches, not mangroves. Leaflets usually 4 or 6, more or less ovate, and usually narrowed to a distinctly pointed apex. Inflorescence often exceeding 8 cm in length. Root system not elaborated, buttresses absent. Bark longitudinally fissured. Fruit about the size of an orange .... *X. moluccensis* (Lamk.) Roem.

2A. Fruit up to 25 cm in diameter, to the size of a melon. Trunk surface smooth, pale, blotched greenish or yellowish, and peeling in patches (Fig. B.39A). Trunk base often enlarged, with well-developed buttresses continued outward as narrow undulating ribbonlike extensions of the surface root system (Fig. B.39B). Inflorescence up to 6 m long, spreading and rather irregularly branched, without a well-developed main axis .... *X. granatum* Koenig

2B. Fruit small, about the size of an orange and not exceeding 12 cm in diameter. Trunk surface rough, dark brown, fissured, and peeling in narrow strips (Fig. B.40). Buttresses absent or very short, horizontal roots developing blunt pneumatophores up to 20 cm high and 3–4 cm in diameter (see Figs. 5.7, B.41). Inflorescence up to 8 cm or longer, with a distinct main axis and regular lateral cymes .... *X. mekongensis* Pierre

From this it is clear that the 3 recognized species should easily be distinguishable in the field on the basis of habitat, trunk, fruit size, and root elaboration. Other distinguishing features, as found in herbarium specimens, are often less easy to apply. The leaflet number varies within the individual's ontogeny; seedling leaves are simple, with a rapid transition to compound leaves, soon with 6 leaflets in all species; this number is reduced as the plant matures. A leaf with 6 leaflets therefore does not necessarily indicate *X. moluccensis*. Leaflet shape is not always so distinct as Noamesi suggests, since both *X. granatum* and *X. mekongensis* can have quite pointed leaves. Noamesi implied differences in bud morphology determined by the arrangement of bud scales, but I have not been able to find any difference. His description of the inflorescence is not very complete and his reference to the inflorescence of *X. granatum* as "usually dichotomously branched" is inaccurate. According to Noamesi, *X. granatum* may have the larger flowers. In my experience of the 2 species growing close to each other, they cannot be distinguished by floral features.

Unfortunately interpretation and description of floral morphology in *Xylocarpus* are deficient because it has not been recognized until recently that flowers are unisexual. The 2 kinds of flowers can be recognized by careful examination, but differences are not likely to be observed in dried specimens. The male flowers have a more slender ovary and stamens with functional pollen; the female flowers have a swollen ovary and empty stamens. Ovules are fewer and larger in female flowers. Noamesi (1958) may have recognized the difference in his diagnosis of *X. mekongensis* when he refers to the ovary as "conical, set on a conspicuous disc, or globose with a moderate to obsolete disc around the base," but he comments no further on the significance of this difference.

A

Figure B.39. *Xylocarpus granatum* (Meliaceae). (A) Flaking bark pattern distinctive for the species. (B) Ribbonlike plank roots (together with aerial roots of *Rhizophora* and *Bruguiera*). Hinchinbrook Island, Queensland.

B

Figure B.40. *Xylocarpus mekongensis* (Meliaceae). Peeling bark distinctive for the species. Magnetic Island, Queensland.

Figure B.41. *Xylocarpus mekongensis* (Meliaceae). Root knees at low tide (together with those of *Avicennia*). Darwin, northern Australia. (From a transparency by A. M. Gill)

Xylocarpus granatum König 1784   (Figs. B.39, B.42, B.43)
[*Naturforscher* 20:2]

Carapa obovata Blume 1825 [*Bijdr.* 179]
X. obovatus (Blume) Juss. 1830 [*Mem. Mus. Paris* 19:244]
X. benadirensis Mattei 1908 [*Boll. Ort. Bot. Palermo* 7:99]

Vegetative growth

Germination is initiated by extension of the hypocotyl so that the radicle is extruded from the seed (Fig. B.42d). The cotyledons stay within the seed, which may remain attached to the seedling for a long time (Fig. B.42g). The plumule grows upward through the cleft between the paired cotyledons, which are fused within the seed. The plumule extends rapidly and it reaches a height of 50 cm in a few weeks (Fig. B.42g). It bears a series of spirally arranged scale leaves, which are progressively more distant. At the same time the radicle extends and branches to form, at least initially, a distinct taproot.

The first leaves are simple, but there is a fairly rapid transition to compound leaves with 2, 4, or 6 leaflets. Subsequently the number of leaflets is reduced to 4 or even 2 on the older parts. Leaves are paripinnate, the rachis being golden brown and continuing beyond the insertion of the distal pair of leaflets as a minute appendage. Leaflets have a short corky petiolule, the same color as the rachis; this petiolule affects the primary positioning of the leaflet but there are no obvious later periodic leaflet movements. Leaflets vary somewhat in shape and size, up to a maximum of 12 cm long and 6 cm wide at the broadest part, but they may be less than half these dimensions in depauperate specimens. Typically the apex is rounded or blunt and sometimes even slightly emarginate, but leaflets with a more or less acute apex are not uncommon. The leaflet thickness varies considerably according to the situation and is often quite succulent, up to 1.5 mm thick.

Growth of the shoot system is rhythmic and corresponds precisely to Rauh's model (Hallé et al. 1978). The resting bud is enclosed in short, thick, brown bud scales, which may retain a vestigial leaf blade. The bud scales leave a series of close-set scars so that the successive increments are clearly recognized. Trees may retain leaves throughout the year, even in seasonal climates, but Chai (1982) reports this species (and *X. mekongensis*) as deciduous in the nonseasonal climate of Sarawak.

Branching follows a period of seedling growth and is the result of the proleptic development of axillary buds at the end of the previous growth flush. Reserve buds are produced in abundance, since all leaves, including most of the bud scales, subtend a dormant bud. Few of these expand during the normal development of the tree. Saplings recover rapidly after damage, and depauperate plants may develop several trunks. The persistence of these reserve axillary buds probably accounts for the marked ability of older trees to sucker basally when they are damaged. Suckers essentially repeat the morphology of seedlings but have only a short series of basal scale leaves.

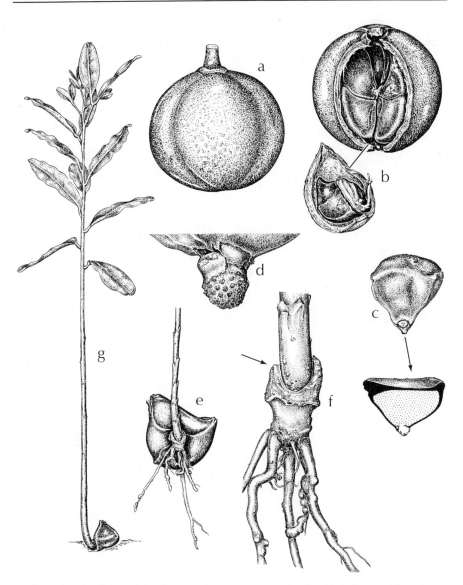

Figure B.42. *Xylocarpus granatum* (Meliaceae) fruit and seedling. (a) Fruit (× ¼); (b) fruit with one valve removed (× ¼); (c) seed, external view and L.S. (× ¼); (d) detail of radicle from recently dispersed seed (× 1); (e) base of seedling (× ¼), cotyledons attached, seedling axis with scale-leaf scars; (f) detail of cotyledonary node, cotyledons attached (× ½), arrow: level of insertion of cotyledons; (g) seedling (× ⅛) with juvenile and transitional leaves. (Fruits from Missionary Bay, Hinchinbrook Island, Queensland; seedlings cultivated at Fairchild Tropical Garden, Miami, Florida)

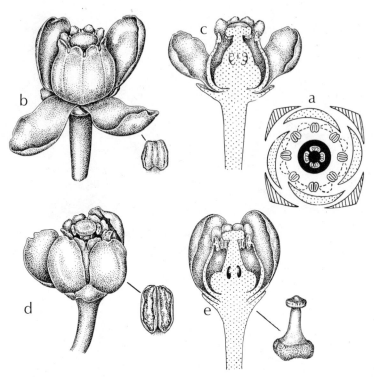

Figure B.43. *Xylocarpus granatum* (Meliaceae) flowers. (a) Floral diagram, scales on stamen tube dotted; (b) female flower from side (× 4), inset: nonfunctional stamen; (c) female flower in L.S. (× 4); (d) male flower from side (× 4), inset: functional stamen; (e) male flower in L.S. (× 4), inset: nonfunctional ovary (× 3). (Material from Semetan, Sarawak. P. B. Tomlinson 19.7.82)

Normally trees remain single trunked. The bark is very characteristic, smooth, pale greenish or yellowish brown, and peeling in irregular patches so that the trunk is blotchy. Lenticels are not conspicuous. Buttress development appears fairly late and is usually preceded by the first evidence of the platelike surface roots in trees about 5 m high. These have not been studied in detail but are commented upon by all authors who have seen the tree in situ.

Reproductive biology

In the absence of extended measurements on phenology, little precise information about the periodicity of shoot extension and flowering is available. Noamesi records this species as flowering throughout the year on the basis of herbarium specimens, but this is probably not generally true because of the pronounced rhythmic growth and synchrony in shoot development over the whole plant. Chai (1982) records quite precise flowering and fruiting periods for both species. Inflorescences appear mainly from the axils of leaves at the base of the current shoot increment, but they also develop directly from older previously dormant buds or on woody shoots. In this situation they may appear to be terminal. This habit may be important in view of the large fruit, which requires a stout twig to support it. The tree cannot be described as ramiflorous, however.

Inflorescences consist of a central axis supporting spirally arranged inconspicuous and ephemeral bracts. Each bract usually subtends a 3-flowered cyme with even smaller bracteoles. The irregular inflorescence structure in this species, emphasized by Noamesi, is the result of zigzag development of the axis and the frequent abortion of flowers in the cymes. In a regular cyme the terminal flower seems most often to be female, the lateral (or distal flower of the whole inflorescence) to be male. Flowers expand rather irregularly (usually the terminal flower of a cyme opens first) and are quite strongly but pleasantly scented. Bees are recorded as flower visitors, and one could expect flowers to be pollinated by short-tongued insects that insert their mouth parts between the stigmatic disc and the stamen tube in search of nectar produced by the ovary disc. In withdrawing this organ, they would retain some pollen (male flowers) or deposit existing pollen on the undersurface of the stigma (female flowers). Minute drops of nectar originate on each wart of the disc and accumulate in the floral cup.

Fruit develops rapidly and normally only 1 fruit appears from a single inflorescence. As the fruit enlarges, a short massive axis is produced, becoming pendulous with increasing weight. The mature fruit is the size of a good melon and weighs 2 or 3 kg. The 4 sutures of the future woody valves are evident as 4 shallow equidistant furrows. The fruit mainly shatters on impact after its fall from the tree, with the valves breaking away from the fruit stalk releasing the several seeds. Fruits are always well filled, as if there is no abortion of ovules, and Noamesi records up to 20 seeds, but 8 to 10 is a more usual number.

Seeds are angular and fitted completely into the fruit cavity; the outer convex surface is pressed against the inner fruit wall, and the angular, often pyramidal internal facets are fitted together in a completely close-packed arrangement. Seeds have a thick corky testa; the embryo has an undifferentized cotyledonary mass with the plumule and radicle sometimes on one internally facing facet but more usually on the outer convex face. The site of the radicle is evident as a small boss. Seeds may start to germinate by extrusion of the radicle while still floating. No information is available about the longevity of seeds. Seedlings are quite common in *Xylocarpus* communities so that maintenance of populations as well as dispersal seems effective.

Xylocarpus mekongensis Pierre 1897   (Figs. 5.7, B.40, B.41)
[*Fl. For. Cochinch*. plate 359B]

Carapa moluccensis Watson 1928 [*Malay. For*. Rec. 6:70, 75, tabs. 34 and 35. non Lam]
X. gangeticus (Prain) Parkinson 1934 [*Indian For*. 60:140]
X. australasicus Ridley 1938 [*Kew Bull*. 1938:291; orig. as *X. australasica*]

This species is probably the "*Xylocarpus moluccensis*" of many nontaxonomic works.

Apart from the diagnostic details outlined earlier, this species is morphologically and biologically very similar to *X. granatum*. The most striking differences are the bark texture and the presence of well-developed peglike pneumatophores. The name *X. australasicus* has been used in several recent accounts for plants from Western Australia and New Guinea, presumably with the acceptance of Ridley's species.

The species is more consistently deciduous than other species, and in Queensland and New Guinea, for example, it characteristically loses its leaves for 1 to 2 months in the cool, drier winter months.

The erect, conical pneumatophores up to 30 cm tall (Fig. B.41) originate from the extended horizontal "cable roots," which grow away from the base of the trunk. They are entirely secondary and represent a localized outgrowth of the vascular cambium (Fig. 5.7). They can be forked or clustered. They may originate in slight loops where the apex of the parent root is displaced upward. This explains the concavity in their base. The cable roots in addition produce branch roots, which may descend or grow horizontally to proliferate the root system. New horizontal roots originate (presumably adventitiously) from the base of the older pneumatophores. The ultimate branches of the descending roots are very fine and lack root hairs.

The surfaces of older pneumatophores become rough with numerous lenticels.

Xylocarpus moluccensis (Lamk.) Roem. 1846
[*Syn. Hesper. Fasc.* 1:124]

This is regarded as a nonmangrove species, although its habitat may frequently impinge on mangrove swamps. I have had no opportunity to investigate this plant. It seems largely to be distinguished by its usually broadly ovate pointed leaflets. The bark is described by Parkinson (1934) as "grey, rough with longitudinal fissures and peeling in flakes."

Family: Myristicaceae

A tropical woody family of 18 genera and 300 species characterized by the dioecious condition, simple entire, alternate, exstipulate leaves, the 10 stamens united into a central column in male flowers, and the ovary of a single superior uniovulate carpel in female flowers. The 1-seeded fruits are fleshy but dehisce to release the large arillate seed, which has ruminate endosperm. The family includes the commercial nutmeg, *Myristica fragrans* L., native to the Moluccas.

Myristica Gronov. 1755
[*Fl. Or.*:141]

A genus of understorey trees with about 120 species widely distributed in the Asian tropics. The following species occurs in back mangal with a distribution entirely restricted to the islands of New Guinea and New Britain. The account is taken largely from Sinclair (1968). In the field this tree is recognized by its basal aerial stilt roots, alternate oblong leaves, copious red exudate, and nutmeglike fruits.

Myristica hollrungii Warb. 1897
[*Monogr. Myrist.* 490, t: 19, f. 1-2]

An evergreen dioecious tree growing to 36 m high with horizontal whorled branches suggesting Massart's model. Aerial roots forming basal stilts. Bark dark grayish brown, finely fissured vertically and flaking. Sap red, exuding copiously from cut surfaces. Leaves spirally arranged on the orthotropic (trunk) axis, 2-ranked on the plagiotropic (branch) axes. Stems with distant leaves and a continuous line on each side of the stem along the internode connecting successive leaves. Leaves glabrous, thinly coriaceous, 20 to 35 by 5 to 13 cm, with a short slender petiole 1.5 to 2 cm long, the blade oblong or oblong-lanceolate, the apex acute or acuminate, the base rounded, to subcordate. Veins 16 to 22 on each side, parallel and equidistant, slightly prominent beneath. Inflorescences axillary; axis short, thick, producing flowers distally and acropetally over an extended period. Male flowers 5 mm long, shortly stalked, subglobose with a pointed apex, with a single lateral bracteole 1 to 4 mm long. Perianth cupular, with minute hairs, yellow outside, white within, shortly 3-lobed apically at anthesis. Stamens 10 united in a central column with a short apiculus and a short basal, sterile, hairy stalk. Female flower similar but on a shorter pedicel, the perianth lobes very short, ovary ovoid with a single basal anatropous ovule and a short bilobed stigma enclosing the receptive surface. Fruit 3 to 4 by 2 to 3 cm, on a short thick stalk and with a pointed apex. Pericarp firm, orange glabrous at maturity, seed brown, surrounded by the laciniate aril that is orange-red at maturity.

Distribution and ecology
Sinclair (1968) describes the tree as one of the most common wild nutmegs in New Guinea. Within its range it is regarded by several authors as a characteristic back-mangal constituent (e.g., Percival and Womersley 1975), but it is recorded from wet places inland and has been collected at altitudes up to 900 m where the stilt roots are little developed or even absent.

Growth and reproductive biology
There is no information in the literature on the biological aspects of this tree.

Systematics and nomenclature

A full description and discussion of the nomenclature of this species are provided by Sinclair (1968) and are not repeated here. There has been some confusion with related species, which was compounded by the destruction of many types in the Berlin Herbarium in World War II, but problems have been resolved by Sinclair.

Sinclair distinguishes *M. hollrungii* from the related *M. subalulata* by its total lack of ant-inhabited swellings on the shoot and the absence of a distinct internodal wing. Where these characters are not clear in herbarium specimens, the two species have sometimes been confused.

Family: Myrsinaceae

A large family of over 1000 species in about 30 genera distributed throughout the tropics and subtropics. The family is characterized by free-central placentation and translucent, often elongated glands.

The following taxa occur in, or in association with, mangroves:

    1A. Fruit elongated, not fleshy, enclosing the single elongated seed............

    ....*Aegiceras* spp.

    1B. Fruit globose, fleshy, ripening black with 1 or more minute seeds 2

    2A. Anthers multilocellate, calyx lobes rounded .... *Ardisia elliptica* Thunb.

    2B. Anthers not multilocellate, calyx lobes pointed .. *Myrsine umbellulata* ADC

Aegiceras Gaertner 1788

[*Fruct. et Sem.* 1:216, pl. 46; based on *A. majus* Gaertn. (not *A. minus*)]

A genus of 2 species restricted to mangrove communities in the Asian tropics, with a distribution from India and Ceylon to South China and Hong Kong, through Malesia to the Philippines, New Guinea, and tropical Australia but not known for the Pacific islands.

The genus has long been recognized as distinctive within the family by virtue of its elongated (not rounded) capsular, dehiscent (not fleshy, indehiscent) fruit, and elongated seed without (not with) endosperm. It has distinctive multilocellate anthers, but these also occur in at least one species of *Ardisia* (*A. elliptica* Thunb.). The discreteness of the genus has been long recognized, and de Candolle (1844, pp. 141–3) even included it as a separate "family," Aegiceraceae, in his *Prodromus*. However, these distinctive features largely relate to its unusual fruit and seed biology, which in turn may be related to its habitat. In other characters it resembles the other Myrsinaceae closely (e.g., in the resinous glands), and it has the placenta with numerous embedded ovules in several series that characterizes the tribe Ardiseae, which is where it has been placed traditionally (e.g., Mez 1902).

Two species are distinguished as follows (see Backer and Bakhuizen van den Brink 1965):

1A. Inflorescence consistently an umbel, with a short peduncle at most 5 mm long, flowers all on first-order branches. Pedicels 8–12 mm long, corolla tube 5–6 mm long, the mouth with a dense weft of fine hairs. Adult fruit strongly curved. Leaves large, up to 11 by 6 cm . . . . *A. corniculatum* (L.) Blanco

1B. Larger inflorescences racemose, with an extended peduncle up to 2 cm long, with the flowers on second-order branches. Pedicels 4–6 mm long, corolla tube about 4 mm long with a thin weft of hairs at the mouth. Adult fruit only slightly curved. Leaves smaller, not more than 6 by 3 cm . . . . *A. floridum* Roem. & Schult.

Backer and Bakhuizen van den Brink record that the species differ in flower odor; *A. corniculatum* has sweet-scented flowers (hence the synonym *A. fragrans* König) and *A. floridum* has sour-smelling flowers. There is no information on floral biology or insect visitors that might account for these differences.

Aegiceras corniculatum (L.) Blanco 1837   (Figs. B.44, B.45)
[*Fl. Filip.*:79]

A. *majus* Gaertn. 1788 [*Fruct. et Sem.* 1:126, pl. 46, Fig. 1]
A. *fragrans* König 1805, [in König and Sims, *Ann. Bot.* 1, 131, Fig. 3]

Low evergreen tree or shrub growing to a height of 6 m. Bark smooth, dark gray. Shoots initially monopodial with articulate construction, forming resting terminal buds enclosed by bud scales, or later sympodial below terminal inflorescences. Leaves alternate, spirally arranged (rarely subopposite), stipules absent; petiole short (0.5 to 1.0 cm), terete but slightly 2-keeled laterally; blade coriaceous, 4 to 8(-11) by 3 to 4(-6) cm, entire, elliptic to obovate, cuneate at the base, apex rounded to slightly emarginate. Surface glabrous but with minute pustules turning black with age and reddish punctate glands especially on or near the slightly ribbed margin. Midrib prominent below, often slightly red. Inflorescences as simple umbels, either terminating long shoots or on short leafy or leafless lateral shoots in the axils of foliage leaves, the older inflorescences resembling short shoots with numerous overlapping conspicuous flower scars. Bracts minute (1 to 3 mm), ephemeral; bracteoles absent. Flowers fragrant, perfect, pentamerous, pointed in bud with a slender pedicel 1 to 2 cm long. Calyx with 5 free imbricate contorted blunt and asymmetric lobes with surface glands like those of the leaves, remaining erect. Petals 5, white, pointed, contorted, and always twisted to the left, fused basally to form a short tube 5 to 6 mm long, reflexed at maturity, with a dense series of hairs in the mouth of the corolla tube and shorter, capitate hairs at the base. Stamens 5, opposite the corolla lobes; filaments about 3 mm long, united below into a short tube with a ring of internal and external hairs at the level of the mouth of the corolla

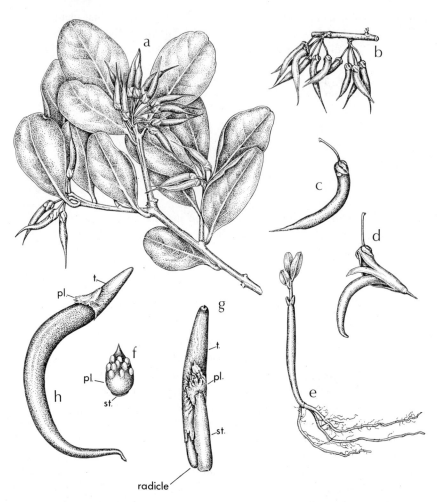

Figure B.44. *Aegiceras corniculatum* (Myrsinaceae) shoot, fruits, and seedling. (a) Habit with fruit (× ½); (b) older axis with fruiting clusters on short shoots (× ½); (c) single mature capsule (× ¾); (d) dehiscent capsule extruding single seed (× ⅜); (e) germinating seedling (× ½); (f) placenta (pl., × 9) showing numerous ovules and short placental stalk (st.); (g) developing embryo (× 3) with extended placental stalk (st.), flattened placenta (pl.), and embryonic radicle protruding from seed coat (t.); (h) mature embryo (× ½) with extruded hypocotyl and remains of seed coat (t.) attached to placental remains (pl.). (Material from various sources: a, Semetan, Sarawak, P. B. Tomlinson 19.7.82; b–d, Botuma, Papua New Guinea, P. B. Tomlinson, 31.10.74D; e–h, Missionary Bay, Queensland, P. B. Tomlinson, 11.3.76B)

tube; the facing surfaces of the stamen and corolla tube glandular basally. Anthers medifixed and versatile, at maturity extended horizontally above the reflexed petals, multilocellate, but dehiscing by longitudinal slits. Ovary 8 mm long, conical, with a single loculus, extended into a long solid, simple style beyond the mouth of the

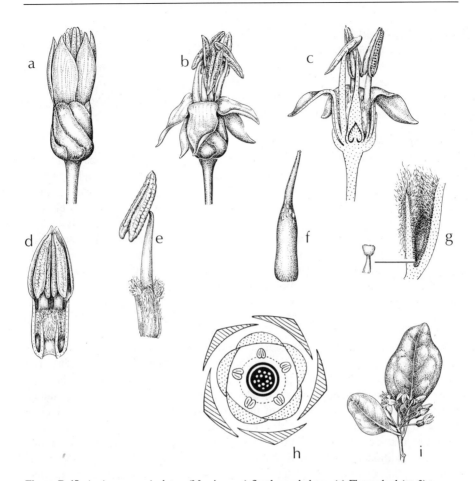

Figure B.45. *Aegiceras corniculatum* (Myrsinaceae) floral morphology. (a) Flower bud (× 3) to show aestivation; (b) open flower (× 3), petals reflexed, stamens reflexing and dehiscing; (c) flower in L.S. (× 3), stamen to left completely reflexed; (d) corolla from flower bud in L.S. (× 4) to show corolla tube and attached stamen tube; (e) single stamen (× 4) dissected from stamen tube in bud position prior to becoming reflexed; (f) ovary (× 4); (g) details of base of corolla tube (× 8) with inset of single capitate hair from base of tube (× 70); (h) floral diagram; (i) flowering axis (× ½) with terminal umbel. (Material from Missionary Bay, Hinchinbrook Island, Queensland. P. B. Tomlinson, 16.12.75)

corolla tube, with pedicellate glands at the base of the style and a nectariferous area at the base of the ovary. Placentation free-central, with numerous flattened ovules embedded in the rounded, but apically conical, somewhat fleshy, and shortly stalked placenta. Fruit 5 to 8 cm long, curved, with a persistent calyx, a 1-seeded capsule pointed apically and filled at maturity by the embryo with extended radicle and attached laterally by a long funicule-like structure. Germination immediately on release epigeal (*Rhizophora* type).

Distribution

This species has a wide distribution from Ceylon (but not India) to South China, through the Malay Archipelago to Polynesia and northeastern Australia as far south as New South Wales. It seems most characteristic of the outer (seaward) mangal fringe and occurs typically as an isolated low shrub, never forming a conspicuous part of the community.

Growth and reproduction

The tree is initially monopodial in its construction, with articulate growth resulting from periodic rest of the terminal buds, which become enveloped by a series of reduced leaves as bud scales. Upon later regrowth these leave a distinct series of scars when they fall. Branching is also rhythmic and seems loosely correlated with periodicity of extension growth so that tiers of branches may be formed. Each leaf subtends a single axillary meristem, which may grow out proleptically with a close series of basal bud-scale scars. Shoots may eventually terminate in umbels so that substitution growth occurs, with one branch eventually overtopping all others. Most inflorescences, however, terminate longer or shorter lateral shoots. Where these are leafless, they are recognized as axillary inflorescences.

Floral biology

Both the calyx and petals are contorted, always seemingly with the same direction of twist, which is usually described as ''left-handed'' (Fig. B.45a). The petals soon exceed the calyx lobes, expand, and become reflexed. The anthers are more or less versatile and become extruded as much by the extension of the stamen tube as by that of the free portion of the filament. The anthers become horizontal and bend toward the style, which they more or less enclose. Pollen grains are large. The hairs that close the corolla tube are short and stiff, some are gland-tipped, and there are additional sessile glands on the outside of the stamen tube and the inside of the corolla tube. One may speculate that these function as osmophores in view of the conspicuous floral fragrance. The flowers secrete nectar, apparently from the ring of tissue at the base of the ovary.

Clearly the flowers are pollinated by insects, but no records of the kind of flower visitors are available. There is considerable synchrony of flower development in one umbel, but an acropetal succession of flowers extends the flowering period. There is no evidence for an outbreeding mechanism, but the seed set seems high.

Fruit and seed biology

Aegiceras is distinct in its fruit and seed morphology within the Myrsinaceae, although it has the same placentation as the related *Ardisia*. Alphonse de Candolle (1841) first clearly described seed development and resolved the discrepancies in Gaertner's original description to which König first drew attention. Carey and Fraser (1932) provide a more recent account.

The placenta is central, shortly stalked, and somewhat conical apically. The sides of the placenta (Fig. B.44f) are covered by numerous semianatropous ovules, but in later development the functional ovule is best described as campylotropous. In fertilized flowers after the corolla falls, one ovule enlarges and the others remain undeveloped. The developing ovule displaces the placenta laterally and the young seed elongates. An unusual feature is the elongation of the placental stalk, which has the same function as a funicle (Fig. B.44g) but is topographically distinct from the ovular stalk. De Candolle makes an analogy between floral pedicel and peduncle in describing this structure. The structure had been referred to by Gaertner and several subsequent authors (e.g., König) as an "umbilical cord" or "umbilicle." The base of the fertilized ovule itself resembles a flattened aril. The fruit and seed develop further by intercalary growth, which is slightly unequal on opposite sides, resulting in the characteristic curvature of these organs. Initially the seed coat covers the embryo, but this is pierced by the elongating radicle in the lower (micropylar) region. This elongation of the embryo is parallel to the direction of extension of the umbilicle.

The seed coat remains complete around the distal (cotyledonary) end of the embryo (Fig. B.44h). Although de Candolle compared ovule development in *Aegiceras* and *Rhizophora*, the major difference is in ovule orientation and the fact that the radicle of *Aegiceras* does not penetrate the fruit wall, as in *Rhizophora*. The propagule consists of the pedicellate fruit, which dehisces early to expose the green radicle, which curves away from the capsule wall. On the ground the radicle penetrates the substrate and elongates to lift the plumule. The plumule itself extends through the fruit wall.

Aegiceras floridum Roemer and Schultes 1819
[*Syst. Veg.* 4:512]

A. *nigricans* A. Rich 1834 [*Voy. Austrol. Bot.* 2:57, Fig. 21]

Differing from *A. corniculatum* in the characters cited in the diagnostic key, this species tends to be somewhat diminutive in all its parts. The 2 species have not always been clearly distinguished, especially in the herbarium, but dried specimens of *A. floridum* can be recognized in the sterile state by the smaller leaves, which are characteristically dark beneath with a narrow light margin that is somewhat inrolled. The species has a much more restricted geographic distribution and is recorded from northern Borneo, Java, the Moluccas, and Celebes throughout the Philippines to Indochina. It is said to be uncommon and, since it may be overlooked, the range may be wider. Its ecological status in relation to the more common *A. corniculatum* has not been described, and it is not known whether the 2 either co-occur or exclude each other.

Ardisia SW. 1788
[*Prod. Veg. Ind. Occ.* 3:48]

A more or less pantropical genus of some 250 species. The following is recorded in Southeast Asia as a mangrove associate in communities only occasionally inundated by the highest tides.

Ardisia elliptica Thunberg 1798
[*Nov. Gen. Pl.* 7:119, see also Trimen 1895 *Fl. Ceylon* 3:74]

This species is distributed widely from Ceylon and South India, Indochina, to Java, Borneo, the Philippines, and Sulawesi.

A shrub growing to 5 m high, with spirally arranged entire elliptical leaves, about 10 by 4 cm, narrowed basally to a short (1 cm) petiole. The twigs are swollen at the base and easily detached. Flowers pentamerous, perfect, in axillary umbels or condensed racemes, about 1 cm in diameter. Calyx with 5 rounded, imbricate lobes; petals 5, white or pink, pointed; stamens 5, the anthers multilocellate. Ovary globose with a simple style, placentation free central. Fruit about 5 mm in diameter, with a globose few-seeded berry ripening black.

Myrsine L. 1753
[*Syst. Sp. Pl.*:196]

A pantropical genus including some 10 species when considered in the strict sense, that is, distinct from *Rapanea*. The following species is recorded in Sarawak as a fringe mangal species, especially in regions transitional to heath forest; it has a limited distribution elsewhere in similar situations.

Myrsine umbellulata A.DC. 1834
[*Trans. Linn. Soc.* 17:135; see also 1844 *Prodromus* 8:95]

A small tree growing to 7 m with spirally arranged ovate-elliptic leaves, about 8 by 3 cm, tapered to a short petiole (scarcely 5 mm long); apex rounded, flowers pentamerous, perfect, small, scarcely 3 mm in diameter, in sessile, condensed racemes, or umbels, on the older wood. Calyx cupulate, with 5 pointed lobes. Petals 5, white, ephemeral; stamens 5. Ovary globose with a simple style and free central placentation. Fruits globose, usually 1-seeded berries, clustered on the older wood well below the leaf-bearing portion of the shoot, ripening black.

Family: Myrtaceae

This large, essentially tropical family of 80 genera and 3000 species includes 1 isolated species associated with mangrove communities. The family is divided into 2 major subfamilies; Leptospermoideae with capsular fruits and Myrtoideae with fleshy fruits.

Figure B.46. *Osbornia octodonta* (Myrtaceae). Distinctive fibrous bark. Townsville, Queensland.

Osbornia F. Mueller 1862
[*Fragmenta* 3:30]

A monotypic genus ranging from tropical Australia to the Philippines. The species is recognized by its opposite, rather small, obovate, and aromatic leaves with pellucid dots, and the spongy gray bark.

Osbornia octodonta F. Muell. loc. cit.   (Fig. B.47)

Small evergreen, hermaphroditic trees or more usually low shrubs, with a gray, soft, spongy flaking and quite thick bark (Fig. B.46), usually on more exposed sites in association with mangroves, never within the dense mangrove community itself. Plants usually with several irregular slender trunks, not forming a regular shape. Leaves opposite, decussate on 4-angled twigs; leaf obovate, usually 2 to 4 by 1.5 cm, but larger on more vigorous shoots; margin entire, apex rounded or

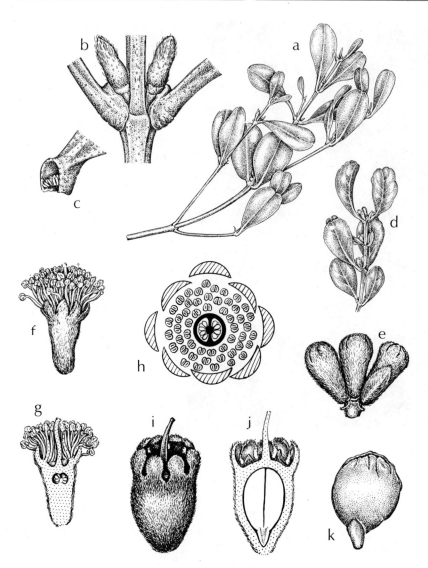

Figure B.47. *Osbornia octodonta* (Myrtaceae). (a) Leafy shoot (× ½); (b) node with axillary flower buds (× 4); (c) detail of notched leaf base with colleters (× 4); (d) shoot with flower buds (× ½); (e) dichasium of flower buds (× 4); (f) open flower (× 3); (g) flower in L.S. (× 3); (h) floral diagram; (i) fruit (× 3); (j) fruit in L.S. (× 3); (k) embryo (× 4). (Material from Kuching River, Sarawak. P. B. Tomlinson, 22.7.82)

sometimes slightly emarginate. Petiole 1 to 3 mm, reddish, grooved adaxially, with an adaxial ridge closing the groove basally. Colleters present as an inconspicuous palisade within the basal groove. Stipules absent. Terminal bud scales absent; axillary and terminal buds protected within pocket formed by grooved petioles of subtending leaf or terminal leaf pair, the pocket closed by the adaxial ridges. Leaves aromatic when crushed, with numerous scattered pellucid or translucent dots. Flowers axillary, either solitary or in 3-flowered cymes (dichasia); dichasia otherwise appearing to be terminal, sometimes then with a pair of reduced leaves immediately below. Inflorescence or flower with a basal pair of somewhat woody prophylls 1 to 2 mm long, subsequent bracteoles 3 mm long, scarious, ephemeral. Flowers about 5 mm long with a well-developed calyx cup 2 mm deep ending in 8 equivalent rounded lobes, the calyx densely silver hairy within and without, tapered below, appearing (and often described as) white. Petals absent. Stamens yellow, numerous (up to 48), protruding as a cluster at maturity, each slender and 2 to 3 mm long, inserted on the mouth of the calyx cup; anther minute (0.5 mm), the connective extended into a blunt appendage. Ovary inferior, apparently unilocular with several (up to 15) ovules on 2 basal, obscurely axile placentae. Style simple, tapered, 3 to 4 mm long, and protruding to the same height as the extended stamens. Fruit a globose, leathery berry about 5 by 7 (-10) mm, somewhat larger than and included within the calyx tube. Seeds 1 (-2). Endosperm absent. Embryo straight, top shaped, cotyledons hemispherical, much exeeding the short hypocotyl. Germination epigeal.

Growth
The habit of the plant is irregular, leading to the development of a low shrub or bushy tree with a dense, rounded crown. Individual shoots are infrequently but irregularly branched with frequent sympodial development as a result of abortion of the terminal bud or, in older plants, the development of terminal flowers. The shoots grow rhythmically, but since bud scales are not formed, there are no obvious articulations. Branching may be correlated with resting periods. Frequently there are 2 buds at a node, with the lower becoming reproductive and the upper remaining vegetative.

Reproductive biology
There is no information about pollination. The flowers are aromatic and made conspicuous by their silvery indumentum and expanded stamens; insect pollination is likely. The glandular appendage on the stamen may function as an osmophore. The fruits apparently do not dehisce, but are detached and float as a unit.

Root system
This remains unspecialized; the main roots may become exposed by erosion of the substrate and develop obscure buttresses. Larger-diameter and exposed roots develop a spongy bark, like that of the stem. Percival and Womersley (1975) record

the presence of pneumatophores; the aerial roots tend to loop much as in *Lumnitzera*, to which they probably refer.

Systematic and other notes

The position of *Osbornia* within the Myrtaceae is disputed, but clearly it is isolated. There is even some uncertainty into which of the 2 subfamilies it should be placed. Briggs and Johnson (1979) include it as a monogeneric group "*Osbornia alliance*" within the subfamily Myrtoideae and indicate this in their phyletic diagram by showing it attached basally, originating more or less directly within the ancestral Myrtoid stock. Schmid (1980) includes *Osbornia* within the subfamily Leptospermoideae because of its nonfleshy fruit, which has been the more traditional approach. Wood anatomy, pollen morphology, and floral anatomy are said to support this generally accepted assignation.

This historical problem resides in the texture of the fruit, which is leathery but not fleshy as in the typical Myrtoid fashion. At the same time the fruit is not dehiscent as is usual (with some exceptions) in the Leptospermoid group. Nevertheless, *Osbornia* was included in the Leptospermoideae (tribe Backhousinae) by the early monographer Niedenzu, a tradition followed subsequently by other authors until this opinion was challenged by Briggs and Johnson. They indicate that features of the leaves, flowers, and embryo suggest Myrtoideae without elaborating details. They argue for an ancestral succulent fruit, as in the Myrtoideae with the fleshy condition being lost in the change from presumed animal dispersal to dispersal by floating in water. Also the fruit is most commonly 1-seeded, with the embryo filling the cavity. The pericarp expands to accommodate this, but increases little in thickness.

Floral morphology

Although the flower of *Osbornia* is usually described as apetalous, with 8 calyx lobes, Briggs and Johnson refer to this octamery as "false"; they consider that the calyx and corolla are scarcely distinguishable and simply simulate a single whorl. This opinion is reiterated by Schmid (1980); neither statement is substantiated in detail. In material that I have examined the lobes overlap in a single spiral, all are about the same size, and there is no textural difference between them. Schmid also comments that *Osbornia* has an incompletely divided ovary with the ovules morphologically axile but basally located on the septum; this configuration can be seen in a transverse section just above the base of the loculus.

Geographic distribution

The distribution map by van Steenis (1936) shows an apparent discontinuity in the distribution of this plant between a western area (Philippines, North Borneo, Sulawesi, and eastern Java) and an eastern area (north Queensland and adjacent Papua New Guinea).

Family: Palmae (Arecaceae)

One of the larger, most natural, and economically important monocotyledonous families with over 200 genera and about 2600 species; typically plants with aerially unbranched, woody trunks and large compound leaves (palmately or pinnately dissected) in a terminal crown. It is divided into 13 subgroups of varying sizes (Moore 1973). *Nypa* is a distinctive mangrove element in the Old World; a number of species may occur as mangrove associates. The taxa treated here may be distinguished as follows:

1A. Unarmed, rhizomatous palm, massive axis buried in estuarine mud; leaves pinnately erect to a height of 7 m; fruiting head globose with numerous fibrous fruits (Fig. B.48) ....*Nypa fruticans* (Thunb.) Wurmb.

1B. Spiny, trees or lianes, leaves 1–3 m at most; fruiting heads paniculate ... 2

2A. Lianes, supported by extended grapnellike structures; stem diameter 3–4 cm; fruits with numerous reflexed, regularly overlapping scales............
.... *Calamus erinaceus* (Becc.) Dransfield

2B. Stems self-supported (Fig. B.49A), clustered, axes to about 10 cm in diameter, armed with spines that become erected after leaf fall (Fig. B.49B); fruits not scaly .... *Oncosperma tigillarium* (Jack) Ridl.

Additional palms that may be encountered, usually as outliers of swamp communities, include *Phoenix reclinata* (tropical Africa), species of *Euterpe, Manicaria*, and *Mauritia* (tropical America), and species of *Raphia* (Central America and tropical Africa). Such palms may have an appreciable tolerance of salt water (e.g., *Phoenix paludosa* Roxb. in Asia).

Nypa Steck 1757
[*De Sagu* 15; see Moore 1962 *Taxon* 11:164–6]

Nipa Thunberg 1782 [*Kongl. Vet. Acad. Nya Hanal.* 3:231]

Nypa fruticans (Thunb.) Wurmb. 1781   (Figs B.48, B.50, B.51)
[*Verh. Batav. Genootsch*

Nipa fruticans Thunberg 1782 [loc. cit. 3:234]

A monoecious palm with a thick, subterranean, somewhat fleshy, dichotomously branched, rhizomatous axis up to 70 cm in diameter, with oblique raised leaf scars; axis and leaf bases usually submerged in mud and inundated at high tide, rarely exposed by erosion. Each terminal shoot supporting a cluster of erect paripinnate leaves to 7 m long. Leaf with a terete petiole 1 to 2 m long and a somewhat bulbous leaf base, closed at the insertion. Leaflets reduplicately folded, numerous (30 to 40), regularly arranged in the grooved rachis, and each lanceolate, to 70 cm long, with a single prominent adaxial midrib. Indumentum absent except for frequent scattered abaxial scales, up to 1.5 cm long, below the midrib. Inflorescence (Uhl 1972) on a long peduncle, axillary, arising above the water surface

Figure B.48. *Nypa fruticans* (Palmae). Mature fruiting heads are pendulous and supported by the tide. Labutale Lagoon, Papua New Guinea.

at high tide; bicarinate prophyll and initial series of spirally arranged bracts sterile and enclosed by the leaf base, distally 7 to 9 similar bracts fertile and subtending lateral branches, the branches themselves branched to 3 or even 4 further orders. Branches usually somewhat adnate to parent axis but without a conspicuous pro-phyll. Main axis of inflorescence ending in a globose cluster of congested female flowers at first enclosed by a series of short sterile bracts; all lateral axes ending in a club-shaped spike of densely arranged male flowers. Flowers dimorphic and sessile, each subtended by an inconspicuous, even filiform, bract. Male flower with 6 short tepals 4 to 5 mm long, in 2 whorls, the outer tepals somewhat widened above; stamens 3, united to form a central column, the anthers dehiscent extrorsely. Pollen spinous. Female flowers with 6 tepals much as in the male flower, but exceeded by the 3 (or 4) separate angular carpels about 1 cm high, the stigmas sessile, funnel shaped. Carpel unilocular with a single basal anatropous ovule. Fruiting head developing into an aggregate of fertile and partly developed but sterile carpels, the individual fruit an angular drupe with a smooth, dark brown epicarp, mesocarp fibrous, endocarp thick, the seed grooved adaxially. Endosperm usually with a central hollow; embryo small, basal (Fig. B.51a). Germination incipiently viviparous (Fig. B.51b–d), essentially hypogeal, but initiated on the fruiting head with the plumule protruding before the fruit is released (Fig. B.52).

A   B

Figure B.49. *Oncosperma tigillarium* (Palmae). (A) A characteristic back-mangal species in Bako National Park, Malaysia. (B) Spiny trunk; stems are used in construction after the spines are shaved off.

Distribution and ecology

Widely distributed from Ceylon, the Ganges Delta, Burma, and the Malay Peninsula through Indonesia to New Guinea and the Solomon Islands, northward to the Philippines and Ryukyu Islands, and southward to north Queensland. *Nypa* has had a much wider distribution outside its present range because fossil fruits are known from the Paleocene in Brazil and the Eocene in Europe, Africa, and India. Its pollen is one of the oldest known identifiable angiospermous pollens referable to a modern species, being recorded in sediments as early as the Upper Cretaceous.

In its present range it is a palm of quiet estuaries or shallow lagoons into which fresh water may run (Fig. 4.3). It makes extensive pure stands by virtue of its rhizomatous dichotomizing habit (Fig. 4.4). It seems to favor brackish waters, often forming a wide border beyond the fringe of adjacent mangroves or swamp forest (Fig. 4.6). It does not occur on shores exposed to much wave action and never in hypersaline conditions.

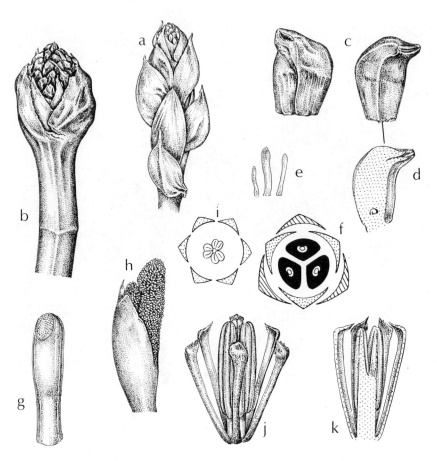

Figure B.50. *Nypa fruticans* (Palmae-Nypoideae) inflorescence and flower. (a) Distal part of inflorescence at female stage (× ⅕; see Fig. B.53). (b) Female flower head with lower bracts removed (× ½). (c) Single carpels from different aspects (× ⅔). (d) Carpel in L.S. (× ⅔). (e) Tepals, one outer, two inner, from a single female flower (× 3). (f) Floral diagram of female flower. (g) Young male spike still enclosed by bract (× ½). (h) Male spike at anthesis (× ½); apex of next higher order spike protruding on left. (i) Floral diagram of male flower. (j) Male flower just before anthesis (× 9). (k) Male flower in L.S. (× 9); staminal column unextended. Note: The perianth has been described as 2 whorls (e.g., Moore 1973); this is most obvious in the female flower (e) and least so in the male (j). Staminal column (k) extends about 2-fold at anthesis, with the stamens standing above the tepals. (Material from Kuching River, Sarawak. P. B. Tomlinson, 28.7.82A)

This plant is an unusual mangrove because of its colonial habit. The axis is normally buried, but when occasionally exposed resembles a series of overlapping cow plats (Fig. 4.5). The stem dichotomy is then evident. Isolated stems that are washed out cannot reroot, but Chai (1982) suggests that large clumps that retain much of their original root system may become reestablished, the only example of vegetative dispersal in mangroves.

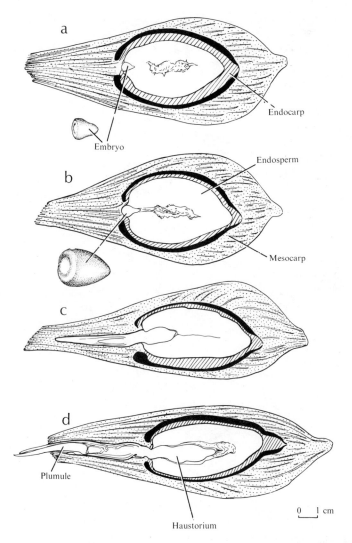

Figure B .51. *Nypa fruticans* (Palmae). Vivipary. (a–d) Attached fruits of increasing age to show development of viviparous embryo. (d) Corresponds to stage when fruit is released. (After Tomlinson 1971)

Nomenclature, taxonomy, and phylogeny

The original spelling *Nipa* of Thunberg used in older literature is predated by *Nypa* of Steck as adopted by van Wurmberg in 1781 and so is considered invalid by Moore. *Nypa* has an isolated position in the Palmae and is placed by Moore (1973) in a monotypic group, the Nypoid palms. It is considered to represent an independent line of specialization within the family, and because of its numerous

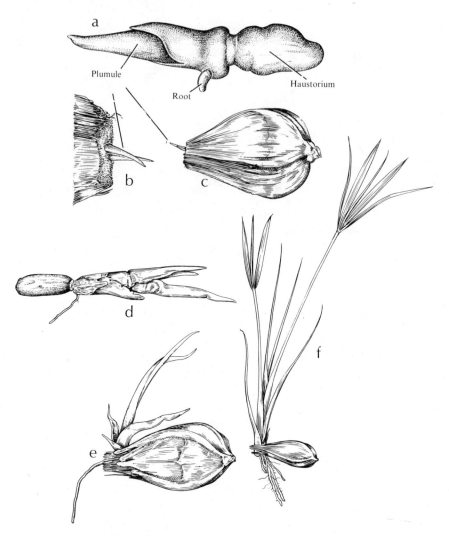

Figure B.52. *Nypa fruticans* (Palmae) seedlings.(a) Isolated embryo ($\times$ ³⁄₂), corresponding approximately to stage of development in Figure B.51c; (b) detail of protruding plumule ($\times$ 1); (c) fruit at time of release ($\times$ ½); (d) isolated embryo ($\times$ ½) from detached, floating seed; (e) established seedling ($\times$ ½); (f) older seedling ($\times$ ⅛). (After Tomlinson 1971)

distinctive features has no obvious close relatives. Relationships with the Pandanaceae have been suggested on the basis of superficial similarities of the fruiting head, but the basic morphology of the 2 forms is quite dissimilar. In its vegetative anatomy it shows all the diagnostic features of the Palmae (Tomlinson 1961); any attempt to segregate it as a distinct family Nypaceae should be discounted.

Reproductive biology

Growth of the axis seems to be continuous, and disinterred stems show no morphological discontinuities except for the branch forks. The leaves are spirally arranged on horizontal axes but are brought into an erect position by unequal growth of the leaf base (Fig. 4.4). The leaf sheath is a closed tube, as in all palms, but the tube is short and extends as a short groove into the base of the terete petiole. Branching as described by Tomlinson (1971) occurs at regular intervals by an equal dichotomy, which is evident in the leaf immediately behind the fork, which has a characteristic double groove and always occupies the same position relative to the fork (Fig. 4.4a,b). Architecturally the palm may be referred to Schoute's model (Hallé et al. 1978), but with plagiotropic rather than orthotropic axes (see *Hyphaene*). The result is an axis that spreads by regular forking of monomorphic axes (Fig. 4.4c,d). No vegetative axillary meristems are produced.

Most leaves are capable of subtending an axillary inflorescence, but not all develop completely. Each inflorescence is protogynous (Fig. B.53); the stigmas of the carpellate clusters become receptive as they are exposed by the separation of the enveloping bracts. Receptivity is indicated by the stickiness of the stigmatic surface. The pollen is released subsequently as the individual male spikes are exposed in turn, elongating beyond the enveloping bracts (Fig. B.54). In each flower the staminal column is extended, carrying the pollen as a mass onto the surface of the spike, since flowers are more or less synchronous in anthesis.

Essig (1973) records 2 kinds of insects visiting *Nypa* inflorescences "in significant numbers." First, there are small bees of the genus *Trigona*; second, small flies of the family Drosophilidae. I can confirm these observations. Apparently, the flies complete their life cycle in the somewhat fleshy axes of the branches of the male inflorescence, since these become extensively damaged by larvae that are presumed to be mainly of the same flies. Essig suggests that the bees are not important pollinators because they visit almost exclusively male flowers. The flies visit both types of flower and pollen becomes attached to them.

Fruits may fill a single head completely, but isolated fruits commonly occur in a mass of partly developed but sterile carpels. The mature fruits are dispersed by water; their detachment is assisted by the plumule, which extends precociously before the seed is loosened, often penetrating the fibrous tissue of the inflorescence axis. The peduncle typically becomes pendulous under the weight of the fruiting head, but it is supported by water at high tide (Fig. B.48). Movement caused by the tides aids in the detachment of mature fruits. The stranded fruit continues to develop producing 2 or 3 scale leaves before the first bladed leaf (eophyll), which is pinnate (Fig. B.52f). The seedling axis is horizontal from the start.

Economic uses

Nypa is one of the more commercially valuable plants of the mangroves on a sustained yield basis (Brown and Merrill 1920). The leaves make a high-quality thatch (attap), the immature endosperm provides a jellylike sweetmeat in

Figure B.53. *Nypa fruticans* (Palmae). Inflorescence at the stage of stigmatic receptivity (''female''); terminal carpellate flowers are exposed and capable of receiving pollen. Staminate flowers on lower branches are still enclosed by overlapping bracts. Kuching River, Sarawak.

Figure B.54. *Nypa fruticans* (Palmae). Inflorescence at male stage; exposed male spikes on lateral branches below the terminal female cluster. (Specimen cultivated at Fairchild Tropical Garden, Miami)

Malaya, and in the Philippines industrial alcohol is extracted by distillation after fermenting the sugar tapped from the inflorescence, in the manner of toddy. Details of the process are not published.

Calamus L.
[Sp. Pl:325]
[See Beccari 1908 *Ann. R. Bot. Gard. Calcutta* 11 and Appendix 1913]

A genus of almost 400 species, mainly dioecious, spiny, climbing palms with whiplike grapnels that are either extensions of the leaf tip (cirrus) or sterile adnate inflorescences (flagellum). Leaves pinnate. Fruit covered with numerous reflexed scales.

Calamus erinaceus (Becc.) Dransfield 1978
[*Kew Bull. 32:484; see also 1979* Malay. For. Rec. 29:130]

A multiple-stemmed rattan forming thickets on the landward edge of mangal or behind coastal sand bars. Stems climbing to 15 m, 2 to 3.5 cm in diameter but up to 6 cm wide including enclosed sheaths. Sheaths orange to yellowish green, very densely armed with horizontal or oblique whorls of spines. Petiole at insertion on sheath with a stout knee. Leaf about 4.5 m long with numerous leaflets up to 40 by 2 cm, leaf axis extended into a cirrus about 2 m long. Inflorescences as lateral panicles from axils of uppermost leaves, with numerous tubular overlapping bracts, ultimate axes (rachillae) with distichous series of flowers, the male much more dense than the female rachillae. Fruit globular, about 1 cm in diameter, covered with about 12 vertical series of triangular scales; embryo minute, endosperm homogenous.

The canes of this species have little commercial use. This species is probably the one referred to as *"Daemonorops leptopus"* in Watson (1928).

Oncosperma Bl. 1838
[Bull. Sci. Phys. Nat. Neerl. 1:64]   (Fig. B.49A,B)

Oncosperma tigillarium (Jack) Ridl. 1900
[*J. R. Asiat. Soc. Straits Branch* 33:173]

O. filamentosa Bl. 1838–9 [*Rumphia* 2:97]

A tall, many-stemmed palm up to 25 m, but the stems rather slender (scarcely 10 cm in diameter) with a crown of pinnate leaves to 4 m long, each leaf with characteristic drooping leaflets. Stems armed with flat, sharp, black spines. Both Watson (1928) and Chai (1982) record this palm as marking the inner limit of the mangrove that are found in regions transitional to nonlittoral forest. Watson indicates that it grows inland and suggests it is likely to be killed by salt water.

Family: Pellicieraceae

A monotypic family; the genus *Pelliciera* has traditionally been placed in the Theaceae or Ternstroemiaceae, usually as an isolated tribe Pelliciereae.* Current opinion tends to place it in a separate family, Pellicieraceae (Planch. and Tri.) Beauvisage, and emphasize its affinities with more remote groups such as Marcgraviaceae. It is distinguished by its regularly pentamerous flowers enclosed by a large pair of bracteoles, mangal habitat, one-seeded indehiscent fruit with corky pericarp, and the large, naked embryo.

Pelliciera Planchon and Triana 1862
[*Benth. Hook. Gen. Pl.* 1:186]

> *Pelliciera rhizophoreae* Triana and Planchon 1862 (Fig. B.56)
> [*Ann. Sci. Nat.* Series 4, 17:381]
>
> Including *P. rhizophoreae* var. *benthamii* Triana and Planchon 1862 [loc. cit.]
> For a detailed citation, see Kobuski (1951).

Small trees, 5 to 10 (but up to 18) m high with somewhat tiered, distally arcuate branches. Trunk swollen and markedly fluted below (Fig. B.55), the ridges each originating as an acropetally developed series of short aerial roots. Bark dark, rough fissured. Stems with conspicuous circular leaf scars and stubs of stalks of fallen fruits at intervals. Leaves spirally arranged, with regular 2/5 phyllotaxis, each sessile, asymmetric oblong-lanceolate and broadest at the middle, 10 to 12 by 2.5 to 4 cm; blade glabrous, leathery, dark glossy green; apex bluntly rounded, margin initially with a series of prominent but ephemeral glands (presumed salt glands) on the wider inrolled side, becoming entire with age. Leaf base abruptly narrowed to the insertion with a pair of glands (extrafloral nectaries). Young leaves involute in bud. Stipules and bud scales absent.

Flowers large, showy, pentamerous up to 12 cm wide at anthesis, solitary in the 1 to 3 leaf axils immediately below the resting terminal bud, the leaves subtending the flowers broadened at the base. Flowers at first enclosed by a pair of narrow whitish or reddish bracteoles about 5 by 0.5 to 0.8 cm. Calyx of 5 short, free, imbricate, white, concave lobes; sometimes reddish without and with numerous small glands internally at the base. Petals 5, free, white, or rose pink with a white midvein, about 6 cm long, 1.5 cm wide at the base, tapered distally to a blunt point, initially erect but reflexed at anthesis, ephemeral and falling with the sepals and stamens. Stamens 5 free, 6 cm long, lying within the alternate grooves of the ovary, anthers narrow, 2.5 cm long, sagittate basally, pointed distally, dehiscence

---

* Also called Pelliceriaceae and *Pelliceria*; see under "History and Systematics."

Figure B.55. *Pelliciera rhizophoreae* (Pellicieraceae). Base of trunk with fluted buttresses. (From a color transparency by A. M. Juncosa)

extrorse by longitudinal slits. Ovary narrow, cylindrical, woody, about 6 cm long with 10 longitudinal grooves, tapering to a narrow style with a bifid stigma; locules 2, each containing a single axile anatropous ovule, only one ovule developing or one locule sterile. Fruit top shaped, 8 to 12 cm long and about as wide, somewhat flattened, the style persisting as a pointed woody but brittle beak, at first green but (with age) becoming reddish brown and developing resinous pustules, the surface with many ridges. Fruit wall about 6 mm thick at the base, leathery externally but spongy within. Seed single, without endosperm, and including a naked cordate embryo, testa at maturity represented by ribbonlike fragments, the hypocotyl extending into the stylar beak, plumule reddish, long, slender and hooked, enclosed by the 2 fleshy cotyledons (Fig. B.57). Germination semiepigeal, with rapid separation of the cotyledons and radicle and straightening of the plumule (Fig. B.58).

Geographic and ecological distribution

This species presently has a limited distribution on the Pacific coast of central and northern South America, from Buenaventura Bay in western Colombia to Costa Rica. Recent isolated collections on the Atlantic coast (Winograd 1983,

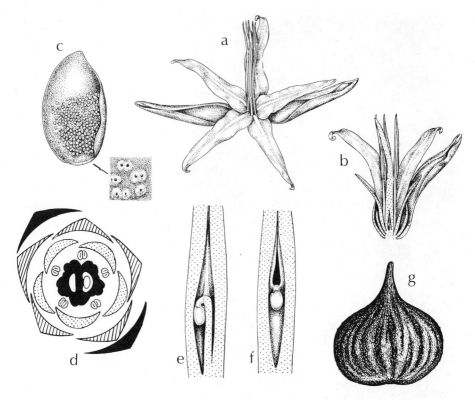

Figure B.56. *Pelliciera rhizophorae* (Pellicieraceae) flowers and fruit. (a) Flower from above
($\times$ ½), stamens not yet reflexed; (b) flower in L.S. ($\times$ ½); (c) sepal from within ($\times$ ³⁄₂), with inset
detail showing adaxial glands ($\times$ 8); (d) floral diagram; (e) ovary in L.S., transverse view ($\times$ 3); (f)
ovary in L.S., dorsiventral view ($\times$ 3); (g) fruit ($\times$ ¼). (Material from Rio Tempisque near Puerto
Humo, Costa Rica. A. M. Juncosa, *s.n.*)

Jiménez 1984) are relicts of its ancient, wider range. Evidence from its occurrence
as a microfossil (pollen) shows that it once had a circum-Caribbean distribution
and extended to Mexico (Graham 1977) and it is found in Eocene deposits in
Panama (Fig. 3.2). Its progressive range restriction is attributed to climatic and sea-
level changes and the increasing competition from *Rhizophora*. *Pelliciera* provides
the best evidence for the mangrove "refugium" theory of Gentry (1982). It occupies
the intertidal zone, commonly in association with *Rhizophora*, but with a much
more localized distribution, typically in sheltered sites such as estuarine banks or
protected beaches.

Growth and reproduction

Plants are evergreen and seem to develop resting terminal buds (which are
not protected by bud scales) only in association with flowering, which occurs from

Figure B.57. *Pelliciera rhizophoreae* (Pellicieraceae). Fruit dissected to show embryo morphology; two embryos each with a single cotyledon removed to show the characteristically curled plumule. (From a transparency by A. M. Juncosa)

Figure B.58. *Pelliciera rhizophoreae* (Pellicieraceae). Seedling with submerged fleshy cotyledons. (From a color transparency by A. M. Juncosa)

the axils of one or sometimes more of the subterminal leaves so that the flowers, because of their size, superficially appear terminal and overtop the displaced apex. Flowers appear on trees only about 1 m high (Howe 1911), a distinctive feature of pioneering species. Flowering is said to be over an extended season in winter and early spring (November to February), probably due to nonsynchronous flowering of different shoots, but there is said to be a peak of heavy fruiting late in the year (October). The seasonality shown by individual shoots is reflected in the fairly regular series of stubs of fruit stalks on each shoot separated by a long sequence of nonflowering nodes.

Each leaf subtends a single vegetative bud (except at flowering nodes), but most buds remain undeveloped and may have a limited life span because prolepsis and suckering are rare. Branching is therefore mainly by syllepsis and can result in a series of sylleptic shoots on vigorous specimens. Branching on the trunk is discontinuous (Attims's model) and thus results in tiers of plagiotropic axes that are themselves little branched but curved upward and bearing the leaves in terminal clusters (hence the superficial similarity with *Rhizophora* commented upon by several observers). Leaves are asymmetric, with the wider, glandular half rolled innermost in bud. Leaves on one shoot seem always to retain the same symmetry, which may change on branches.

The floral parts surrounding the ovary are ephemeral and tend to fall together. There is no information about the floral mechanism or pollination biology, but Howe (1911) suggests that flowers are scented. The fruits fall and usually release the seeds, which are dispersed in the water, germinating immediately they become stranded. Howe suggests that the beak of the fruit is an adaptation for anchoring or even planting, implying that the point is forced into the mud when the fruit falls. This is contradicted by Johnston (1949, p. 207), who points out that the heavy end of the fruit is not the pointed end and that fallen fruits tend to land on their side or the blunt end. He suggests that broad dispersal by ocean currents is unlikely, since the protection of the seed by the fruit wall is not only poor but also of short duration. Seeds in water swell after only a few hours and sink (Collins et al. 1977).

Root system

The short radicle may persist or abort, but its presence is not important in the development of the root system. The first adventitious roots arise as a series at the base of the hypocotyl; as in *Rhizophora* they are differentiated in the embryo. They are replaced by wider aerial roots on the plumular axis. The anatomy of aboveground and subterranean roots is sharply contrasted. With age several orthostichies of aerial roots are formed, each younger root above the next older root in the series. These roots do not immediately penetrate the stem surface but grow down inside the bark closely appressed to older roots. This raises the cortical tissue (which seems to be supplemented by some kind of secondary growth) so that the surface remains unbroken, except for numerous conspicuous lenticels. These vertical series of adventitious roots below a covering layer of bark form the fluted buttresses. The roots become secondarily thickened and branch within the bark. Roots emerge below the substrate with the submerged type of anatomy; they may be seen exposed in specimens on the banks of creeks. A clear picture of the system can be obtained from dead specimens, from which the rotted bark is easily removed; the stem proper is shortly obconical. The whole buttress system in this species, with its numerous lenticels, seems to function as one giant pneumatophore.

Ant protection

Collins et al. (1977) drew attention to the frequent presence of an aggressive species of *Azteca* that is suggested to provide protection against some insect predators that the ants attack vigorously. The association may be mutualistic, since the ants may colonize *Pelliciera* and derive some nutrients from the pair of extrafloral nectaries at the base of each leaf blade and also from the insects they attack. The association seems nonobligatory, however, because healthy trees of *Pelliciera* may lack ants and trees with foraging ants may be extensively damaged by insects. In some parts of its range *Pelliciera* lacks both nectaries and ants.

History and systematics

There are relatively few illustrations of the plant that show diagnostic details because Triana (who first collected the plant) produced only scanty specimens, presumably used as the basis for Baillon's illustrations (1875, p. 245). The only extensive figures are provided by Hemsley (1879–88) and reproduced by both Szyszylowicz (1893) and Beauvisage (1918); Howe's excellent account is illustrated by photographs. Beauvisage, in his description of the genus, repeats the earlier statement that the ovary is 5-locular; this is incorrect, although it refers to the type collection (from Buenaventura, Colombia). Later collections have all shown 2-celled ovaries, including specimens collected by Sutton-Hayes on which Hemsley's description and illustrations were based (see Howe 1911). The reportedly 5-celled ovary of the more southern form is the basis for var. *benthamii* of Triana and Planchon, but the existence of such forms is doubtful and was admittedly based on incomplete specimens.

The place of this genus within the angiosperms is somewhat problematic. Bentham and Hooker (1862) included it between the genera *Gordonia* and *Pyrenaria* in the tribe Gordonieae of their Ternstroemiaceae. Baillon (1875) included it within the same family as an isolated tribe Pellicerieae, close to the Marcgravieae. Beauvisage (1918), emphasizing anatomical features, drew attention to its similarities with certain Marcgraviaceae, especially the genus *Souroubea*, and effected the compromise of creating the family Pelliciéracées intermediate between Ternstroemiaceae and Marcgraviaceae, largely in deference to Baillon. Beauvisage used the spelling *Pelliciera* throughout his account, and it is presumably on this basis that this spelling has been accepted by most subsequent authors. This is appropriate, since it commemorates the naturalist Bishop Pellicier of Montpellier; *Pelliceria* as originally used by Triana and Planchon seems to be an error (Kobuski 1951).

Family: Plumbaginaceae

A cosmopolitan family, mainly of herbs or low shrubs, frequently in saline habitats and including one mangrove representative (van Steenis 1948). The genus *Aegialitis* (2 species) is sometimes segregated as the family Aegialitidaceae from the Plumbaginaceae because of its numerous distinctive features, including anomalous secondary thickening, abundant sclereids, and incipiently viviparous seeds. Pollen morphology is distinctive, but the greatest resemblance is with the tribe Plumbagineae and particularly with *Plumbago* (Weber-El Ghobary 1984); separate subtribal status is suggested.

Aegialitis R. Brown 1810
[*Prodromus Florae Novae Hollandae*:118, 426]

> The name is sometimes, but incorrectly, rendered *Aegialites*.

A low-growing treelet or small shrub with perfect flowers, to 3 m high, with a basally swollen, fluted axis; typically occurring in exposed, often rocky or sandy sites. Anomalous secondary thickening producing included phloem. Bark dark, smooth, fissured or flaking with age. Buds mucilaginous. Twigs with conspicuous annular leaf scars. Leaves to 15 cm long, somewhat fleshy, spirally arranged, erect, and clustered terminally on the shoots. Petiole to 8 cm long, grooved adaxially and extended basally into a tubular leaf sheath with a completely encircling insertion. Blade broadly ovate, 6 to 8 by 2 to 5 cm, glabrous, margin entire, apex rounded or bluntly acuminate.

Flowers pentamerous, in many-flowered terminal, irregularly one-sided cymes (modified dichasia), with pairs of opposite linear bracteoles. Pedicel smooth, long (to 2 cm), up to three-fourths the length of the unexpanded flower bud. Calyx 7 to 13 mm long, tubular, somewhat inflated, fluted externally, 5-lobed apically, the lobes about 3 mm high and minutely but bluntly apiculate by extension of median vein. Petals 5, 8 to 10 mm long, white, imbricate, free above with bluntly rounded lobes, shortly fused basally to form a corolla tube 2 to 3 mm long. Stamens 5, about 10 mm long, inserted on the corolla tube alternately with the petals, filaments 6 to 7 mm long, slender but inflated close to their attachment, anthers 2 to 3 mm long, basifixed but extended horizontally at the time of pollen release; dehiscence extrorse by longitudinal slits; pollen sparse with large grains.

Ovary superior, grooved or angular below, narrowed above, with each lobe extending into 1 of the 5 free styles, each style 6 to 8 mm long with an extended oblique peltate stigma initially facing inward. Ovary unilocular with a single basally attached anatropous ovule, the funicle recurved to orientate the micropyle apically. Fruit an elongated, bluntly pointed, longitudinally dehiscent capsule, 4 to 5 cm long by 3 to 5 mm wide, enveloped basally by persistent calyx. Pericarp thin, thickened somewhat distally. Single seed with a thick testa; endosperm absent.

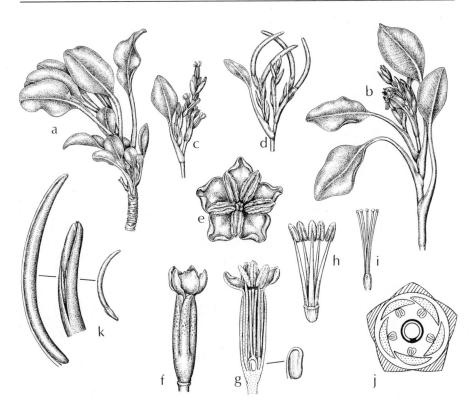

Figure B.59. *Aegialitis annulata* (Plumbaginaceae) vegetative, flowering, and fruiting shoots. (a) Vegetative shoot with cogested internodes (× ½); (b) flowering shoot with extended internodes (× ½); (c) inflorescence (× ½); (d) infructescence (× ½); (e) flower from above (× 6); (f) flower from side (× 3); (g) flower in L.S. (× 3), inset: single ovule and recurved funicle (× 12); (h) stamens (× 3); (i) ovary (× 3); (j) floral diagram; (k) fruit (× ½), inset: details of apical part of capsule dehiscing to expose seed (center × ³⁄₂) and detached seed (left × ³⁄₂). (Material from Three Mile Creek, Townsville, Queensland. P. B. Tomlinson, 26.6.77)

Embryo elongated, with an extended hypocotyl up to 4 cm long; cotyledons short, bluntly pointed, enclosing plumular leaves within a profuse mucilage. Germination immediate, epigeal.

Two very similar species are known, distinguished as follows:

1. Leaves dull above, calyx 7–8 mm long. Corolla with 9 or 10 lobes....... *A. annulata* R. Br.
2. Leaves shining above, calyx 13 mm long. Corolla with 12 lobes .... *A. rotundifolia* Roxb.

Aegialitis annulata R. Brown 1810 (Fig. B.59)
[Prodromus Flora Novae Hollandae:426]

This species is recorded for northern Australia and east Malesia.

Aegialitis rotundifolia Roxburgh 1824
[*Fl. Ind.* 2:111]

This species is recorded in Burma, Bengal, and the Andaman Islands.

Geographic and ecological distribution

The genus is used by van Steenis (1949) as an example of vicarious distribution, since the 2 species together occupy a range somewhat comparable to that of many other Asiatic mangroves, but their individual ranges do not overlap. Fossil pollen of *Aegialitis* is known from the intermediate area, however.

Aegialitis is a characteristic mangrove associate but does not itself occur within closed mangrove communities, since it prefers or even requires exposed sites. At the same time, it withstands wave and tidal action. Consequently it is most characteristic of open rocky shores or exposed beaches, sometimes as the outermost zone of a narrow mangrove belt. It is less common in back mangal, although it can become established in highly saline soils.

Anatomy

Aegialitis is unusual in the large number of trichosclereids that are produced in all parts except the flowers. It also has anomalous secondary thickening from successive discontinuous cambia. Derivatives of these cambia may themselves differentiate as sclereids, an unusual, if not unique, condition.

Family: Pteridaceae

A large and diverse family of the true ferns with about 35 genera and over 1000 species. It is divided into 6 tribes (Tryon and Tryon 1982). The genus *Acrostichum* is often described as the "mangrove fern," although it is not restricted to mangal; it occurs in the back mangal but seems particularly suited to disturbed sites. *Acrostichum* is included in the small tribe Pterideae as a somewhat anomalous member with the sporangia covering the whole undersurface of the fertile pinna and not aggregated into sori. Fossil records indicate the occurrence of *Acrostichum* as early as the Eocene.

Acrostichum L. 1753
[*Sp. Pl.* 2:1067, 73]

Rhizomatous fern, common and often dominant in the understorey of the back mangal (sometimes inland in fresh-water sites) and opportunistic in disturbed estuarine sites. Axis horizontal or erect, irregularly branched with a terminal cluster of erect, once-pinnate leaves up to 3 m long, with a terminal leaflet. Pinnae up to 40 by 6 cm, stalked (almost sessile distally), entire, narrowly oblong or lanceolate, the apex narrowly acute to abruptly acuminate or rounded. Basal pinnae sometimes reduced to distant spines along the petiole. Venation reticulate, with uniform elongate areoles diverging from the thickened midrib, without free vein endings. Scales broad, restricted to base of frond, each growing to 1 cm long, with a thickened multicellular middle, margin entire. Fertile fronds with all or only a few distal pinnae fertile.

Fertile pinnae without sori or indusia, the lower surface uniformly covered with sporangia mixed with capitate paraphyses, sporangia large; spores large, tetrahedral.

A pantropical genus; in the subsequent description 3 species are recognized, as is traditional, but careful scrutiny of the genus throughout its range may reveal more. Individuals are very plastic in their development, and there are few reliable diagnostic characters. The species in the New World tropics seem much more clearly circumscribed than they are in the Asian tropics.

The following key is only tentative:

1A. Leaf usually at least 1 m (to 3 m long), apex of sterile pinnae rounded or truncate, at most abruptly acuminate; juvenile leaves with an oblong, blunt blade 2

1B. Leaf commonly less than 1 m long; apex of sterile pinnae narrowly acuminate; juvenile simple leaves with a lanceolate, pointed blade....... *A. speciosum* Willd. (restricted to tropical Asia)

2A. Fertile fronds with all or most of the pinnae fertile (rarely only a few upper pinnae fertile). Pinnae many (40–60), closely set, overlapping and regularly arranged, often subopposite, the lowest pinnae relatively short stalked (less than 2 cm). Areoles next to the midrib broad, never more than 3 times longer than wide. Rachis with several shallow grooves below, flat or scarcely grooved above, the margin of the groove blunt. Basal spines absent. Scales on petiole base leaving prominent scars. Scales also a little way up the petiole. Paraphyses with the terminal cell horizontally extended, often eccentric and with a smooth or little-lobed outline. Plant axis little branched, often erect..........*A. danaeifolium* Langsd. & Fisch. (restricted to tropical America)

2B. Fertile fronds with only upper pinnae (up to 5 pairs plus the terminal pinna) fertile. Pinnae few, not more than 30, rather distant and often irregularly distributed, usually not overlapping, the lowest pinnae always distant, long stalked (up to 3 cm). Areoles next to the midrib narrow, always 3 times longer than wide. Rachis rounded and smooth below, decidedly grooved above with the margin of the groove acute. Spines (the midrib of aborted pinnae) frequent but distant on the lower part of the petiole, black. Scales on petiole base not leaving a prominent scar, no scales up the leaf axis. Paraphyses with the terminal cell unextended, symmetric, the outline irregular and much lobed. Plant axis branched, horizontal .... *A. aureum* L. (pantropical)

Tryon and Tryon (1982) indicate that the spores of *Acrostichum* represent a discrete type within the pteroid ferns; there is some variation in spore morphology within individual species and slight differences between species.

Acrostichum aureum Linnaeus 1753   (Fig. B.60)
[*Sp. Pl.* 2:1069, 73]

Chrysodium aureum Mett. 1856 [*Fil. Hort. Lips.*:21]
See Holttum 1954, *Flora of Malaya*, vol. 2, *Ferns of Malay*:409.

This species is recognized by the few distal fertile pinnae on fertile fronds and the shape of the paraphyses. In tropical America it is readily distinguished from *A. danaeifolium* (Adams and Tomlinson 1979). The description in the key is largely based on American plants. In Southeast Asia the young leaves have a characteristic crimson color.

Acrostichum danaeifolium Langsdorff and Fischer 1810   (Fig. B.61)
[*Ic. Fil.* 5 T1]

This species is readily recognized. Fertile fronds usually have all pinnae fertile; the pinnae are crowded, often overlapping, and relatively long stalked. It commonly occurs inland.

Acrostichum speciosum Willdenow 1810
[*Sp. Pl.* 5:117]

See also Troll (1933b).

The distribution of this species is not well known, since it is imperfectly circumscribed from and commonly confused with *A. aureum*. It is described as being smaller in all its parts, with the sterile pinnae gradually tapering to a narrow point. Troll (1933b) emphasizes the difference between the early simple leaves of this species, which have a lanceolate, relatively short blade, and the long, oblong blade of *A. aureum*.

Ecological and geographic distribution
The dimensions of individuals vary considerably according to the locality and there may be locally depauperate races. J.-M. Veillon (personal communication) refers to the fern on the tiny atoll of Ilot Matthew, southeast of New Caledonia, as being covered by a population (presumably of *A. aureum*) of plants all shorter than 25 cm. An autecological study of the genus, especially in relation to its taxonomy, needs to be done.

Although the fern is a characteristic element of back mangroves and associated tidally influenced estuarine communities, it is by no means restricted to them. Adams and Tomlinson (1979) comment that in South Florida, for example, *A. danaeifolium* extends into fresh-water swamps and is commonly found inland in sink holes within hammocks (islands of forest). When *A. aureum* and *A. danaeifolium* grow together, the latter is more common and dominant.

Acrostichum seems to have strong weedy tendencies and can be very aggressive in disturbed sites. In Vietnam it is reported to suppress the regeneration of mangroves

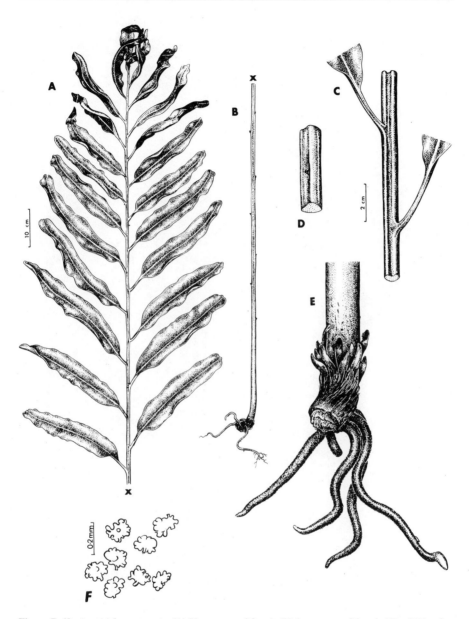

Figure B.60. *Acrostichum aureum*. (A) Upper part of frond; (B) lower part of frond; (C) middle of frond showing pinnule attachment; (D) part of leaf axis with a marginal spine; (E) leaf base with adventitious roots and scales; (F) paraphyses. (After Adams and Tomlinson 1979)

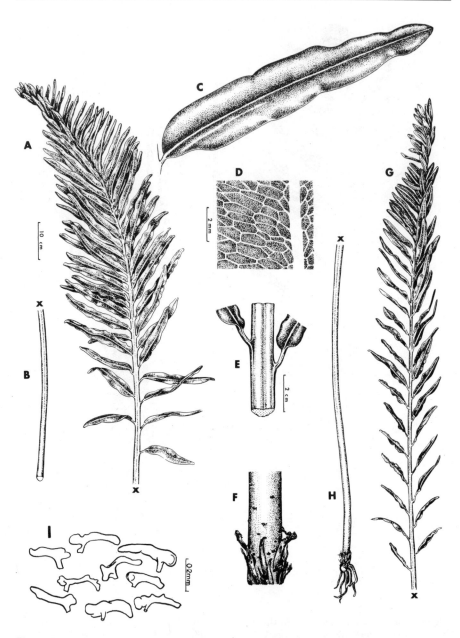

Figure B.61. *Acrostichum danaeifolium.* (A) Upper part of sterile frond; (B) unarmed axis of sterile frond; (C) single pinna; (D) areoles in midrib region; (E) middle of frond showing pinnule attachment; (F) base of leaf axis with persistent scale scars; (G) upper part of fertile frond; (H) lower part of fertile frond; (I) paraphyses. (After Adams and Tomlinson 1979)

that were destroyed by military operations. These tendencies seem related to its high reproductive capacity and its light-tolerant or even light-demanding propensities. Chai (1982), Watson (1928), and others comment on its characteristic location on lobster mounds so that it usually grows beyond tidal influence though within tidal limits.

In the New World, *A. aureum* and *A. danaeifolium* have a somewhat similar range, although *A. danaeifolium* appears to have a wider distribution (e.g., it ranges farther north in South Florida, apparently in keeping with its larger stature and vigor). The 2 species often occur together, apparently without hybridization, although Garcia de Lopez (1978) reports putative hybrids between them in the Dominican Republic.

In tropical Asia the situation is reversed. *Acrostichum aureum* is reported to be larger than *A. speciosum* (with fronds up to 4 m according to Holttum), and has the wider geographic distribution and apparently greater ecological range. However, *A. speciosum* is apparently more restricted to saline environments. *A. aureum* can survive without regular tidal inundation.

Morphology and establishment

Acrostichum offers an unusual opportunity to study the population dynamics of plants in mangal, since it is rhizomatous and has a capacity for vegetative persistence and propagation different from that of all other members of the community except *Nypa*, whereas it is sexually established from gametophytes via widely dispersed spores. The rhizome of *Acrostichum* has an extensive proliferative capacity because each leaf supports a branch meristem (Fig. 4.10); in fact, few of these develop so that rhizomes are rather infrequently branched and vegetative spread may be restricted. This restriction is clear in situations where the plant grows on a localized, specialized substrate, as on lobster mounds in the Malaysian mangroves. The extent of vegetative spread has not been investigated.

New individuals are readily established from spores via gametophytes, especially in disturbed sites. This success is due in part to an appreciable salt tolerance by gametophytes (A. Juncosa, personal communication). Germination of spores seems most successful in fresh water, however. Sex-organ (antheridia and archegonia) ontogeny occurs in sequence so that outcrossing is promoted, but genetic analysis reveals frequent selfing (Lloyd 1980).

Family: Rhizophoraceae

A small pantropical family of about 16 genera and 120 species of trees and shrubs, mainly in the Old World. Traditionally it has been included in the Myrtales; more recent critical evaluation has tended to emphasize its isolation (e.g., Tobe and

Raven 1983), and it is now frequently placed in a separate order, Rhizophorales, either still related to the Myrtales (Dahlgren 1980) or closer to the Lecythidales of the subclass Dilleniidae (e.g., Smith 1981, p. 601). Within the family a number of tribes are recognized. Of them, the Rhizophoreae Blume, which includes only the 4 exclusively mangrove genera *Bruguiera, Ceriops, Kandelia*, and *Rhizophora*, is universally recognized as a natural and discrete taxon; the evidence is derived from a diversity of disciplines (e.g., van Vliet 1976). It is convenient therefore to refer to the tribe as the "mangrove Rhizophoraceae." The uniformly viviparous condition is particularly distinctive. All reported chromosome counts are $n = 18$ (Sidhu 1968).

The following account is based partly on that of Ding Hou (1958).

Rhizophoraceae (tribe Rhizophoreae): key to genera based on technical characters of the flowers

1A. Flowers tetramerous, calyx 4-lobed; petals 4, lanceolate, entire, without distal appendages, not enclosing pairs of stamens. Anthers sessile, multilocellate, dehiscing by an adaxial valve. . . . . *Rhizophora*

1B. Flowers 5- or more (up to 16-)-merous, calyx with 5 or more lobes; petals deeply emarginate, with apical appendages, sometimes enclosing pairs of stamens. Anthers with long (short in *Ceriops decandra*) filaments, 4-locular, dehiscing by slits. 2

2A. Stamens numerous, more than twice as many as calyx lobes, petals bilobed with a long seta in the sinus, each lobe multifid, without marginal hairs. Petals not enclosing a stamen pair. Hypocotyl of seedling slender, gradually narrowed and distally pointed. . . . . *Kandelia*

2B. Stamens twice as many as calyx lobes. Petals with marginal hairs, each enclosing a pair of stamens (except *Ceriops decandra*). Hypocotyl of seedling blunt or abruptly pointed distally. 3

3A. Calyx 5- or 6-lobed, lobes blunt, 2–4 mm long; petals less than 0.5 cm long. Stamens 10 or 12. Hypocotyl of seedling more or less ridged. . . . .
. . . . *Ceriops*

3B. Calyx 8–16-lobed, lobes lanceolate, pointed, up to 15 mm long; petals 0.5–2 cm long. Stamens 16–32. Hypocotyl of seedling not or scarcely ridged. . . .
. . . . *Bruguiera*

Rhizophoraceae (tribe Rhizophoreae): Key to genera based on vegetative characters

1A. Trees with extensively developed aerial stilt roots forming a looping or pendulous complex; leaves black dotted below (at least in dried material). (Trichosclereids present). . . . . *Rhizophora*

1B. Trees without extensively developed aerial stilt roots, leaves without black dots. (Trichosclereids absent) 2

2A. Trees usually without pneumatophores, at most with inconspicuous, slender, negatively geotropic lateral roots . . . . *Kandelia*

2B. Trees with kneelike pneumatophores developed by looping of horizontal roots, the pneumatophore rounded and knobby with age. 3

3A. Leaves usually more than 10 cm long, acute at apex, terminal buds not flattened . . . . *Bruguiera*

3B. Leaves usually less than 10 cm long, rounded at apex, terminal buds
flattened... *Ceriops*

General features of Rhizophoreae

Phyllotaxis

Although the phyllotaxis of this group is usually described as decussate (i.e., strictly with successive leaf pairs at 90° to each other), the pairs of leaves are not strictly at right angles to each other, as is evident from the arrangement of leaf scars on shoots with congested internodes. Tomlinson and Wheat (1979) have shown that the arrangement seen on mature shoots is primary (see Fig. 4.8); that is, it represents the position of the initiation of leaf pairs by the apical meristem and is not the result of secondary twisting of the internode. The phyllotaxis is therefore strictly bijugate; that is, the pairs of leaves diverge from each other at an angle less than 90°. This arrangement has important architectural consequences, since mutual shading is reduced (see Fig. 4.9) and branches diverge at diverse angles. Bijugate phyllotaxis may, in fact, characterize most members of the family Rhizophoraceae.

Stipular and bud morphology

The arrangement of leaves in the buds in the Rhizophoreae (and many terrestrial Rhizophoraceae) is constant and characteristic. Each leaf pair has an associated pair of stipules that stand above the leaf insertion so that the stipules do not enclose the associated leaf but rather the younger appendages, including axillary appendages at the same node (inflorescences, branches, buds). Each stipule is a tubular structure open on the ventral side, but encircling the node at its insertion and rolled distally (see Fig. 8.5). Adjacent margins of opposite stipules overlap the same way, so that the left side of one is overlapped by the right side of the other; that is, only one side of each stipule is exposed. The leaf blade, which is more tightly rolled than the stipule, overwraps in the same way. Within the base of each stipule is a series of glandular colleters, in 2 or 3 ranks. These seem to be the source of the sticky material that exudes within the stipules and covers the surface of the organs enclosed by the stipules. The color and consistency of the exudate are variable and somewhat diagnostic for different species. It can be a somewhat milky fluid in *Rhizophora*, but a more resinous material that dries to a varnish in *Ceriops*. Primack and Tomlinson (1978) have observed honey eaters licking this fluid exudate from the buds of *Rhizophora stylosa* and suggest that this encourages the visits of the birds, which also eat predatory insects.

The stipule color is somewhat variable, even within a species; it may be red in *Rhizophora* (e.g., *R. apiculata*) and especially in *Bruguiera gymnorrhiza*. The stipules are always ephemeral and abscise cleanly as soon as the enclosed organs begin to expand. This leaves an annular stipule scar, which in older shoots stands above the petiole scar but below the scar of branch structures or dormant buds. In shoots with short internodes the helical arrangement of the petiole scars, which results from the bijugate phyllotaxis, is very clear. Ding Hou (1960) and Kenneally et al. (1978) have commented independently on the following diagnostic differences between *Rhizophora* and *Bruguiera* based on the different arrangement of the vascular bundles that appear on the petiole scar;

    1A. Leaf scars with several vascular bundles, arranged in 2 rows ... *Rhizophora*
    1B. Leaf scars with 1 series of 3 bundles .... *Bruguiera*

This difference is the result of the earlier dispersion of the traces in *Rhizophora* because the nodal anatomy is similar in all members of the Rhizophoraceae.

Bud construction in the Rhizophoreae is further characterized by the small and constant number of appendages within the exposed pair of enveloping stipules, to a maximum of 3 nodal sets. As a pair of leaves expands out of the terminal bud, it is replaced by a new pair initiated by the shoot apex. The rate of expansion fluctuates considerably, however, and buds can remain relatively inactive for extended periods, neither initiating new primordia nor expanding older ones. The bud retains the capability to develop and expand new appendages at any time depending on exogenous and endogenous factors. There is no periodic accumulation of appendage primordia to form a resting bud and their subsequent rapid expansion in a pronounced "flush" of growth, as occurs in many tropical trees.

The size and shape of buds vary in different species but are somewhat diagnostic and largely dependent on stipule size. Stipules may be as long as 8 cm in *Rhizophora apiculata*, but less than 3 cm long in *Ceriops* and the small-flowered species of *Bruguiera*. *Ceriops* has characteristically flattened buds with the dorsal sides almost angular. A feature of the group that is particularly well shown in *Ceriops* is that the developing primordia are not close packed and there is ample free space to a large extent occupied by colleter secretions.

Architecture and crown shape

Although the Rhizophoreae have a basically uniform shoot construction, they are somewhat diverse architecturally, have some differences in reiterative responses, and consequently show different crown shapes. Constant features include continuous growth (in the broad sense), since there are no obvious resting terminal buds and unbranched shoots lack obvious discontinuities. Continuous growth can result in continuous branching, but more usually branching is discontinuous.

Sapling growth is characterized by an initial unbranched phase followed by diffuse branching. The extent of the unbranched phase varies and seems dependent on general vigor. In dense shade, plants up to 3 m tall with short internodes can develop. In the branched phase, intermittent tiers of branches develop, sometimes rather regularly. Branching is always by syllepsis; reserve buds are produced but they seem never to contribute to crown development. In *Rhizophora mangle*, for example, reserve buds are known to last for 3 years at a maximum, so the tree has no capability for sprouting from older wood. This seems a general property of *Rhizophora*, which has a rather uniform physiognomy with no specific architectural variation. The inherent deterministic features of crown shape nevertheless allow the development of a considerable variety of crown shapes (Fig. B.62).

.A

B

Figure B.62. *Rhizophora mangle.* Contrasted growth forms to show architectural plasticity. (A) Scrub form with dense crown and cone of aerial roots from landward margin of the mangroves in the Florida Everglades (see Fig. 1.11). (B) Old exposed specimen on the shoreline, Biscayne Bay, Florida. The original trunk is lost and replaced by numerous reiterated crowns arising from old branches, which are, in turn, supported by an extensive inverted "canopy" of aerial roots. (From negatives by A. M. Gill)

321

In the tribe Rhizophoreae there are 2 architectural models (following the system of Hallé et al. 1978), but the physiognomy of the crown has 3 fairly distinct types.

1. Attims's model *(Rhizophora)*. In the definition of this model the lateral axes are, at least initially, orthotropic, whereas growth and branching are continuous or diffuse. In *Rhizophora* sylleptic branches are produced, usually in tiers of 1 to 3 pairs at successive nodes, separated by about the same number of branch-free nodes. The orthotropic nature of the branches is at first apparent and they tend to repeat the branching pattern of the parent tree, but as the branch ages and extends, it reclines, with the branches becoming restricted to the lower side and plagiotropy gradually imposed. Progressively, in axes of 2 or 3 successive orders, the distal portion of the axis develops short internodes and eventually functions as a terminal short shoot, which is more or less abruptly erect. In this way plagiotropy by apposition is gradually achieved, leading to a branch complex that has much the same physiognomy as that characteristic of Aubreville's model.

2. Attims's model (*Kandelia, Ceriops*, **small-flowered** *Bruguiera* **species**). These correspond to the preceding growth pattern, but the axes remain much more strongly orthotropic, and plagiotropy by apposition is little developed.

3. Aubréville's model (large-flowered *Bruguiera* **species)**. In the definition of this model the trunk axis shows continuous or rhythmic growth, but always with the production of regular tiers of branches that show plagiotropy by apposition, with regular and pronounced sympodial development. The plagiotropy in *Bruguiera* is not as precise as the clear pattern described for *Terminalia*, but the same physiognomy results, with erect terminal short shoots substituted by extended lateral axes, each of which in turn repeats the process. Proliferation of the system results when 2, instead of 1, lateral renewal shoots develop. The greater precision of this pattern in *Bruguiera* compared with *Rhizophora* is best appreciated when the young leader shoot complexes of the 2 types are compared; the abrupt plagiotropy of the former is then contrasted with the gradual plagiotropy of the latter.

Reiteration

Bruguiera more regularly conforms to its architectural model because reiteration seems difficult and results in shoots in which a direct substitution of the leader is most common. Consequently the trunk remains single and the crown shape narrow and conical. In *Rhizophora* the leader is readily replaced by one or more existing laterals because of the strong orthotropy. Several leader complexes can be developed; the crown becomes open, broad, and irregular. The reiterative ability of *Rhizophora* is best revealed in low scrubby plants in marginal habitats with numerous erect axes and asymmetry of the crown. Prolepsis plays no role in crown development.

Differences between the 2 contrasted tree forms are inherently subtle, but the resultant crown shapes are readily contrasted where the trees grow together. Quantification of the differences has not been attempted, but one can speculate that the more plastic method of organization in *Rhizophora* accounts, in part, for its greater success. The difference in root systems between *Rhizophora* and other members of the tribe may aid in this differential success.

Floral biology

The floral biology of mangrove Rhizophoraceae is complex and is best understood in a comparative context, summarized in Table 7.2. The subject has been dealt with in some detail by Tomlinson et al. (1979), but this includes only an account of the different kinds of floral mechanisms and the kinds of flower visitors with some reasonable inference about the likely method of pollen transfer. There is very little information about incompatibility and other isolating mechanisms because no experimental work has been carried out.

The contrasted pollination mechanisms are summarized in the following key, with additional explanatory notes (see also Table 7.2):

1A. *Pollination predominantly by wind* (anthers dehiscent in the flower bud)........... 2
  2A. Mature flowers within the leafy cluster............. *Rhizophora* (most species)
  2B. Mature flowers below the leafy cluster......................... *R. apiculata*
1B. *Pollination by animals* (only the usual and presumed most effective pollinator is included; flowers may receive a diversity of visitors) 3
  3A. *Pollination mechanism explosive* (anthers dehiscent in the flower bud)........ 4
    4A. Pollination by daytime visitors...................................... 5
      5A. Pollination by birds .... large-flowered *Bruguiera* species (i.e., *B. exaristata, B. gymnorrhiza, B. sexangula*)
      5B. Pollination by butterflies....small-flowered *Bruguiera* species (i.e., *B. cylindrica, B. hainesii, B. parviflora*)
    4B. Pollintation by nighttime visitors.......................................
      6. Pollination by moths .... *Ceriops tagal*
  3B. *Pollination mechanism not explosive* (anthers not dehiscent in the flower bud).
    7A. Pollination presumably by small, short-tongued animals (insects) .. *Ceriops decandra*
    7B. Pollination presumably by fairly large, long-tongued animals (insects) .... *Kandelia candel*

The simplest expression of differences in floral mechanisms is the difference in flower size, shown in Figure B.63.

Wind pollination in *Rhizophora*

Evidence for this mechanism comes from the floral syndrome and from the observation that *Rhizophora* has an appreciably larger pollen/ovule ratio (the ratio of the total number of pollen grains to the total number of ovules per flower). When this value is high, it has been used as an indication that a plant is pollinated by wind, since the method of pollen transfer is very wasteful of pollen (Cruden 1977). High pollen/ovule ratios on a relative scale are characteristic of familiar wind-pollinated taxa. For example, *Rhizophora* has a ratio as much as an order of magnitude higher than the ratio in other genera.

Other characters that are indicative are the brief pollen presentation time and the short functional life of the flower. The anthers dehisce before the flower bud opens

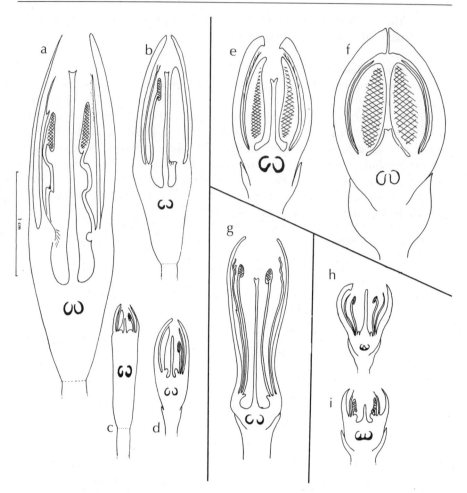

Figure B.63. Rhizophoraceae flower size. Longitudinal sections of flower buds show relative sizes as follows: (a–d) *Bruguiera*: (a) *B. gymnorrhiza*; (b) *B. exaristata*; (c) *B. parviflora*; (d) *B. cylindrica*. (e–f) *Rhizophora*: (e) *R. stylosa*; (f) *R. apiculata*. (g) *Kandelia candel*. (h–i) *Ceriops*: (h) *C. tagal*; (i) *C. decandra*. Extent of pollen-producing tissue indicated by crosshatching. (From Tomlinson et al. 1979)

and, in all species except *R. apiculata*, are drawn out with the marginal hairs of the petals as they expand. The petals fall within a day. Flowers are also typically pendulous. In *R. apiculata*, which lacks marginal petal hairs, the flowers point downward at maturity and also mature below the leafy crown, so that the foliage does not hinder pollen dispersal. There is no copious secretion of nectar in any species of *Rhizophora*.

The style of *Rhizophora* is not elaborated in the manner typical of wind-pollinated flowers, however. Furthermore, flowers are visited by insects (e.g., bees), although these may only seek pollen, whereas thrips are common in even freshly opened flowers. It is possible that the pollination mechanism is being transformed from an original animal-pollinated ancestral condition (which persists in all other genera of the tribe) to a wind-pollinated condition. In this way *Rhizophora* escapes the competition for animal pollination that otherwise predominates in mangrove communities, as we have discussed (see Chapter 7).

The abundance of *Rhizophora* pollen in marine sediments indicates the productivity of the tree in this respect and does not conflict with the idea of the plant as a wind-pollinated species.

Unspecified pollination by insects

This is characteristic of *Kandelia* and *Ceriops decandra*. *Kandelia* has a generalized flower but a well-developed calyx cup into which nectar is secreted. *Ceriops decandra* differs from *C. tagal* because it lacks the elaborate pollen-discharge mechanism of that species. It is unusual to have such contrasted pollination devices in one genus.

Explosive pollen discharge and animal pollination

This group includes *C. tagal* and all 6 species of *Bruguiera*. The petals enclose pairs of stamens; the antisepalous set has a characteristic basal twist of the filament to allow this. Opposite margins of the petal are held together by a set of interlocking hairs. The anthers dehisce precociously so that at the time the flower opens, each petal pouch includes loose pollen. The petals are under tension, generated during floral expansion; the tension is released suddenly by a slight touch to the base of each petal, such as is provided by the probing mouth or beak of a flower visitor searching for nectar in the floral cup. This explodes the petal, which unzips instantly, scattering a cloud of pollen on the head of the visitor.

Individual petals work independently so that pollen discharge can involve more than one visitor; the presentation time, in the absence of a visitor, is several days.

It could be argued that the mechanism is designed simply to disperse pollen freely for wind pollination, but this seems unlikely in view of the nectar, which is copious in larger flowers, the elaborateness of the mechanism, and the usual inability of petals to explode in the absence of precise triggering by a visitor. Scented flowers also attract visitors.

Although the mechanism applies to all 7 species that have this structure, individual species are designed to attract a given class of visitor because of their flower size and orientation and time of anthesis. The large-flowered species of *Bruguiera* (*B. exaristata*, *B. gymnorrhiza*, and *B. sexangula*) are pollinated by birds, and have downward-pointed flowers, the smaller-flowered species (*B. cylindrica*, *B. hainesii*, and *B. parviflora*) are pollinated mainly by day-flying insects (e.g., butterflies) and have erect or divergent inflorescences. *Ceriops tagal* is pollinated by small night-flying moths. The clear impression is that competition is minimized by this division of labor.

Isolating mechanism

Some degree of dichogamy, involving weak protandry, seems characteristic of most species, although the time of stigma receptivity has not been established precisely. There is evidence that *Rhizophora* is self-compatible from the ability of isolated greenhouse plants to set fruit, but no careful and extensive experimental work has been done. On theoretical grounds, Primack and Tomlinson (1980) suggest that mangrove plants should be self-compatible, since this is a characteristic of pioneering species that may become established as isolated individuals in new environments remote from parental sources. The available evidence does not conflict with this view.

Seedling anatomy

In studying the structure of fossilized propagules of *Ceriops cantiensis* Chand. in the London Clay flora (Eocene, Lower Tertiary), Wilkinson (1981) contrasted the hypocotyl anatomy of several species in the tribe Rhizophoreae and summarized older literature. She notes that *Rhizophora* can be distinguished immediately by the spherical masses of stone cells in the outer cortex and the multiradiate sclereids in the cortex and pith (the latter corresponding closely to those found abundantly in other parts of this genus). Differences in epidermal features and cuticle texture are said to distinguish other taxa. That *Ceriops* can be recognized unequivocally from deposits of this age is a further clear indication of the ancestry of mangrove Rhizophoraceae. Wilkinson also comments on the absence of pollen of Rhizophoraceae from the London Clay flora. This could be evidence that the mangrove propagules were delivered to the site as flotsam rather than for the existence of mangal in situ.

Rhizophora L. 1753
[*Sp. Pl.*:443]

See Ding Hou (1958) for an extended synonymy.

A pantropical genus of perhaps 8 species (including 3 putative hybrids), which is very uniform in its vegetative construction.

Evergreen trees growing to 30 m tall, with characteristic aerial stilt roots, perfect flowers; often depauperate, scrubby, and much branched in marginal habitats. Leaves with bijugate phyllotaxis, each pair associated with a pair of lanceolate, interlocked stipules to form a distinctive pointed terminal bud, the stipules with a 2-ranked palisade of glandular colleters internally at their insertion; stipules early caducous to leave an annular scar, the petiole scar itself elliptical and evident below the associated stipule scar. Leaves simple, entire, blade elliptical, margin often somewhat recurved, 10 to 20 by 4 to 10 cm, but often somewhat larger in saplings, the range of size and shape somewhat characteristic for each species but much influenced by the habitat and position of the tree. Petiole short, 2.0 to 4.0 cm by 3 to 5 mm, cylindrical. Leaves and sometimes stipules abruptly yellowing with senescence.

Leaf texture coriaceous, glabrous but with numerous microscopic cork warts on the lower surface visible on older leaves as fine black dots in dried, if not always in fresh specimens; veins evident but not prominent below. Apex with a short persistent and erect mucro 2 to 3 mm long (one group of species) or apex blunt and recurved (a second group of species); base usually acute to attenuate.

Inflorescence axillary either within the leafy crown or, maturing below the leafy crown in *R. apiculata*, with a shorter (1 to 3 cm) or longer (8 to 12 cm) peduncle, the length and thickness fairly diagnostic for each species. Inflorescence cymose, often pendulous in species with longer peduncles. Inflorescence 2 to 4 or more (up to 128) depending on the number of bifurcations, but with the frequent abortion of one branch, or (e.g., in *R. mangle*) often with a trifurcation at the first (rarely subsequent) node, with less precise flower number; flower number or range of flower number usually diagnostic for each species. Each node on inflorescence with a pair of short cupular connate bracts, the bracts either fleshy or even corky (*R. apiculata*), with a pair of similar bracteoles below each flower. Pedicel above bracteoles absent or very short. Flowers perfect, tetramerous; calyx green, yellow, white, or brown, as a shallow tube with 4 thick, fleshy, valvate, acute lobes, persistent in fruit. Petals 4, ephemeral, delicate, white or greenish white, lanceolate and somewhat longer than the calyx lobes, glabrous (*R. apiculata*) or otherwise with a slightly to markedly hairy margin. Stamens 8, 12, or 16 (or some number close to these) inserted on margin or receptacular disc; each sessile (*R. apiculata*) or at most with a very short (to 1mm) filament, acute, multilocellar, dehiscing introrsely by a short adaxial valve. Ovary semi-inferior, the apex obscurely to distinctly conical and extended into a single style 1 to 5 mm long or stigmas almost sessile (*R. mucronata*). Stigma obscurely to evidently bilobed. Ovary 2-locular, with 2 anatropous ovules on an axile placenta in each loculus. Fruit ovoid, somewhat extended apically, with a brown or olive green, somewhat leathery pericarp and including 1 (rarely more) developing seed. Seeds viviparous, germinating by extension of hypocotyl through stylar canal, hypocotyl maturing on tree to a length of 15 to 30 cm (70 cm or more in *R. mucronata*). Cotyledons without stipules and fused to form a tubular collar that remains on the tree after seedling is lost. Cotyledonary collar extended and becoming exposed before abscission of seedling. Propagule a seedling with extending hypocotyl, the plumule protected by a stipule pair of first (aborted) plumular leaves.

The genus is distinguished anatomically from other members of the tribe by the cork warts on the leaves and the abundant sclereids, including characteristic H-shaped trichosclereids, which are especially abundant in the aerial roots and seedling hypocotyl. They can be seen with a hand lens protruding from any broken surface and are a useful diagnostic field character.

Nomenclatural history of *Rhizophora*

Rhizophora as used by Linnaeus was a generic name for all mangrove plants, much as Rumphius (1743, 3:102–20) used the name *Mangium*. The characteristics of the genus were based largely on vivipary and habitat, and Linnaeus listed 7 species (including 2 more in the second edition of *Species plantarum*, 1762). Of Linnaeus's original 7 species, 6 are now assigned elsewhere; 4 to other genera of the mangrove Rhizophoraceae (*Bruguiera, Ceriops*, and *Kandelia*), whereas *Aegiceras* (Myrsinaceae) and *Sonneratia* (Sonneratiaceae) were appropriately recognized and segregated by subsequent workers. The only remaining Linnaean species in the genus is *Rhizophora mangle*, based on the name *Mangle*, which goes back at least as far as Oviedo (1526, *Hist. Gen. Nat. Ind.*) and had been used for the plant by pre-Linnaean authors. A complete history of the origin and use of the name *Rhizophora* is provided by Salvoza (1936).

Rhizophora: Key to Species

1A. Leaf apex with an erect mucro 2–3 mm long....... eastern species (species exclusively of the Old World tropics, ranging from East Africa to the Pacific in the Tonga-Samoan region) 2
1B. Leaf apex without an erect mucro, usually blunt with a recurved margin, at most (*R.* × *selala*) with an irregular recurved (but never prominent) mucro western species (species of the Atlantic coasts of West Africa, tropical America, the Caribbean and Pacific coasts of central and southern tropical America, with an extension into the Pacific as far west as New Caledonia and the New Hebrides)... 5

Eastern Species

2A. Flowers in 2's (rarely in 4's) on stout peduncles (about 10 mm long) shorter than petiole, borne below the leafy crown, that is, in the axil of a leaf scar; each flower sessile with a pair of massive cupular corky bracteoles. Petals glabrous. Stamens usually 12 (sometimes fewer or more), sessile. Stigmas almost sessile. Fruits and seedlings always maturing well below the leaf rosette .... *R. apiculata* Bl.(Indo-Malaya)
2B. Flowers in 4's or higher numbers (rarely in 2's) on stout to somewhat slender peduncles (15–40 mm long), longer than the petiole, borne within the leafy crown, that is, in the axil of an attached foliage leaf; each flower with an evident (even if short) pedicel, bracteoles green, fleshy but never corky. Petals slightly to conspicuously hairy on the margins. Stamens either 8 or about 16 with a short (1–2 mm) filament. Stigmas sessile or ovary extended into a slender style. Fruits and seeds borne within or just below leafy rosette ....3 (remaining eastern species)
3A. Flowers per inflorescence usually in 4's (sometimes 2's), peduncle short (about 15 mm long) and rather stout. Stamens variable in number (8–15-) 16 (up to 22) and some often distorted, aborted, or represented by a filamentous staminode. Petals with inconspicuous marginal hairs. Styles 2–3 mm long. Trees usually sterile and not producing seedlings. Pollen sterile ......*R.* × *lamarckii* Montr. (a possible hybrid *R. apiculata* × *R. stylosa*) (New Caledonia, New Hebrides, Queensland, New Guinea, Solomon Islands)
3B. Flowers 4, 8, or more per inflorescence (rarely 4 or fewer by reduction in depauperate specimens), peduncle usually long (25–40 mm), slender. Stamens almost always 8, rarely aborted. Petals with densely woolly,

conspicuous marginal hairs. Style either long or absent. Trees fertile (able
to develop numerous seedlings). Pollen fertile 4

4A. Stigmas sessile, seedlings warted, 50–70 mm long. Leaf blade broad (to 10
cm) and long (to 20 cm) . *R. mucronata* Lamk. (East Africa to the Western
Pacific, that is, possibly as far as New Hebrides and the Gilberts)

4B. Stigmas on a slender style 4–5(-6) mm long, seedlings smooth, not
exceeding 30 cm. Leaf blade narrow (to 7 cm) and short (to 12 cm) ....
.... *R.stylosa* Griff. (India to Tonga and Samoa)

Western Species

5A. Peduncle commonly trifurcate at first node; flowers per inflorescence 2–9.
Flower buds yellow or yellowish white, somewhat to distinctly angular .. 6

5B. Peduncle bifurcate at all nodes, flowers per inflorescence usually more than
10 (up to 64). Flower buds white or greenish white, not angular 8

6A. Leaf apex with an irregular curved, sometimes deciduous mucro. Peduncle
2.5–3.0 cm long, but often longer, 2.6–3.2 mm wide, often with more
than 2 orders of branching. Flowers 2–9 per inflorescence. Mature flower
buds white, neither sharply angular in cross section nor abruptly narrowed
to a distinct shoulder at the base, 12–14 mm long. Apex of ovary extended
into a distinct style 1–2 mm long. (Plants sterile, lacking fruits and
viviparous seedlings). .... *R.* × *selala* (Salvoza) Tomlinson (a probable
hybrid *R. samoensis* × *R. stylosa*) (New Caledonia, Fiji, and the New
Hebrides)

6B. Leaf apex blunt, recurved, without a mucro. Peduncle 2.0–2.5 cm long (or
longer), 2.0–2.4 mm wide, rarely with more than 2 orders of branching.
Flowers usually 2–5 per inflorescence. Mature flower buds yellow, sharply
angular in cross section and distinctly shouldered at the base, 10–12 mm
long. Apex of ovary steeply conical and without a discrete slender style.
(Plants fertile, capable of developing fruits and seedlings) 7

7A. Bracteoles conspicuous, 1 mm or longer, flower buds pointed at apex......
R. mangle L. (tropical America on the Atlantic and Pacific coasts)

7B. Bracteoles inconspicuous, less than 1 mm long, flower buds rounded at
apex ....
R. samoensis (Hochr.) Salvoza (New Caledonia, New Hebrides, Fiji, and
probably elsewhere in the Western Pacific; possibly occurring on the Pacific
coast of tropical America)

8A. Bracteoles thick, short, rounded; flower buds rounded, flowers often 32–64
per inflorescence....*R. racemosa* G.F.W. Meyer (West Africa and Atlantic
coast of tropical America)

8B. Bracteoles thin, long, pointed; flower buds pointed apically, flowers usually
8–32 per inflorescence.........*R. harrisonii* Leechm. (regarded as a hybrid
between *R. mangle* and *R. racemosa* but reported as occurring outside the
range of one of its putative parents) (West Africa and Atlantic and Pacific
coasts of tropical America)

Further notes on the taxonomy of *Rhizophora* are found in the discussion on the
variation in morphological features that are used diagnostically.

In general, *Rhizophora* species are very similar, and diagnostic features are often
quantitative, referring to a range of sizes or numbers that are readily appreciated
in the field where large samples are available, but not so in the herbarium. Examples

of measurements that show the ranges of some of these features are found in Duke and Bunt (1979) and Tomlinson (1978). Normally where more than one species of *Rhizophora* grow together, the fieldworker soon learns that they can be distinguished easily and completely by the combination of characters in the key, together with more elusive "Gestalt" features, such as leaf color and size, which are unavailable to the herbarium worker. Some field characters, however, may not be permanent; Duke and Bunt (1979) record the absence of leaf dots in their key to distinguish *R. apiculata*. This species does have the cork warts, which are a generic diagnostic feature; they are simply obscure in fresh leaves, but appear in dried or fluid-preserved leaves. It would be misleading to apply this character elsewhere than in the field.

Furthermore, the ranges of variation so far recorded in any detail refer to local populations and may not be applicable throughout the entire range of one species, although in my experience *Rhizophora* species are remarkably uniform in their morphology at the extremes of their ranges.

Field study so far demonstrates that even though the measured ranges of given characters overlap among different species, there is no continuum and species (and even hybrids) remain as clearly recognizable entities.

Species aggregates in *Rhizophora*

The species of *Rhizophora* may be aggregated in various ways. The most morphologically distinctive species is *R. apiculata* in terms of flower number per inflorescence, inflorescence position, bracteole structure, petal indumentum, and stamen number; this is the basis for its segregation as a monotypic section *Aerope* Blume, in contrast to the remaining species forming section *Rhizophora* (sect. *Mangle* Blume). However, representatives of the 2 sections are capable of crossing if we accept *R. × lamarckii* as a hybrid *R. apiculata × R. stylosa*.

Another method of aggregation could recognize 2 groups distinguished by the morphology of the leaf apex; one group with a prominent mucro (*R. apiculata, stylosa, mucronata*, and *× lamarckii*) and the other without (*R. samoensis, mangle, harrisonii*, and *racemosa*). This arrangement is strongly correlated with geographic distribution, the first group constituting the eastern species and the second the western species. Again, however, members of the 2 groups appear capable of crossing as in the production of *R. × selala* (*R. stylosa × samoensis*), which is morphologically intermediate in its leaf apex and occurs in the area of geographic overlap between the 2 groups.

A third method of segregation could contrast species with the capability of trifurcating at the first node of the inflorescence (*R. mangle* and *R. samoensis*) with the remaining species that do not. Once again the 2 groups are interfertile, in the hybrids *R. × selala* (*R. samoensis × R. stylosa*) and *R. × harrisonii* (*R. mangle × R. racemosa*).

These considerations indicate that there is no clear-cut basis for subdividing the genus *Rhizophora*.

Distribution of *Rhizophora* species

Rhizophora is a pantropical genus in climatically rather uniform coastal environments, but with limited extension into the subtropics. Although the range of most species is appreciable, there are considerable differences in longitudinal ranges that are difficult to account for ecologically, especially because some species co-occur. Latitudinal limits are known with some precision in many parts of the world because botanists have been interested in the limits of the climatic tolerance of mangrove species at the margins of their range. The chief restriction is cold; *Rhizophora* is killed by frost and cannot survive extended periods of near-freezing temperatures. Some species (e.g., *R. apiculata* and *R. mucronata*) seem restricted to wetter climates; *R. stylosa* seems tolerant of drier climates. The limits of the latitudinal range are not precise because populations may be killed in exceptional winters but are restored subsequently by seedlings brought some distance via ocean currents, as occurs in central Florida.

The following information summarizes the distribution. In many instances precise distributional limits cannot be stated, but the illustrative maps (Figs. B.64, B.65) are based on herbarium records all examined and determined by myself, together with limited field experience, in the South Pacific, Malesia, and the Caribbean. Breteler (1969, 1977) has provided some information for the western species, and the decreasing distribution of eastern species eastward in the Pacific has been documented by Fosberg (1980). Recent experience shows that despite the frequency and wide distribution of *Rhizophora*, significant new records are easily made; *R. stylosa* was first collected on Singapore, a botanically accessible island, by H. M. Burkill in 1960.

The species are discussed in decreasing order of their range.

Eastern Species   (Fig. B.64)

Rhizophora mucronata

Ranges from East Africa to the Western Pacific. It has a southern limit in Africa at about the latitude of Durban; its limit in the Persian Gulf, where it is poorly developed by virtue of the dry climate, is not precisely known. In much of the Indian Ocean areas, it is the sole representative of the genus but does coexist with *Bruguiera* and *Ceriops*. Further eastward in the Indo-Malayan region it grows with *R. apiculata* and *R. stylosa* but becomes progressively a less conspicuous element of mangrove floras as one moves eastward. It is thus not known in Western Australia (possibly because of the dry climate) and has a localized distribution in northern Australia, Queensland, and into eastern New Guinea and the Solomon Islands. There is a record for the Gilbert and Ellis Islands. Older records for Fiji and Tonga are based on misidentification and confusion with *R. stylosa*. In the Northern Hemisphere it occurs in the northern Philippines and Indochina, but without precisely ascertained limits.

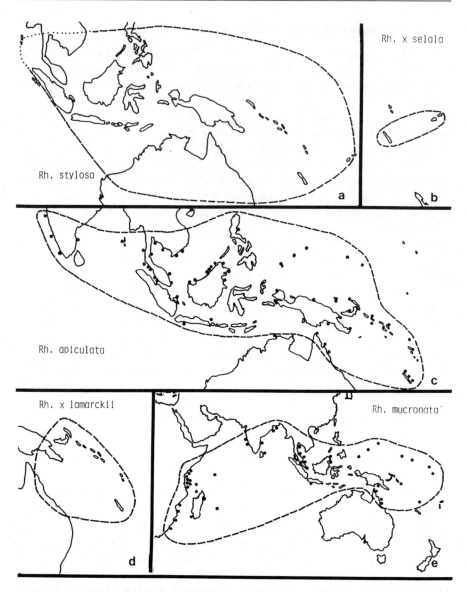

Figure B.64. Distribution of eastern species of *Rhizophora*. The generalized outline is based on field identification or herbarium records checked by the author. These are indicated by the dots in the distribution of *R. apiculata* and *R. mucronata*. (a) *R. stylosa*; (b) *R.* × *selala* in New Caledonia and Fiji; (c) *R. apiculata*; (d) *R.* × *lamarckii*; (e) *R. mucronata*.

Rhizophora stylosa

Ranges from southern India to the Eastern Pacific (Samoa), southward to New South Wales, and westward to Indochina. It is not known from South America. This species becomes a progressively more conspicuous element of mangrove vegetation in an easterly direction (cf. *R. mucronata*). In the Western Pacific this species seems to be particularly vigorous and will survive in extreme habitats (coral reefs, rocky coastlines) though often depauperate. Probably for this reason, it appears to be the sole representative of the genus in Western Australia. Its precise limit in the North Pacific is not certain; here it is largely supplanted by *R. mucronata*.

Rhizophora apiculata

This species has a more restricted range than *R. stylosa*; it is not known east of the Andaman Islands. Otherwise it is a common and even dominant constituent of mangroves in the Malesian region as far west as Queensland and Papua New Guinea in the south and the Philippines in the north. It occurs on New Caledonia (its easternmost limit) but restricted to the east (north) coast of this island, probably because of temperature and rainfall. In the central Pacific its easternmost limit is Ponape (Fosberg 1980).

Rhizophora samoensis

This species is distinguished with difficulty from *R. mangle* by morphology alone and may be regarded as a geographic outlier of that species in the Western Pacific, occurring from New Caledonia and the New Hebrides to Tonga and Samoa.

Rhizophora × lamarckii

This species was first described from, and considered to be endemic to, New Caledonia. More recently its existence in isolated localities in Queensland, New Guinea, the Solomon Islands, and the New Hebrides has been established on the basis of field observation and herbarium records. On abundant circumstantial evidence it is recognized as a hybrid of *R. apiculata* with *R. stylosa* and always coexists with its putative parents. In New Caledonia, for example, it is restricted, like *R. apiculata*, to the east coast. In the type locality, Canala, it occurs with 4 other species of *Rhizophora* (*R. apiculata, samoensis, × selala*, and *stylosa*), representing the greatest concentration of species in the genus and providing a further example of the richness of the New Caledonian flora.

Rhizophora × selala

This species was first recognized by Guppy (1906) in Fiji, who suggested that it was a hybrid between *R. stylosa* (which he incorrectly called *R. mucronata*) and *R. samoensis* (which he called, more appropriately, *R. mangle*). Salvoza accepted the taxon but did not provide a formal diagnosis, which was first done by Tomlinson (1978), who subsequently demonstrated its existence in New Caledonia and the New Hebrides. It is likely to be found elsewhere in places where its parents coexist.

Western species   (Fig. B.65)

Rhizophora mangle

Ranges from western Africa to the Pacific coast of tropical America. In Africa its precise latitudinal limits are not certain, but it is recorded as far south as Angola and as far north as Mauritania. In the Americas it has, on the Atlantic side, a wide distribution to about latitude 25°N in Florida and to eastern Brazil, and on the Pacific side from Mexico to northern Chile, where its southern range is limited by the cold, dry climate. A better understanding of its possible co-occurrence with *R. samoensis* in Pacific South America is desired.

Rhizophora × *harrisonii*

There is circumstantial evidence that the species is a hybrid between *R. mangle* and *R. racemosa* but, if so, its geographic distribution is peculiar, being greater than one of its putative parents. It co-occurs with its 2 parents in West Africa (Keay 1953) and the Atlantic coast of tropical America. However, Breteler (1977) claims that it occurs on the Pacific coast of tropical America in the absence of *R. racemosa*. Since the Panama Isthmus was last open over 3 million years ago, it seems that the situation needs reexamining, because it is unlikely that a species could exist in a vegetative condition for this length of time if *R. harrisonii* is infertile. However, Prance (1975) has since recorded *R. racemosa* on the Pacific coast of Colombia.

Rhizophora racemosa

Ranges from western Africa to northern South America, but the precise limits of its distribution are uncertain, especially its occurrence on the Pacific coast.

Summary

This brief summary shows that the broad facts of *Rhizophora* phytogeography are known but that even quite basic information has recently been derived. Without a more complete and precise factual knowledge of the geography and ecology of the genus, any attempt to account for the distribution of *Rhizophora* on a historical basis is highly speculative. More extensive fieldwork and a more precise diagnosis of taxa are needed.

Brief diagnosis of *Rhizophora* species

A more complete description and diagnosis are found in Ding Hou (1958, 1960) and Salvoza (1936).

1A.(*Rhizophora* sect. *Aerope* Bl. 1849, *Bot. Mus.* 1:134)

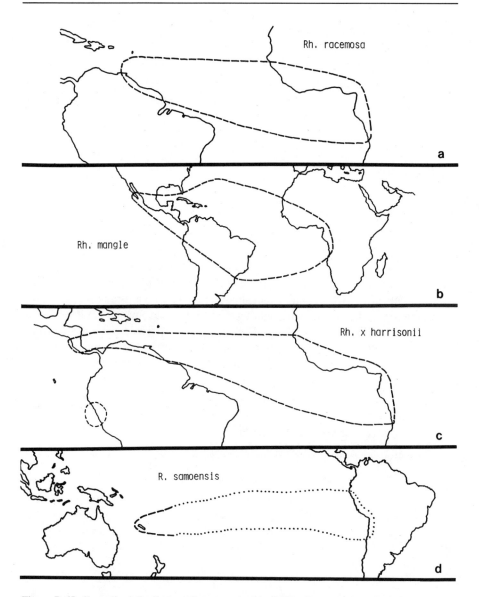

Figure B.65. Generalized distribution of western species of *Rhizophora*. (a) *R. racemosa*; (b) *R. mangle*; (c) *R.* × *harrisonii*; (d) *R. samoensis*, morphologically almost indistinguishable from *R. mangle*, known from New Caledonia and Fiji. The dotted extension of its range to South America is hypothetical.

Rhizophora apiculata BL. 1827   (Fig. B.66)
[*En. Pl. Jav.* 1:91]

R. conjugata Arn. (non L.) 1838, [*Ann. Mag. Nat. Hist.* 1:363]
A more complete citation is found in Ding Hou (1958); see also Salvoza (1936).

This species is perhaps the most distinctive in the genus, with large, dark glossy green leaves and flowers usually in pairs on a stubby axis, maturing below the rosette of leaves.

1B. *Rhizophora* sect. *Mangle* Bl. 1849, *Bot. Mus.* 1:132 [sect. *Rhizophora* Ding Hou 1958, *Fl. Males.* 1(5):450]

Rhizophora mangle L. 1753   (Fig. B.67)
[*Sp. Pl.* ed. 1:443]

This species is distinguished by its usually rather small leaves, with the blade 10 to 12 by 5 cm, the petiole about 2 cm long, pale and extending into the midrib, which is pale green to almost white below, and particularly by the blunt apex. Stipules are 6 to 8 cm long. Older shoots are light brown. Fruits are elongated, 2 to 3 cm long, often asymmetric and tapered to the apex, which is attenuated to the persistent style 2 to 3 mm long. The fruit is characteristic olive green with a roughened surface. Peduncles in more robust shoots may be 6 cm long but are usually shorter, the pedicels about one-third this length. Seedlings are 15 to 20 cm long, blunt, and without conspicuous lenticels.

Rhizophora samoensis (Hochr.) Salvoza 1936
[*Bull. Nat. Appl. Sci. Univ. Philipp.* 5:220, pl. 6]

Rhizophora mangle Guppy 1906 (non L.) [*Obs. Nat. Pacif.* 2:441, frontisp., as "*mangle chico*"]
Rhizophora mangle var. *samoensis* Hoch. 1925 [*Gand.* 2:447]

This species is scarcely distinguishable in its morphology from *R. mangle*, but it has blunt (not pointed) flower buds and obscure (not obvious) bracteoles. Geographic isolation alone is not sufficient reason for maintaining a separate species, but it is more than useful to retain the name for populations in New Caledonia, Fiji, Samoa, and Tonga.

Rhizophora racemosa Meyer 1818   (Fig. B.68m)
[*Prim. Fl. Esseq.*:185]

R. mangle var. *racemosa* Engler 1872 [*Fl. Bras.* 12(2):427]

This species is distinguished most readily by the much-branched inflorescence, with up to 6 orders of bifurcation, leading to a potential flower number of 128 on any one axillary flowering branch, although numbers are fewer than this

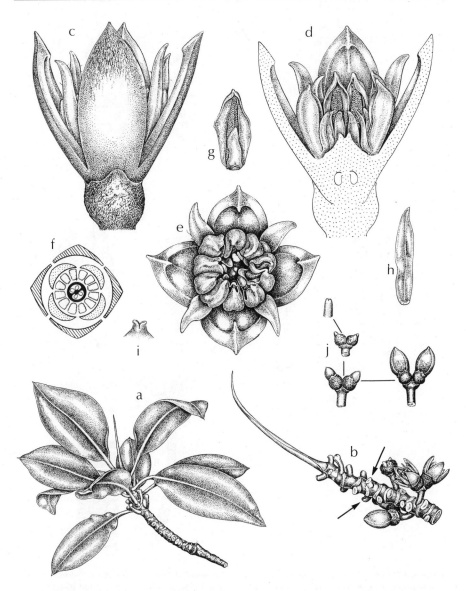

Figure B.66. *Rhizophora apiculata* (Rhizophoraceae) leaves and flowers. (a) Shoot with developing inflorescences (× ¼); (b) flowering shoot with leaves removed (× ¼), stipules enclosing apex; open flowers only below level of insertion of oldest leaves (arrows); (c) flowers from side (× 3); (d) flower in L.S. (× 3); (e) flower from above (× 3); (f) floral diagram, stamen number fairly constant; (g) dehisced stamen (× 3); (h) petal (× 3); (i) stigma at receptive stage (× 3); (j) sequence of developmental stages (× ½) of inflorescence (flower pair). (Material from Singaua, Lae, Papua New Guinea. P. B. Tomlinson, 30.10.74C).

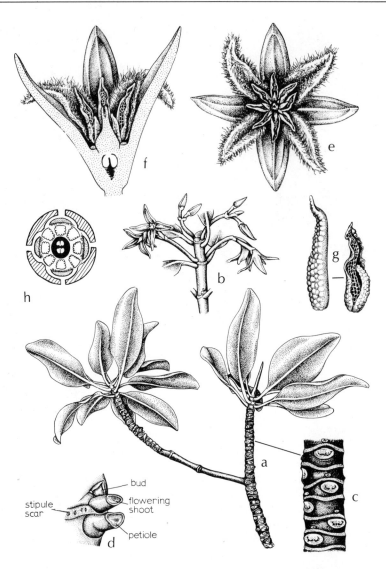

Figure B.67. *Rhizophora mangle* (Rhizophoraceae) leaves and flowers. (a) Shoot (× ⅜) branching by apposition; (b) part of flowering shoot with axillary inflorescences (× ½); (c) detail of scar pattern on terminal short shoot (× 1) (cf. part d); (d) node with leaf and flowering shoot cut off (× 2), vegetative bud above inflorescence; (e) flower from above (× 3); (f) flower in L.S. (× 3); (g) undehisced and dehisced stamen (× 4); (h) floral diagram. (From Tomlinson 1980)

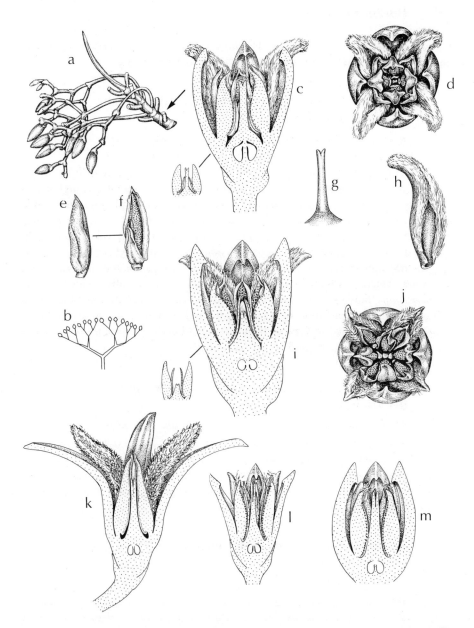

Figure B.68. *Rhizophora* spp. (Rhizophoraceae) floral morphology. (a–j) Indo-Pacific species. (a–h) *Rhizophora stylosa*: (a) shoot with leaves detached (× ½), inflorescences borne within leafy shoot (cf. *R. apiculata*), lowest leaf on shoot at arrow; (b) diagram of 16-flowered (2⁴) inflorescence with branches shown in one plane; (c) L.S. flower (× 3), inset shows relation of style to dehisced stamens; (d) flower from above (× 3); (e,f) undehisced and dehisced stamen (× 6); (g) style (× 3); (h) single petal (× 3). (i–j) *R. mucronata*: (i) flower in L.S. (× 3), inset shows relation of style to dehisced stamens; (j) flower from above (× 3). (k–m) Atlantic-Caribbean species, L.S. flower (× 3): (k) *R. mangle*; (l) *R. harrisonii*; (m) *R. racemosa*. (Material sources: *R. stylosa*: Bootless Bay, Port Moresby, Papua New Guinea, P. B. Tomlinson, 30.10.74C; *R. mucronata*: Semetan, Sarawak, P. Chai, *s.n.*; *R. mangle*: Fairchild Tropical Garden, Miami, Florida, P. B. Tomlinson, *s.n.*; *R. harrisonii*, *R. racemosa*: Monrovia, Liberia, West Africa, D. de May *s.n.*)

because of abortion of some axes. The flower buds are characteristically pointed and the flowers rather small, the sepals being 8 to 10 mm long.

Rhizophora × *harrisonii* Leechman 1918   (Fig. B.68l)
[*Kew Bull.* 1918:8, Fig. A]

R. *brevistyla* Salvoza 1936 [*Bull. Nat. Appl. Sci. Univ. Philipp.* 5:211]

The synonymy follows the interpretations implied in Ding Hou (1960)

This species is, on circumstantial evidence, considered to be a hybrid between *R. mangle* and *R. racemosa* (intermediate in its morphology, high degree of sterility, and overlapping geographic range). It remains morphologically distinct and is recognized by its many-flowered inflorescences with blunt or rounded flower buds, the sepals being 10 mm or longer. Descriptions of its reproductive capacity suggest that it sets abundant seed, unlike the other *Rhizophora* hybrids of the western species. *Rhizophora brevistyla* refers to specimens from the Pacific coast of Central and South America, which Salvoza (1936) considered specifically distinct from *R.* × *harrisonii*. Further field study is needed to see whether his opinion can be substantiated. Differentiation between Pacific and Atlantic populations of *Rhizophora* is to be expected in view of their presumed long separation by formation of the Isthmus of Panama about 3×10^6 years B.P.

Rhizophora mucronata Lamk.1804   (Fig. B.68i–j)
[*Encycl.* 6:189]

R. *mucronata* Lamk. var. *typica* Schimp. 1891 [*Bot. Mitt. Trop.* 3:92]

The flowers of this species differ from those of *R. stylosa* only in the sessile stigma; inflorescences in the 2 species are very similar. The leaves of *R. mucronata* are typically much larger than those of *R. stylosa*, however. Another conspicuous character is the very long seedling hypocotyl with a rough, warty surface (Fig. B.69).

Rhizophora stylosa Griff. 1854   (Fig.B.68a–h)
[*Nat. Pl. As.* 4:665]

R.*mucronata* Lamk. var. *stylosa* Schimp. 1891 [*Bot. Mitt.Trop.* 3:92]

It is interesting that the type specimen of this species is an apparent mixture, only some of which is *R. stylosa*. In the field *R. stylosa* can be distinguished from *R. mucronata* by the characters cited here.

Rhizophora × *lamarckii* Montr. 1860   (Fig. B.70)
[*Mem. Acad. Sci. Lyon* 10:201]

R. *conjugata* var. *lamarckii* (Montr.) Guillaumin 1914 [*Notul. Syst.* 3:56]

This is interpreted as a sterile F_1 hybrid between *R. apiculata* and *R.*

Figure B.69. *Rhizophora mucronata* (Rhizophoraceae). Viviparous seedlings of this species are the largest in the mangrove Rhizophoraceae. Klang, Malaysia.

stylosa by Tomlinson and Womersley (1976). It was originally considered to be endemic to the northeast coast of New Caledonia, and at its type locality at Canala it grows not only with its putative parents but also with *R.* × *selala* and *R. samoensis*, so that here we have the largest concentration of *Rhizophora* anywhere (5 out of a possible 8 taxa). Its wider distribution is now well established (Tomlinson 1978; e.g., Queensland, Papua New Guinea, and the Solomon Islands). It is always consistently intermediate in its morphology between its putative parents (Duke and Bunt 1979, Tomlinson 1978, Tomlinson and Womersley 1976) and occurs only where the 2 grow together. The pollen appears to be sterile and seedlings are rarely produced.

Duke and Bunt (1979) note that in Queensland *R.* × *lamarckii* is multitrunked and rambling, forming communities of gnarled and stunted individuals in higher tidal contours and lower to middle tidal reaches in areas of moderate to high rainfall. The frequently rather battered appearance of trees suggests that once established *R.* × *lamarckii* is very tenacious and plants may be considerably old; unfortunately, precise age cannot be estimated.

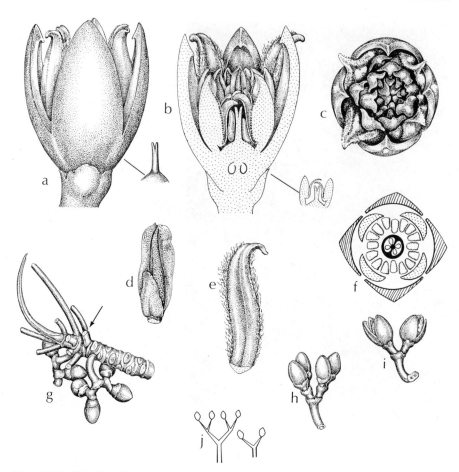

Figure B.70. *Rhizophora lamarckii* (Rhizophoraceae) floral morphology. (a) Flower from the side (× 3), inset: detail of style (× 3); (b) flower in L.S. (× 3), inset: relation of style to dehisced stamens; (c) flower from above (× 3); (d) dehisced stamen (× 3); (e) petal (× 3); (f) floral diagram, stamen number is not constant; (g) shoot with leaves detached (× ½), mature inflorescences just within or below leafy crown (arrow, position of oldest leaf in crown); (h) 4-flowered inflorescence; (i) 2-flowered inflorescence; (j) diagram of 2- and 4-flowered inflorescences with branches shown in one plane. (Material from Barune, Port Moresby, Papua New Guinea. P. B. Tomlinson, 31.10.74X; from Tomlinson and Womersley 1976)

Rhizophora × *selala* (Salvoza) Tomlinson 1978
[*Arnold Arbor*. 59:159]

"*Selala*" Guppy 1906 [*Obs. Nat. Pacif.* 2:445, 487, frontisp.]
R. mucronata Lamk. var. *selala* Salvoza 1936 [*Bull. Nat. Appl. Sci. Univ. Philipp.* 5:219]
R. selala Tomlinson and Womersley 1976 [*Contrib. Herb. Aust.* 19:9]

Interpreted as a sterile F_1 hybrid between *R. stylosa* and *R. samoensis*. This taxon was first recognized by Guppy in Fiji. It is characterized by its distinct but recurved or deciduous leaf tip, the inflorescence with 2 to 9 flowers commonly branched beyond 2 orders, the white and scarcely angled flower buds, and its sterile condition. In my experience it always co-occurs with its putative parents from which it is readily distinguished, at least in the field, in its consistent intermediate morphology.

Bruguiera Lamarck 1793–7

[*Tabl. Enc. Meth.* t. 397; see also *Encycl. Meth. Bot.* 4:696, 1797–8]

For a full synonymy, see Ding Hou (1957, 1958).

Trees growing to 40 m high, evergreen, with continuous growth, hermaphroditic; short aerial roots of sapling developing into stout buttresses of adult trees; pneumatophores kneelike. Leaves opposite, bijugate; petiolate, entire; blade glabrous, usually ovate to oblong-ovate, with a cuneate to acuminate base, apex usually bluntly pointed. Stipules in pairs, 1 to 4 cm long, sometimes reddish; orientation as in *Rhizophora*. Flowers either solitary or in shortly stalked, bifurcate, 2 to 6 flowered cymes. Bracteoles never enclosing base of flower. Flowers either large (2 to 4 cm long) or smaller (1 to 1.5 cm long) with either about 13 (large flowered) or about 8 (small flowered) pointed, initially valvate, fleshy calyx lobes. Petals delicate, as many as and alternating with the calyx lobes, bilobed and with marginal, interlocking hairs, the lobes commonly with appendages distally and usually with a rigid filament between them. Stamens with slender, somewhat contorted filaments, twice as many as the sepals and enclosed in pairs by the pouched petals, dehiscing precociously, the pouches exploding when triggered. Ovary inferior, obscurely 2 to 4-locular, with 2 ovules in each loculus. Floral disc cup shaped; style slender with an obscurely 3-lobed stigma. Fruit 1-seeded with persistent, erect, or reflexed lobes. Hypocotyl terete or obscurely ribbed, cigar shaped, blunt apically. Fruit falling with the seedling, the cotyledonary collar not extending.

This genus is distinguished from other mangrove Rhizophoraceae in its numerous (more than 6) calyx lobes and the propagule, which consists of both seedling and fruit, unlike the other genera in which the seedling is detached by abscission of the cotyledonary node so that cotyledons and fruit remain on the tree, abscissing independently after the seedling is detached. *Bruguiera* is also consistent in its explosive mechanism of pollen release; this is triggered by a visiting pollinator, in which only *Ceriops tagal* of the other Rhizophoreae resembles it.

Nomenclature

The genus is somewhat imperfectly segregated into 2 groups of species; the large, solitary-flowered group (e.g., *B. gymnorrhiza*) and the small, many-flowered group (e.g., *B. parviflora*). The difference in flower size is correlated with differences in leaf size and fruit size. This distinction was the basis for the recognition of 2 genera: *Bruguiera* (sensu stricto) and *Kanilia*, representing the first and second group, respectively, by Blume. Subsequently Miquel [1855, *Fl. Ned.*

Ind. 1(1):585] reduced *Kanilia* to a section of *Bruguiera* and applied the name *Mangium* (from Rumphius) to the other section. However, Ding Hou (1958) pointed out that the distinction between the 2 sections was obscured by the later discovery of the 2 species *B. hainesii* and *B. exaristata*, which in some features are morphologically intermediate. The first species sometimes has solitary flowers; the second sometimes has flowers in pairs and the size difference between flowers is not great. On the other hand, there is a fairly sharp distinction between species on the basis of flower orientation in relation to contrasted classes of pollinators (birds versus insects). The morphological continuum is thus transcended by a functional discontinuity, but this is insufficient justification to revive the sectional names.

The following key provides diagnostic information about the 6 species recognized in *Bruguiera*:

1A. Inflorescence 1 (-2)-flowered; flowers recurved, 2–3 cm long, calyx lobes about 13; sinus between lobes of petals empty or at most occupied by short bristles not exceeding the lobes. Fruit stout, 0.5–2 cm in diameter. Large-flowered species pollinated mainly by birds 2

1B. Inflorescences (1-) 2–6-flowered; flowers oblique or erect, 1–1.5(-2) cm long, calyx lobes about 8(-10); sinus between lobes at petals always occupied by a long bristle exceeding the length of the lobes. Fruit slender, 0.4–0.8 cm in diameter. Small-flowered species pollinated by insects 4

Large-flowered species (Fig. B.71)

2A. Leaves obovate, 5–9 by 3–4 cm. Flowers sometimes in 2's, 2–2.5 cm long, yellowish green. Petals 9–11 mm long, usually without a bristle in the sinus between the petal lobes . . . . *B. exaristata* Ding Hou

2B. Leaves elliptic-oblong, 10–20 by 5–7 cm, usually longer than 10 cm. Flowers 3–4 cm long, reddish. Petals 12–15 mm long, with a bristle in the sinus between the petal lobes 3

3A. Tips of petal lobes blunt, without filamentous appendages . . . . .*B. sexangula* (Lour.) Poir.

3B. Tips of petal lobes acute, each extended into 3 filamentous appendages . *B. gymnorrhiza* (L.) Lamk.

Small-flowered species (Fig. B.72)

4A. Calyx lobes slender, short (less than 3 mm), only one-fourth to one-fifth the length of the calyx, erect or at most slightly spreading in fruit. Petals 1.5–2 mm long . . . . *B. parviflora* Wight & Arn.

4B. Calyx lobes stout, long (more than 3 mm), half the length of the calyx, horizontal or reflexed in fruit. Petals 4–9 mm long 5

5A. Mature flowers 10–12 mm long. Petals 3–4 mm long. Calyx tube at mouth 2 mm in diameter; calyx lobes 8 and completely reflexed in fruit . . . . *B. cylindrica* (L.) Bl.

5B. Mature flowers 18–22 mm long. Petals 7–9 mm long. Calyx tube at mouth 5 mm in diameter; calyx lobes 10 and extended at right angles in fruit . . *B. hainesii* C. G. Rogers

Detailed descriptions are provided for one representative of each group. A complete synonymy is provided by Ding Hou (1958); see also Ding Hou (1957).

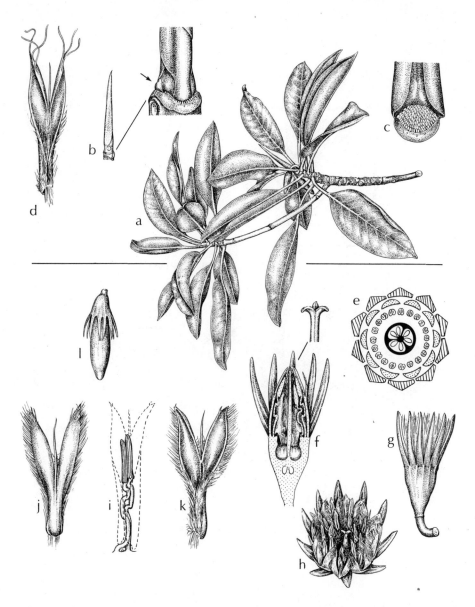

Figure B.71. *Bruguiera* (large-flowered species) (Rhizophoraceae) shoot and flowers. (a–d) *Bruguiera gymnorrhiza*: (a) leafy shoot (× ¼) showing apposition growth; (b) lower: terminal bud (× ½) with expanded leaves removed to show stipule pair; upper: detail (× 3) with stipules removed to show axillary bud (arrow) and palisade of colleters; (c) detail of detached stipule from inside (× 3) to show colleters; (d) petal from inside, partly unfolded (× 3) to show diagnostic features of lobe tips. (e–l) *Bruguiera sexangula*: (e) floral diagram; (f) flower in L.S. (× ⅔), the petals enclose the stamens; inset: detail of stigmas (× 3); (g) flower from side (× ⅔); (h) flower from above (× ⅔), all petals "exploded"; (i) stamen pair (× ⅔) with outline of enclosing petal; (j) petal from abaxial (outer) side (× ⅔); (k) petal from adaxial (inner) side (× ⅔); (l) young fruit (× ½). (*B. gymnorrhiza* cultivated at "The Kampong," Douglas Road, Miami, Florida; *B. sexangula* from Bootless Bay, Port Moresby, Papua New Guinea. P. B. Tomlinson, 30.10.74; from Tomlinson et al. 1979)

345

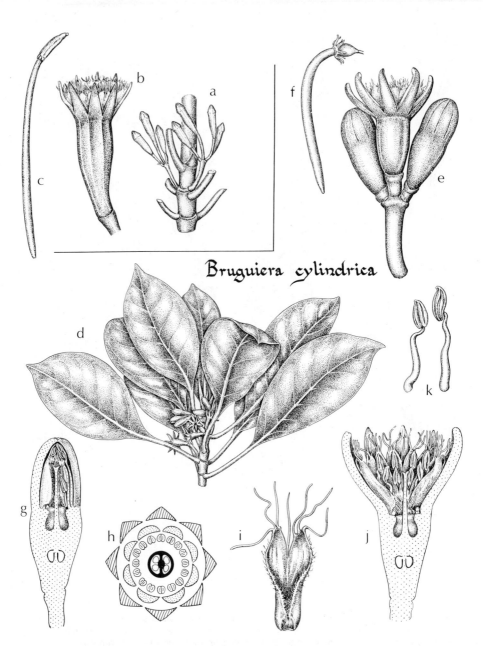

Figure B.72. *Bruguiera* (small-flowered species) (Rhizophoraceae) leaves, flowers, and seedlings. (a–c) *Bruguiera parviflora*: (a) part of shoot with leaves removed (× 1); (b) flower from side (× 3), petals all "exploded"; (c) mature detached seedling (× ½). (d–k) *Bruguiera cylindrica*: (d) flowering shoot (× ½); (e) 3-flowered cymose inflorescence (× 3); (f) mature detached seedling (× ½); (g) unopened flower in L.S. (× 4); (h) floral diagram; (i) petal (× 4) after pollen release; (j) flower in L.S. (× 4) after petals have "exploded"; (k) antepetalous (left) and antesepalous (right) stamens (× 9). (*B. parviflora* from Labu Lagoon, Lae, Papua New Guinea, P. B. Tomlinson 23.10.74; *B. cylindrica* from Tarauma Link Road, Port Moresby, Papua New Guinea, P. B. Tomlinson 5.11.74; from Tomlinson et al. 1979)

346

Figure B.73. *Bruguiera gymnor-rhiza* (Rhizophoraceae). Checkered roughened bark. Hinchinbrook Island, Queensland.

Large-flowered *Bruguiera* species (Fig. B.71)

Bruguiera gymnorrhiza (L.) Lamk. 1797–8　(Fig. B.71a–d)
[*Encycl. Meth. Bot.* 4:696; see also *Tabl. Encycl. Meth.* 2, 1819:517, t. 397, as *Gymnorhiza*]

B. rheedii Blume 1827 [*En. Pl. Jav.*;92]
B. conjugata Merr. 1914 [*Philipp. J. Sci. Bot.* 9:118]

Trees to 30 m or more high with short buttresses. Bark rough, black, fissured in a regular checkered pattern (Fig. B.73). Shoots plagiotropic by apposition, developing terminal short shoots (Aubréville's model). Leaves elliptic-oblong, coriaceous, 8 to 22 by 5 to 10 cm, apex bluntly pointed, base cuneate. Petiole up to 4 cm long, often glaucous with a white wax, as on the young stems, leaves often reddish beneath. Stipules to 4 cm long, often reddish. Flowers solitary, up to 3.5 cm long, reflexed at anthesis, pedicel naked, up to 2.5 cm long. Calyx usually reddish with 12 to 14 lobes that remain erect in fruit. Petals to 15 mm long,

2-lobed with the deep median sinus occupied by a slender bristle, additional 3 to 4 bristles on the acute petal lobes. Stamens enclosed in pairs by the petals. Calyx cup deep, style slender, filiform with 3 or 4 stigmatic lobes, calyx tube scarcely elongating in fruit. Hypocotyl to 25 cm long, cigar shaped, blunt apically, slightly angular.

Ecological and geographic distribution
This tree is a characteristic element of the middle mangrove community throughout its range, extending into the transitional landward communities. Ding Hou (1958) describes it as the "largest and probably the longest lived of the mangrove community." It may grow on exposed shores where the mangrove belt is thin or depauperate.

It has the broadest range of the genus and indeed of all mangroves, extending from East Africa (including Madagascar) through Ceylon and the Malay Archipelago into Micronesia and Polynesia (Samoa) northward to the Ryukyu Islands and southward in tropical Australia (Western Australia and Queensland).

Variation
The red to almost scarlet color of the flowers, which is repeated in the reddish stipules and leaf undersurfaces, is not a constant feature of this species. I have seen scattered individuals without this color among populations of the plant in New Caledonia; the contrast renders them very striking but they do not seem to differ in any morphological feature.

Architecture
Bruguiera gymnorrhiza offers a good example of Aubréville's model and may be contrasted with *Rhizophora* (Attims's model), although there are some close resemblances. The difference between these 2 architectural models leads to important differences in crown shape, however, and particularly in crown plasticity. The sapling axis is orthotropic and with radial symmetry as a result of the bijugate phyllotaxis. It goes through a period during which it remains unbranched, with the length of time depending on the vigor of the sapling. This vigor is reflected by the spacing of the leaves; axes with numerous, close-set leaf scars suggest slow growth; axes with few distant leaf scars suggest rapid growth.

Branching is by syllepsis, as is usual in the family. The branches are markedly plagiotropic and sympodial by apposition. After a limited time of horizontal growth, the axis turns erect and continues its activity as an erect short shoot, producing a rosette of leaves as a consequence of the bijugate phyllotaxis. A renewal shoot arises in the axil of a leaf at the base of the erect shoot; this extends horizontally to repeat the developmental process (see Fig. 4.7). The system is proliferated when 2 instead of 1 renewal shoots develop, the 2 renewal shoots usually diverging at an angle of about 60° to form an irregular hexagonal pattern.

This arrangement is not as regular as in *Terminalia*, which has the same architecture, but one can appreciate that the overall pattern of the branch system (as in *Terminalia*) minimizes the path length and maximizes the effective photosynthetic area. The regularity is not achieved in *Bruguiera* because the phyllotaxis does not permit a precise repetition of leaf positions and the units of extension are not of such uniform length, so that the number of internodes between successive branchings and consequently their spacing are rather variable.

The net result, however, is a young tree with regular tiers of precisely oriented, almost pagodalike branches. This regularity is lost in older trees as units are damaged or broken off. In the process of reiteration the trunk axis is replaced by an existing orthotropic branch axis, so that the mature tree characteristically retains a single trunk. This should be contrasted with *Rhizophora*, which because of its imposed instead of obligate plagiotropy, responds to damage by "dedifferentiation" in the sense of Hallé et al (1978), and several substitution shoots can develop. The crown readily loses its regular shape, and in stressed individuals the process results in shrubby trees, a condition not usual in *Bruguiera*.

The obligate versus imposed plagiotropy in these 2 genera is best appreciated by examining the leader of a vigorous sapling. The pronounced orthotropy of young branches found in *Rhizophora* contrasts with the plagiotropy of *Bruguiera*. This simple difference, reflecting contrasted architectures, leads to a limited plasticity of crown development in *Bruguiera* but a considerable plasticity in *Rhizophora*.

Bruguiera sexangula (Lour.) Poir. 1816　(Fig. B.71e–l)
[*Lamk. Encycl.* 4:262]

B. eriopetala Wight and Arnold 1838 ex Arn. [*Ann. Mag. Nat. Hist.* 1:368]
B. malabarica F.-Vill. (non Arn.) 1880 [*Nov. Appl.*:79]

This species resembles *B. gymnorrhiza* closely, but differs in the somewhat smaller, thinner leaves and most clearly in the absence of long appendages on the blunt petal lobes. Wyatt-Smith (1953a) suggests that the calyx is smaller and more distinctly ribbed. The twigs and petioles lack the white waxy covering that is often characteristic of *B. gymnorrhiza*; unlike that species, the calyx is never a conspicuous scarlet, although it is occasionally reddish. The aerial roots at the base of the trunk, which appear in the sapling, may be quite well developed.

Bruguiera exaristata Ding Hou 1957
[*Nova Guinea n.s.* 8:166, f. 1–2]

It seems remarkable that this distinctive species should have remained overlooked until so recently, but the reason is probably its range (northern Australia, Timor, and south New Guinea) in areas that are poorly explored botanically.

It is distinguished from the 2 other species of the large, solitary-flower group in its appreciably smaller, yellowish green flowers (never reddish), with an 8(-10)-lobed calyx, smaller leaves, and limited stature (not exceeding 20 m). In Queensland it is a common constituent of the landward part of mangal. The most frequent flower visitors are honey eaters.

Small-flowered *Bruguiera* species   (Fig. B.72)

Bruguiera parviflora Wight and Arnold ex Griffith 1936
(Fig. B.72a–c)
[*Trans. Med. Phys. Soc. Cal.* 8:10]

This species is representative of the small, many-flowered group. It is distributed throughout Southeast Asia, from the Malay Peninsula to tropical Australia, the New Hebrides, and the Solomon Islands, but it is not recorded for New Caledonia. It occurs on the inner mangrove fringe and river banks and has somewhat the characteristics of a pioneering species. Good field characters are the light green stipules that form a slender tube and the delicate light green foliage with slender petioles scarcely 2 mm wide.

Tree growing to 24 m with smooth, gray, obscurely lenticellate bark. Leaves 7 to 13 by 2 to 4 cm, elliptic, apex bluntly pointed, base narrowly cuneate. Stipules 4 to 6 cm long. Inflorescence 3- to 4-flowered, peduncle 2 cm long with obscure bracteoles. Flowers yellowish green, erect at anthesis, pedicels to 13 mm long. Calyx tube ridged, to 9 mm long; calyx lobes usually 8, short, that is, less than one-fourth the length of the calyx tube. Petals to 2 mm long, each petal lobe with bristles; each petal enclosing 2 stamens. Style 1 to 1.5 mm long, stigmas 2 or 3. Fruit with calyx tube enlarged 2- to 3-fold, calyx lobes remaining erect. Hypocotyl smooth, to 15 cm long, truncate at the apex, scarcely 5 mm in diameter, at first green and erect, becoming brown and pendulous.

Floral biology
In this species the flowers remain erect and so point distally, that is, outward in the crown. They are visited by small insects including butterflies, which seem capable of the delicate triggering of the explosive pollen release and are probably effective pollinators.

Seedling establishment
The hypocotyl becomes arched a few days after falling to the ground so that its radicular end is usually in contact with the mud (see Fig. 8.7). The cortex at this end becomes stellately expanded and aids in fixing the hypocotyl, facilitating the penetration of the extending radicle, which becomes a taproot.

Growth
This species conforms more closely to Attims's model in contrast to the clear-cut Aubréville's model shown in *B. gymnorrhiza*. Shoots have a strong tendency to remain orthotropic, and the crown of saplings is rather narrow and conical.

Root system

The looping aerial roots that produce pneumatophores are rather poorly or at most inconspicuously developed in this species.

Bruguiera cylindrica (L.) Bl. 1827   (Fig. B.72d–k)
[*En. Pl. Jav.* 1:93]

Rhizophora caryophylloides Burm.f. 1768 [*Fl. Ind.*:109 (excl. var. B)]

B. malabarica Arnold 1838 [*Ann. Mag. Nat. Hist. 1:369*]

Tree to about 20 m high with short buttresses, bark grayish. Leaves elliptic, 7 to 17 by 2 to 8 cm, with a bluntly pointed apex and cuneate base, petiole to 4 cm long, often reddish. Stipules to 3 cm long. Inflorescence usually 3-flowered, peduncle about 1 cm long. Flowers greenish, erect at anthesis. Calyx tube 4 to 6 mm long, 2 mm in diameter, with 8 lobes about as long as the tube. Petals shortly bilobed with 2 or 3 bristles at the apex of each lobe; each petal enclosing a pair of stamens. Calyx tube somewhat enlarged in fruit, the lobes reflexed, detached with the seedling. Hypocotyl up to 15 cm long at maturity, grooved or angled.

Ecological and geographic distribution

The plant usually grows like a small tree in the inner mangroves but is described as specifically characteristic of newly established substrates, being also somewhat susceptible to prolonged submersion. Because of its pioneering propensities, it may form pure stands, but it is also said to be very slow growing.

It is distributed through the Malay Archipelago as far as New Guinea and north Queensland.

Growth and floral biology

This species has the same habit as *B. parviflora*, with which it may occasionally grow. The floral mechanism is also the same and involves explosive pollen discharge after triggering by small insect visitors.

Bruguiera hainesii C. G. Rogers 1919
[*Kew Bull.*:225]

B. eriopetala Wight of Watson 1928 [*Malay. For. Rec.* 6:109]

This species resembles the other small-flowered *Bruguieras* but is distinguished by its larger flowers and the usually 10-lobed calyx, which stands out at right angles to the calyx tube in the fruit. It is somewhat intermediate between the 2 flower types in the genus. The inflorescences are usually 2- or 3-flowered, but solitary flowers occur. In its erect flower position it suggests that it is suited to small insects that are the putative pollinators of *B. parviflora* and *B. cylindrica*, but there are no records of flower visitors.

This species has a wide distribution from south Burma and Thailand through the Malay Archipelago to the Port Moresby region of Papua New Guinea, but it has not been recorded in Queensland. It is rather infrequent in the inland side of mangroves in regions that are not regularly inundated by normal tides.

Ceriops Arnold 1838
[*Ann. Nat. Hist.* 1:363]

> For a full synonymy and citation, see Ding Hou (1958).

Evergreen, small to moderately tall (up to 20 m high) tree with perfect flowers, typically with a distinct trunk and short basal buttresses apparently orig-inating as short basal stilt roots. Pneumatophores sometimes developed as looped surface roots, as in *Bruguiera*. Sclereids absent from all parts. Bark smooth or slightly fissured, pale grayish or white, often with a red tinge, peeling in thick strips from the buttressed portion. Twigs jointed with swollen nodes. Leaves op-posite, blade on the order of 7 to 12 by 3 to 5 cm, with a well-developed terete petiole 1 to 2 cm long; blade ovate to slightly obovate or elliptic-oblong, glabrous, without prominent veins, apex rounded or slightly emarginate, never apiculate, primordia including minute protruding hydathodes on lower surface. Cork warts lacking. Buds 1.5 to 4 cm long, characteristically spear shaped, flattened, the individual stipules thickened dorsally and including copious mucilage drying to a varnishlike consistency, both within the free space of the bud and covering young expanded portions to make them shiny. Inflorescence solitary, axillary with a short (5 mm) or moderately long (1.5 cm) axis and few (2 to 8), rarely many, flowers in compact bifurcating cymes, the flowers and axes protected by thick enveloping bracts and bracteoles. Flowers small (5 mm long) each enclosed immediately in a cup-shaped pair of partially fused bracteoles. Calyx with 5(-6) pointed valvate lobes. Petals 5(-6) free, keeled, 3 mm long with apical appendages and marginal hairs, at first white but becoming brown with age. Stamens twice as many as the petals, either with long filaments and at first enclosed in pairs by the petals (*C. tagal*), or with short filaments and uniformly spaced (*C. decandra*). Disc within the stamen ring well developed, lobes enclosing the base of the stamens. Ovary semi-inferior, calyx tube short; locules 3 with 2 ovules per loculus; style slender and usually with 3 obscure stigmas. Fruits conical mainly by extrusion of the upper part of the ovary, surface brown and roughened, calyx tube warty, lobes persistent. Seedling and ultimately the cotyledonary tube penetrating the fruit. Hypocotyl up to 20 cm long, 5 mm at the widest part, terete or ridged, detached from the cotyledon at the moment of release. In *C. tagal* $n = 18$ (*Taxon* 27:29, Kilwa Rd, Dar-es-Salaam, Tanzania).

Geographic and ecological distribution

There are 2 species widely distributed from East Africa and Madagascar throughout tropical Asia and Queensland, to Melanesia (New Caledonia and the Solomon Islands) and through Micronesia north to Hong Kong. The precise eastern limit is not known. In Melanesia *C. tagal* is recorded only for the very north of New Caledonia (Diohot, e.g., *Schmid 1594* 21.9.66, as *C. timoriensis*) and only from Winton Bay, Malekula (*Gillison RSNH 3556*, 9.10.71), New Hebrides. The species are typical constituents of the inner mangrove, often forming pure stands on better drained sites, becoming stunted in exposed and highly saline sites, but rarely losing their simple trunk. They are easily distinguished from *Bruguiera* and *Rhizophora* by their glossy green, rounded (never pointed or apiculate) leaves, flattened buds, slender and rather short hypocotyls, and pale-barked trunk with restricted flaky-barked buttresses. The clusters of small shiny flower buds or short peduncles encased in varnishlike material are also characteristic.

Species and varieties

Although similar vegetatively, the 2 species can be distinguished in technical details as follows:

1A. Inflorescence axis relatively long and uniformly slender (10–20 by 2 mm). Petals enclosing the stamens in pairs at anthesis and opening explosively, the apex of the petal with 3 clavate appendages. Stamens with long, slender filaments, much exceeding the blunt anthers. Hypocotyl terete or slightly ridged . . . .*C. tagal* (Perr.) C.B. Rob.

1B. Inflorescence axis short and thick, expanded distally (10 mm or less by 3–4 mm). Petals not enclosing stamens at anthesis, the stamens in a single series not opening explosively, the apex of the slightly keeled petals with a fringe of filamentous appendages. Stamens with a short filament, equaled or exceeded by the anther, which ends in a longer or shorter appendage. Hypocotyl sharply ridged . . . .*C. decandra* (Griff.) Ding Hou

Fruit and seedling characters have been used to separate the species as follows:

1A. Hypocotyl long, up to 25 cm or more, warty throughout; calyx lobes in fruit usually widely spreading or reflexed,* calyx smooth. Fruit pointed apically, 15–20 mm long at maturity . . . .*C. tagal*

1B. Hypocotyl not exceeding 15 cm long, warty toward apex; calyx lobes in fruit more or less erect,* calyx lenticulate or warty. Fruit blunt apically, 10–15 mm long at maturity . . . . *C. decandra*

These characters are not always easy to apply because they depend in part on the development of the fruit and seedling. There seem to be no constant vegetative features that distinguish the species. Wilkinson (1981) lists anatomical features of seedlings in which the 2 species differ.

Ceriops tagal (Perr.) C. B. Robinson 1908   (Fig. B.74)
[*Philipp. J. Sci. Bot.* 3:306]

C. candolleana Arn. 1838 [*Ann. Mag. Nat. Hist.* 1:364]
C. timoriensis Domin 1928 [*Bibl. Bot.* 89:444]
C. boiviniana Tulasne 1856 [*Ann. Sci. Nat.* 6:112, see *Fl. Veg. Madagas.*:589, 1974]

Although several specific names have been recognized, current opinion relegates them all to one rather variable entity. The distinguishing features are

---

* In a morphological sense, calyx orientation with respect to gravity varies considerably.

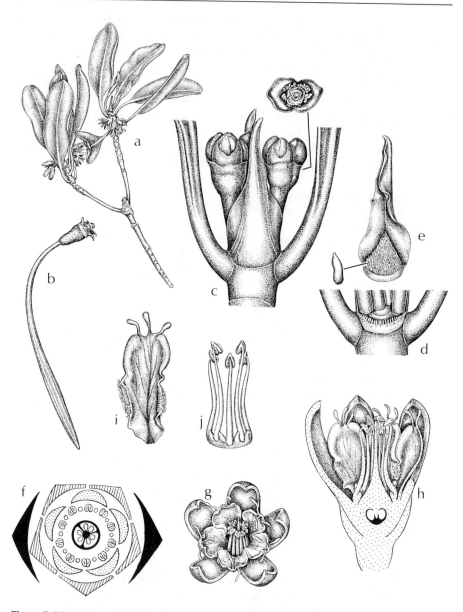

Figure B.74. *Ceriops tagal* (Rhizophoraceae) flower, fruit, and seedling. (a) Flowering shoot
($\times$ ½); (b) mature detached seedling ($\times$ ½); (c) details of terminal bud with expanding inflorescences ($\times$ 3); inset: cupular bracteoles with flower removed ($\times$ 4) to show colleters; (d) detail of
terminal bud ($\times$ 3) as in part a but with stipules removed to show palisade of colleters; (e) stipules
from within ($\times$ 3); inset: single colleter ($\times$ 20); (f) floral diagram; (g) flower from above ($\times$ 6),
petals all ''exploded''; (h) flower in L.S. ($\times$ 6); (i) petal ($\times$ 10); (j) stamens and presumed nectaries
($\times$ 10). (Material from Barune, Port Moresby, Papua New Guinea. P. B. Tomlinson, 21.3.76; from
Tomlinson et al. 1979)

usually flower size and number per inflorescence, which are variable in any population and may relate to the vigor of the tree. In Queensland 2 varieties are recognized by White (1926): *C. tagal* (Perr.) C. B. Rob. var. *tagal* for plants with distinctly fluted seedling hypocotyls and *C. tagal* (Perr.) C. B. Rob. var. *australis* for plants with terete, not fluted, seedling hypocotyls. There seem to be no other distinguishing features. Jones (1971) indicates that var. *australis* has a more southerly range than var. *tagal*.

Ceriops decandra (Griff.) Ding Hou 1958
[*Fl. Males.* Series 1, vol. 5(4):471]

C. roxburghiana Arn. 1838 [*Ann. Mus. Nat. Hist.* 1:364]

Ecological distribution

Ceriops tagal is typically a plant of the inner mangroves, apparently on better drained soils; flooding with salt water is infrequent but produces high soil salinity by evaporation. Under these circumstances it will form pure stands. *Ceriops decandra* more often grows within the tidal zone mixed with other Rhizophoraceae. The 2 species differ in the major features of floral biology, as is discussed later; otherwise, they are sufficiently similar that a general account of vegetative features illustrates both. In Queensland *C. decandra* seems to have larger and darker leaves, but this may be a consequence of habitat differences.

Seedling and sapling

Dispersal and establishment of seedlings follow the pattern usual for the Rhizophoreae. The sapling is initially unbranched, but once branching begins it may be continuous or intermittent, depending on the vigor of the plant. Branching is by syllepsis, with the lateral axes developing in pairs within the terminal bud. This pattern is repeated on successive orders of branches but without any marked tendency to develop plagiotropy by apposition, as in *Rhizophora* and *Bruguiera*. Consequently the crown remains relatively narrow. Vegetative branching becomes less frequent on distal axes. This minimizes the congestion of axes.

The terminal buds of *Ceriops* are distinctive in their construction. The individual stipules, which are shorter than those in the other Rhizophoreae, have a marked dorsal thickening so that the bud as a whole is flattened. There is much free space even within this flattened bud, since only a single younger pair of leaf primordia or stipules is present at any one time. In addition the fluid, presumably secreted by the stipular colleters, dries to form a semisolid plug of material that occupies the free space within the bud. The same material coats the surface of the newly extended internodes and young appendages in the manner of a varnish that flakes and falls off as the organs expand.

With the onset of flowering, which occurs in quite small saplings, the full complement of lateral appendages developed by the terminal bud is present so that each leaf may subtend the following:

1. A single resting bud (the most usual condition)
2. A single sylleptic branch, which is commonly associated with a supernumerary bud between branch and petiole
3. An inflorescence, which also may be associated with a supernumerary bud

Internodes on distal shoots are shortened, characteristically in relation to flowering.

Architecture

Ceriops has an architecture close to that shown by Attims's model, since axes, at least in early stages of development, show the potential for continuous growth. Rauh's model is suggested by the peripheral axes, however, which branch infrequently and intermittently rather than rhythmically because the growth of shoots is essentially continuous. In early stages of crown development, there is much replacement of damaged axes by the direct substitution of an existing lateral (sylleptic) shoot for the damaged leader, but prolepsis can occur, though seemingly as a minor mechanism of repair. There is no basal sprouting.

Phenology

In seasonal climates, as in north Queensland, there is a clear restriction of flowering to summer; the flower buds develop the previous winter. Fruiting follows from these flowers in the next winter, with seedlings on the tree in late summer (February–March). This suggests that seedling development takes about a year, but some fruits form in late summer over winter. This extends the seedling season, since embryos of 2 summers mature at the same time.

Less evident periodicity of leaf development occurs simply by fluctuations in the time of leaf production according to climate (temperature). Otherwise leaf expansion can occur at any time of the year.

Inflorescence structure

The lateral inflorescence is a modified cyme with regular dichotomy of axes; the plane of orientation changes by 180° at each successive fork. Axes are enclosed by thick bracts, and each flower, which is the ultimate product of branching, is at first enclosed by a pair of bracteoles. Irregularities occur in the system by the abortion of axes or flowers. Flower number is low (2 to 8); *C. decandra* usually has fewer flowers, since its slightly larger flowers are congested on a short inflorescence axis. However, populations of *C. decandra* in north Queensland may have larger flower clusters with up to 40 flowers per cluster.

Floral biology

The 2 species differ strikingly in their floral mechanism. Most information about pollination is available for *C. tagal*.

In *C. tagal*, the flower buds appear to open mainly in the evening, emitting a faint but fragrant odor, and they may remain open the following day. Anther dehiscence occurs in bud, and at anthesis the petals are closed, enveloping the stamens in pairs exactly as in *Bruguiera*. Pollination is probably mainly by night-flying insects; moths have been observed visiting the flowers, presumably for the small quantity of nectar secreted by the disc, but bees may be daytime visitors. Pollen release is explosive and is triggered by a delicate touch of the petals. This would presumably throw pollen onto a suitably positioned flower visitor. The tension that sets this mechanism is generated by the enclosed stamen pair held back by the pouched petal.

In *C. decandra*, the stamens are not enclosed at anthesis, and they retain their antesepalous or antepetalous position without displacement. Antepetalous stamens are appressed to the subtending petal, but there is no explosive release of pollen; the short filament is not suitable for developing tension. Details of flower visitors are lacking for this species, but the floral mechanism is apparently much less specialized than that of *C. tagal*.

Root architecture

Development of an above-ground system is less conspicuous in *Ceriops* than in *Bruguiera* but follows the same pattern. Buttresses and aerial roots are restricted to the base of the trunk; the aerial roots appear first on seedlings, becoming the buttresses of the mature tree (Fig. B.75A). These in turn may continue to develop short descending aerial roots so that the base of the trunk comes to be supported by a series of narrow flanges that form a conical structure enclosing a central hollow into which further aerial roots develop (Fig. B.75B). On more exposed and presumably better drained sites, this may be the sole elaboration of above-ground roots, and the relative instability of the system is demonstrated because dead trees are easily pushed over. On wetter sites, where trees are presumably more vigorous, pneumatophores of the *Bruguiera* type are developed and well-developed projecting knobs can be seen, initiated as loops in the horizontal root system. This plasticity of the *Ceriops* root system has not been appreciated by several workers so that there are reports of it lacking pneumatophores.

Kandelia Wight and Arnold 1834
[*Prodromus*:310]

A monotypic genus, ranging from the Ganges Delta, Burma, through Southeast Asia to south China, the Ryukyu Islands, and south Japan. In the southern part of its range it is confined to northeast Sumatra and northern Borneo, an unusually restricted southern limit for mangrove Rhizophoreae. It is distinguished from related genera in the Rhizophoreae by the slender calyx lobes, the numerous, indefinite number of stamens with long slender filaments, the multifid, glabrous petals, the essentially unilocular ovaries, and the absence of either buttresses or specialized aerial roots. The hypocotyl of the seedling is characteristically slender and tapered at each end.

A

B

Figure B.75. *Ceriops* cf. *tagal* (Rhizophoraceae). (A) Development of a stilt-root system in the sapling. Arrow indicates level of original hypocotyl, showing that most roots are adventitious. (B) Mature root system. Dead plant at left shows the flanged, flattened aerial roots. Townsville, Queensland. (A from a color transparency by A. M. Gill)

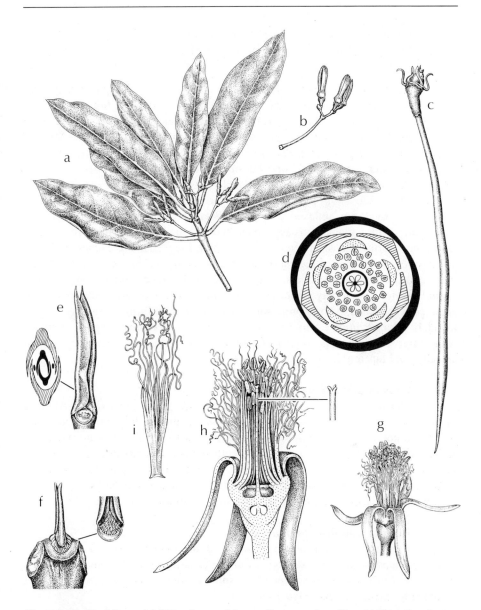

Figure B.76. *Kandelia candel* (Rhizophoraceae) leaves, flowers, and seedling. (a) Flowering shoot (× ⅓); (b) inflorescence (× ½); (c) mature detached seedling (× ⅓); (d) floral diagram; (e) terminal bud represented by enveloping stipule pair (× 3/2), inset: vernation of vegetative parts at one node; stipules crosshatched, leaves and branches solid; (f) bud in part e with stipules removed (× 3), inset: detail of detached stipule from within to show colleters; (g) flower from side (× 3/2); (h) flower in L.S. (× 3), inset: detail of stigma (× 6); (i) petal (× 4). (Material from Semetan, Sarawak. P. Chai *s.n.*; from Tomlinson et al. 1979)

Kandelia candel (L.) Druce 1914   (Fig. B.76)
[*Rep. Bot. Exch. Club Brit. Isles* 1913(3):420]

K. rheedii Wight and Arnold 1834 [*Prodromus*:311]

Small understorey tree growing to 7 m. Buttresses and pneumatophores absent. Bark smooth, grayish or reddish brown. Leaves opposite, bijugate, 6 to 13 by 2.5 to 6 cm, oblong-elliptic to narrowly elliptic-lanceolate, apex obtuse, base cuneate, margin entire, somewhat reflexed. Petiole 1 to 1.5 cm long, terete. Stipules to 2 cm long, overlapping as in *Rhizophora*. Inflorescence axillary, bifurcating with 4 or more flowers, peduncles 2 or 3 but up to 5 cm long, slender. Bracteoles cup shaped, the pair below the flower either enclosing the calyx cup or separated from it by a pedicel 2 to 3 mm long. Flowers 1.5 to 2 cm long, white; calyx lobes 5(-6) linear, 15 by 2 mm, reflexed at anthesis. Petals 5 (-6) to 15 mm long, scarcely 1 mm wide, with a filamentous appendage between the lobes, the lobes themselves with 3 or 4 further linear appendages 5 to 6 mm long. Stamens numerous (30 to 40) inserted on the rim of the calyx cup, filaments uneven in length, 8 to 15 mm long, anthers minute. Floral disc cup shaped, enclosing the shallow nectar cup. Ovary inferior, essentially unilocular but with evidence of 3 carpels; ovules 6, anatropous, attached to basal projecting placenta. Style filiform, 1 cm long, stigma minutely 3-lobed. Fruit 1.5 to 2 cm long with persistent reflexed sepals, peduncle elongating. Seedling hypocotyl up to 40 cm long at maturity, slender, pointed apically, and tapered at each end.

Growth and reproduction

The architecture in the sapling stage corresponds precisely to Attims's model, with continuous growth and continuous to diffuse branching. The floral biology is not well known but seems to be the least specialized of all mangrove Rhizophoraceae, since the flowers apparently attract a diversity of small flying insects. Petals and stamens are ephemeral. In its floral biology it most resembles *C. decandra*, although its flowers are much larger. The generalized floral construction of this species may represent the ancestral condition in the tribe.

Ecology and distribution

Kandelia occupies a narrow niche in the mangrove forest, since it is nowhere abundant and typically occurs in the back-mangrove communities or on the banks of tidal rivers farther inland. It still has some pioneering propensity, as recorded by Steup (in Ding Hou 1958).

Family: Rubiaceae

A very large, cosmopolitan, and natural family with about 500 genera and over 6000 species represented in the tropics mainly by woody plants. Interpetiolar stipules with associated glandular structures (colleters) are a diagnostic feature; they protect terminal buds and usually persist on mature stems. The family includes several commercially important species, notably coffee.

Scyphiphora is a frequent but minor mangrove constituent in the Old World; a mangrove associate from the New World is also described.

Scyphiphora Gaertn.f. 1805
[*Fruct. Sem. Pl.* 3:91, t. 196, Fig. 2]

A monotypic genus with a range from south India and Ceylon, Indochina and Hainan, through the Malay Archipelago and Philippines to tropical Australia and New Caledonia, northward to the Solomon Islands and Palau. It is an uncommon constituent of mangroves and more exposed coastal sites including beaches.

Scyphiphora hydrophyllacea Gaertn.f. 1805   (Figs. B.77, B.78)
[*Fruct. Sem. Pl.* 3:91]

Ixora manila Blanco 1837 [*Fl. Filip*: 635]

An erect shrub, rarely exceeding 2 m. Leaves opposite, decussate, branching mainly by syllepsis but discontinuous (Attims's model, Hallé et al. 1978). Leaves glabrous, somewhat fleshy or coriaceous, oblong-ovate, 4 to 9 by 2 to 5 cm, acute at base, apex rounded to bluntly pointed or slightly emarginate, margin entire; petiole 1 to 2 cm long. Interpetiolar stipules short (to 3 mm), rounded, minutely hairy on the margin, at first enclosing younger parts; with well-developed colleters internally at the base. Flowers perfect, protandrous, usually tetramerous, in axillary regular, usually condensed cymes with 3 to 7 (up to 13) flowers, bracteoles obscure. Peduncles 2 to 15 mm long; flowers sessile or shortly but obscurely pedicellate. Calyx tube glabrous, 3 to 5 mm long, free portion of calyx scarcely 2 mm long entire with 4 (-5) obscure teeth. Corolla tube 3 to 4 mm long with 4 (-5) at first contorted white (or slightly red), bluntly pointed petal lobes, reflexed at anthesis; throat of corolla tube occluded with dense hairs. Stamens 4 (-5) inserted on the mouth of the corolla tube with a short (1 mm) filament, anthers 2 mm long, medifixed and dehiscing introrsely before the flower opens. Ovary inferior, indistinctly separate from calyx tube and pedicel; 2-locular, with 2 ovules in each cell, 1 above the other, the upper ovules erect, the lower pendulous. Style slender with a club-shaped bilobed stigma, the lobes eventually spreading. Fruit green, glabrous, up to 1 cm long, crowned with persistent calyx, with 8 (-10) longitudinal ridges. Outer layers of fruit fleshy, inner corky. Seeds 4 (-or fewer) separated horizontally

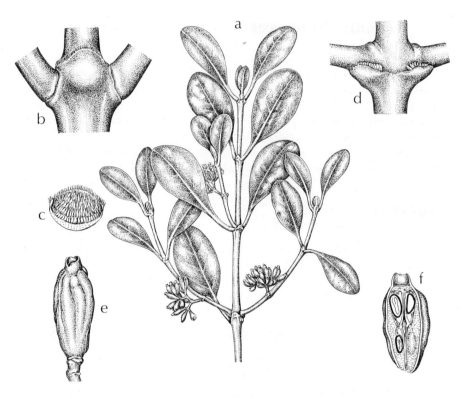

Figure B.77. *Scyphiphora hydrophyllacea*. Shoot and fruit morphology. (a) Distal shoot ($\times$ ½) with flowers and young fruits; (b) detail of node ($\times$ 4) with interpetiolar stipules; (c) stipule ($\times$ 4) detached to show colleters on inner face; (d) first node on inflorescence ($\times$ 4) to show bracts with internal colleters; (e) fruit; (f) fruit in L.S. ($\times$ 3) with 4 seeds (cf. Fig. B.78c). (Material from Barune, Port Moresby, Papua New Guinea. P. B. Tomlinson, 31.10.74B)

by an incomplete septum; each with a straight embryo, endosperm present, testa thin. Germination hypogeal.

Growth and reproduction

Vegetative development. The terminal buds are protected by the hoodlike stipules that secrete a copious varnishlike substance (presumably originating in the palisade of colleters internally at the base of each stipule pair) that forms a plug enclosing the bud. This material also coats the exposed surfaces of younger organs, especially inflorescences. The stipules are persistent. Branches are wide-angled and sylleptic, inserted a little above the node so they characteristically extend from above the mouth of the stipules. Growth is seemingly intermittent but irregular. There are apparently no supernumerary buds so that reserve buds occur only at nodes that do not develop sylleptic branches. This restricts the reiterative capacity of the plant.

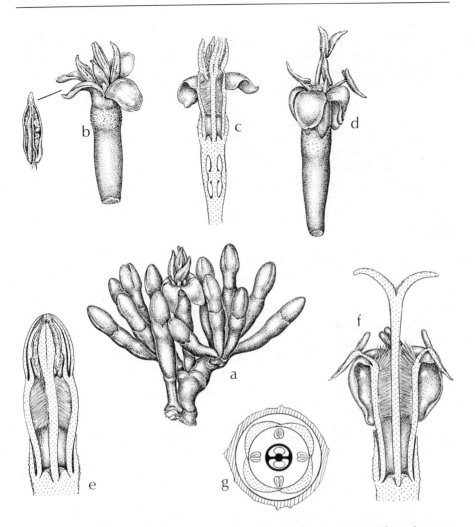

Figure B.78. *Scyphiphora hydrophyllacea* flowers. (a) Flowering branch ($\times$ 4); (b) flower just open-ing ($\times$ 4), inset: detail of anther; (c) flower in L.S. ($\times$ 4); (d) flower at late stage ($\times$ 4), stamens reflexing stigma lobes diverged; (e) flower bud in L.S. ($\times$ 9); (f) flower at late stage ($\times$ 9); stigmas diverged, stamens reflexed; (g) floral diagram. (Material from Barune, Port Moresby, Papua New Guinea. P. B. Tomlinson, 21.3.76V)

Floral biology. The flowers are in fairly regular compound but congested dichasia, but the order of flowering does not follow their age very precisely. Flowers in one head are often synchronous in development. The anthers dehisce introrsely before the flower opens (Fig. B.78e). The large and sticky pollen grains form a mass that is retained between the club-shaped stigmas and the stamens, being prevented from falling into the corolla tube by the weft of stiff hairs at the mouth of the corolla. The petals reflex, exposing the still-closed, club-shaped stigmas;

pollen is further exposed by elongation of the style (Fig. B.78c). The shriveled anthers fall back with the petals and so do not impede access to the corolla tube. Nectar is secreted by the glandular disc at the base of the corolla tube; it is accessible to short-tongued insects. Pollen is picked up either from the stamens themselves or from the back of the stigmas (much as in the related genus *Ixora*). The stigma lobes now diverge (Fig. B.78f) and presumably present their receptive inner surface to a visiting insect transporting pollen from another flower. Eventually the corolla lobes close again and the style falls to one side of the tube. Pollen adherent to the outside of the style may promote selfing.

Fruit biology. Incidence of fruit set is high (almost 100 percent) and this suggests either self-compatibility or a very efficient pollination mechanism. There are reports of a low percentage of seed germination, however. The fruit turns brown at maturity as or before it is shed; it floats because of the spongy layers of the inner fruit.

Nomenclature and taxonomy

The genus has been regarded as isolated since it was erected by Gaertner. It is included by Schumann (1891) in Cinchonoideae-Gardeniinae-Gardenieae but is distinguished by its evergreen habit, cymose inflorescence, 2 ovules per loculus with their characteristic orientation, few-seeded, drupelike fruit with smooth (not hairy) seeds, and marine habitat.

Merrill (1918) suggests that the city of Manila in the Philippines takes its name from this plant, whose vernacular Tagalog name is reported by Blanco as *nilad* or *nilar*, from which he took his specific name *manila*, literally "the place where *nilar* grows."

Gaertner used the spelling *hydrophylacea* in his original description, but *hydrophyllacea* is orthographically correct (from *hydro*, water, and *phyllon*, leaf).

Rustia Klotsch in Hayne 1846
[*Arzn. Gew*. 14, sub. t. 14 and 15]

A genus of 12 species in tropical America and the West Indies.

Rustia occidentalis (Benth.) Hemsl. 1879
[*Biol. Cent. Am. Bot*. 2:14]

This species has a coastal distribution and may be found as a mangrove associate. It is a shrub or small tree growing to 15 m with terminal panicles of pale violet flowers that produce 2-valved capsules. The petiolate leaves are large (15 to 25 by 5 to 15 cm), ovate-lanceolate, in opposite pairs, but with inconspicuous interpetiolar stipules.

Family: Rutaceae

A large, cosmopolitan family with over 1000 species in several genera. Many species are the source of aromatic oils. The family includes the genus *Citrus*.

Merope Roem. 1846
[*Syn. Monogr.* 1:44]

A monotypic genus.

Merope angulata (Willd.) Swingle 1915
[*Bull. Wash. Biol. Soc.* 7:22–23]

A shrubby, low tree or even scrambling plant with axillary, often paired, spines. Leaves alternate, simple, coriaceous, aromatic with pellucid dots; blade 7 to 16 by 3 to 7 cm; petiole short, slender, unwinged. Flowers white, axillary, pentamerous, with 10 stamens; ovary 3-locular, with 4 ovules in each loculus. Fruits triangular in section, 2 to 3 cm long with large long, flattened seeds.

This species seems restricted to back mangal and river banks but is rather scattered. It occurs throughout Malesia. It has a somewhat isolated position within the family, but interest has been expressed in it as a possible salt-tolerant root stock for *Citrus* (Swingle 1943).

Family: Sapindaceae

A large tropical family of about 150 genera and 1500 species of trees, shrubs, and vines, usually with compound, exstipulate leaves. The following species is recorded as a mangrove associate in the Asian tropics.

Allophyllus L. 1747
[*Fl. Zeyl.*:58; *Sp. Pl.* 1753:348]

Allophyllus cobbe (L.) Bl. 1847
[*Rumphia* 3:131]

See also Corner (1939). Gardens Bull. Straits Settlements 10:38.

A species with a wide range from India, Ceylon, and Malesia to New Guinea, but not the Philippines. It is variably recorded as a small tree but more usually as an understorey shrub and commonly as a climber, a common habit in the family. It is recognized by its trifoliate leaves with stout woody petioles about 4 cm long, the leaflets ovate, up to 15 cm long, with incised margins, the apex blunt to pointed. Flowers are small, about 2 mm long, white, in slender axillary spikes 8 mm long. Fruits are 5 mm or more in diameter, in hanging bunches, green turning red and fleshy; they are described as edible.

The species also has a wide range, both ecological and altitudinal, in inland communities and is somewhat variable in its vegetative parts. Radlkofer (1932) divided it into numerous species, but Corner (1939) prefers to treat it as one variable species but still creates a number of varieties, which do not necessarily correspond to Radlkofer's species. Four of Corner's varieties have a coastal distribution, at least in part: var. *limosus* (with lanceolate leaflets 2 to 6.5 cm wide, becoming glabrous with age) is restricted to back mangal; var. *marinus* (with larger ovate glabrous leaflets, 4 to 10 cm wide and whitish twigs) is said to be restricted to rocky and sandy shores. The flowers arise as condensed cushions on the spike and have greenish white sepals; stamens protrude at anthesis and render the flower relatively conspicuous.

Family: Sapotaceae

A mainly tropical, woody family with about 40 not well circumscribed genera and 600 species, commonly with a milky latex.

Pouteria Aubl. 1775
[*Pl. Guia*. 1:85]

A large genus of over 300 species, not clearly distinguished from a number of genera that are often included. The following species, with a wide distribution, is sometimes recorded in back-mangal communities.

Pouteria obovata (R. Br.) Baehni 1942
[*Candollea* 9:423]

Planchonella obovata H. J. Lam 1925 [*Sap.* 209 etc., Fig. 58]
See Baehni for a full synonymy.

Recorded as a tree to 30 m but usually smaller, exuding white latex but not copiously. Leaves simple entire, obovate, to oblong-lanceolate, 8 to 12 by 3 to 5 cm, attenuate basally to a short petiole (2 to 3 cm long), the apex bluntly pointed or rounded. Surface of young twigs, petiole and lower leaf with a reddish brown tomentum. Flowers in axillary clusters, often on older wood; scent fetid. Fruit rounded, about 1 cm long, a 1- (2 to 3) seeded drupe with a brown tomentum.

The species is uncommon in back mangal and has a range from India to south China, the Malay Archipelago, the Philippines, New Guinea, and tropical Australia. In Sarawak it is recorded inland on limestone to altitudes of 250 m.

Family: Sonneratiaceae

A small tropical family of 2 small genera in the Indo-Malayan region. *Sonneratia* is restricted to mangrove communities; it has essentially solitary flowers, numerous stamens, vestigial or no petals, a fruit that does not dehisce regularly, and seed without extended tails. *Duabanga* includes 2 species in lowland tropical rain forest and has a more restricted range. It is distinguished from *Sonneratia* by its several-flowered inflorescence, conspicuous petals, 12 (or more) stamens, and especially the capsular fruit, dehiscing by 4 or more valves to release the numerous tailed seeds; vegetatively it lacks the conspicuous pneumatophores found in *Sonneratia*.

The family is considered close to Punicaceae, Crypteroniaceae, and especially Lythraceae; similarities between seedling *Sonneratia* and species of *Lythrum* are particularly striking. The following account is based on that by Backer and van Steenis (1954), where a complete synonymy is given.

Sonneratia L.f. 1781
[*Suppl. So. Pl.*:38, a conserved name]

A genus of 5 to 6 species, ranging from East Africa through Indo-Malaya to tropical Australia and into Micronesia and Melanesia. It is a typical constituent of mangrove communities throughout its range, often forming a seaward fringe, and it is recognized by its tall conical pneumatophores arising from horizontal roots, its opposite, simple, orbicular leaves, large flowers, and large, globose fruits with persistent calyx. It is sometimes referred to as the "firefly mangrove" because these insects congregate on the trees at night. Trees are relatively uniform in their vegetative features; the following account describes these for one representative species. Where vegetative differences are known, they are included in the key.

Tentative key to Sonneratia species

1A. Calyx 4 (-6) lobed, ovary and calyx not exceeding 2 cm long, ovary 5–8-celled, stigma expanded, mushroomlike; fruit 1.5–2 cm in diameter. Leaves narrow (less than 5 cm wide), but sometimes wider, gradually tapering toward the apex .... *S. apetala* Buch.-Ham.

1B. Calyx usually 6–8-lobed, ovary and calyx 2.5 cm or longer, ovary 14–21-celled, stigma capitate but not expanded; fruit 3–7 cm in diameter. Leaves usually 5 cm wide or wider, abruptly narrowed to the rounded or even emarginate apex 2

2A. Calyx flat, extended horizontally, not enclosing the ripe fruit, at most obscurely ribbed 3

2B. Calyx cup shaped, enclosing the base of the fruit, prominently ribbed 4

3A. Twigs not pendulous. Leaves obovate to suborbicular, petiole scarcely developed, midrib green throughout; veins conspicuous, prominent on the upper blade surface; petals absent; filaments white........ *S. griffithii* Kurz

3B. Twigs slender, pendulous. Leaves elliptic-oblong or oval-obovate to narrowly elliptic, petiole short, midrib often red at base. Veins inconspicuous, not prominent; petals usually present, filaments red below, white distally ....*S. caseolaris* (L.) Engl.

4A. Apex of fruit not depressed at base of style. Tube of the fruiting calyx smooth, lobes reflexed, petals present. Leaves ovate to oblong-ovate, with a short thick petiole.. *S. alba* J. Smith

4B. Apex of fruit depressed at base of style. Tube of the frutiing calyx finely warted, lobes ascending, petals absent. Leaves broadly ovate, as broad as long, with a distinct, narrow petiole ....*S. ovata* Backer

Wyatt-Smith (1953b) gives field characters for the 3 common species in Malaya.

Sonneratia caseolaris (L.) Engler 1897   (Figs. B.79, B.80)
[In Engler and Prantl, *Nachtr.*:261]

S. *acida* L.f. 1781 [*Suppl. Sp. Pl.*:252]
S. *lanceolata* B1. 1851. [*Herb. Lugd. Batav.: 567]*

Trees to 15 m with continuous growth but diffuse branching (Attims's model), branches horizontal or drooping. Leaves glabrous, opposite, without stipules, shortly petiolate to almost sessile. Adult leaves broadly ovate or obovate, usually with a blunt apex, somewhat fleshy with an entire margin, 3 to 5 by 4 to 7 cm, juvenile leaves lanceolate, less fleshy, often with extended, red petioles. Apex in most leaves with a minute recurved spiculus serving apparently as a domatium rather than a hydathode. Nodes (including bracteolar nodes) with 2 lateral pairs of circular glands, 1 pair on each side of the stem. Axis angled, usually almost square in transverse section. Flowers ephemeral, opening in late evening and lasting 1 night. Flowers solitary or in few-flowered dichasia, terminal at the ends of outer twigs or in the axils of distal leaves with at least 1 pair of subtending bracteoles. Flowers with a shallow, green calyx tube with 6 to 8 somewhat longer valvate teeth; petals slender, reddish, alternate with calyx teeth (or absent), inconspicuous, and best seen in flower bud. Stamens numerous, inserted on the inner rim of the calyx tube, filmaents slender, 2 to 3 cm long, white above, reddish below, anther bilocular, medifixed. Ovary globose but flattened or even depressed above, style simple, about twice the length of the stamens, with a capitate stigma; ovary with numerous (up to 20) locules, each with numerous ovules on essentially axile placentas. Fruit a green, somewhat leathery "berry" with persistent subtending calyx;

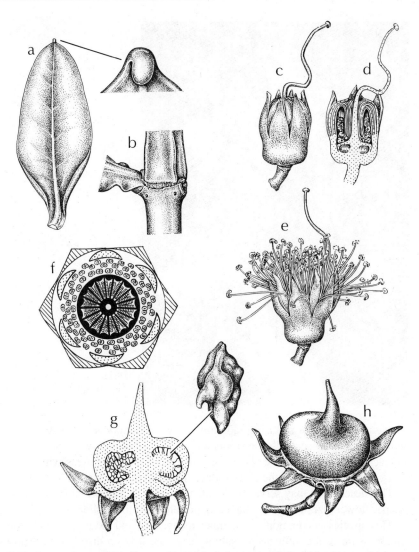

Figure B.79. *Sonneratia caseolaris* (Sonneratiaceae) leaf, flower, and fruit. (a) Leaf incompletely unrolled to show vernation (× ½), inset: detail of recurved leaf tip (× 9); (b) node (× ³⁄₂) with one leaf removed; (c) flower bud (× ½); (d) flower bud in L.S. (× ½); (e) flower at early anthesis (× ½); (f) floral diagram; (g) fruit in L.S. (× ½), inset: detail of seed (× 9); (h) fruit (× ½). (Material from Semetan, Sarawak. P. B. Tomlinson, 19.7.82, and P. Chai, *s.n.*)

seeds numerous in the fleshy pulp of the placenta, the individual seeds irregular and angular. Germination epigeal. Root system including an extended series of cable roots giving rise to narrow, shallowly descending lateral roots and erect pneumatophores. Pneumatophores at first greenish gray with a flaky back, extending

Figure B.80. *Sonneratia caseolaris*
(Sonneratiaceae). Spent and opening
flowers. (From a color transparency
by A. M. Juncosa)

to as much as 2 m at maturity, tapering, woody via secondary thickening and with
numerous narrow second-order roots developing horizontally in the substrate. The
characteristic habitat of the species is river banks and tidal areas with mud banks.
It ascends the mouth of the river farther than other species.

This description may serve for the other species, which are distinguished by leaf
shape, flower size, and fruit shape, as presented in the earlier key.

This species is common in the inner mangroves and extends inland along tidal
creeks as far as the influence of salinity extends. It has a distribution from Ceylon
throughout Southeast Asia, the Malay Archipelago and Philippines, north to Hainan,
east to the Solomon Islands and the New Hebrides (but not New Caledonia), and
into tropical Australia.

This species is distinguished by the large fruits in which the calyx lobes expand
horizontally (not erect) in fruit and the red petals, which contrast with the greenish
or yellowish white of the surface of the sepals. The leaves do not have prominent
veins and are usually pointed at the apex. The seeds, which tend to float and not
sink, are smaller than those of *S. alba* and *S. ovata*.

Soneratia lanceolata has been distinguished from *S. caseolaris* by its narrower
leaves, white stamens and uniformly round (not constricted) flower buds.

Sonneratia alba J. Smith 1819
[*Rees Cycl.* 33(no. 2)]

S. griffithii Watson 1928 [*Malay. For. Rec.* 6:120, 121, f. 24]

This is said to be the most widely distributed species; it ranges from east Africa and Madagascar to Southeast Asia, the Malay Archipelago to the Philippines, and tropical Australia to Micronesia, the New Hebrides, and New Caledonia. According to Chai (1982) it is a pioneer species, colonizing newly formed sandy mud flats in sheltered situations.

It is distinguished by the numerous white petals that are tinged red apically and the cupular calyx under the fruit with smooth (not hairy) lobes that are usually reflexed in a characteristic manner. The leaf is obovate to oval, with the apex broadly rounded and even emarginate. The seeds are falcate and smooth.

Sonneratia apetala Buch.-Ham. 1800
[*Symes Embassy Ava* 3:477]

This species is perhaps the most distinctive in the genus but is the least common, being restricted to southern India and Burma. It is recorded as very rare in Ceylon, with a population of only 6 trees near Muttur in the estuary of the Koddiyar River (MacNae and Fosberg 1981). It is a small to medium-sized tree growing to 15 (-20) m high with narrow leaves gradually tapering toward the apex. The flowers (1.5 to 2 cm) are consistently smaller than those of the other species, and the usually 4 calyx lobes are twice as long as the tube. The ovary is 5- to 8-celled, and the stigma is broad and mushroom shaped at anthesis. The fruit (1 to 2 by 2 cm) is proportionately smaller than in the other species; the calyx lobes remain flat and do not enclose the fruit.

The specific name is not entirely appropriate, since other species of *Sonneratia* are apetalous, whereas apetaly occurs in flowers of populations of species that may otherwise retain petals.

Sonneratia griffithii Kurz 1871
[*J. Asiat. Soc. Bengal* 2:56 in the key]

This species has a restricted distribution along the shores of the Andaman Sea, northward to Bengal and southward to the upper Malay Peninsula. Though described as locally common, it is little collected. It is close to *S. caseolaris*, from which it is said to differ in the prominent leaf veins and absence of petals.

Sonneratia ovata Backer 1929
[*Bull. Jard. Bot. Buitenz.* 3(2):329]

S. alba of several authors [e.g., Watson 1928, *Malay. For. Rec.* 6]

This species has a distribution from Thailand through the Malay Peninsula and Malay Archipelago (but not Borneo) to the Gulf of Papua in New Guinea and into Queensland. It is recorded by Chai (1982) as an inland river-bank species but still within salt-water influence. It lacks petals and is distinguished by the finely warty calyx, the lobes of which are red on the inner side, and by the erect calyx segments in fruit.

General notes on *Sonneratia*

Heteroblasty

Leaf shape in *Sonneratia* is variable but is diagnostically useful. Figure B.81 gives an indication of the range of variation in the 4 most common species. The phenomenon of leaf dimorphism between seedling and adult plants is pronounced in *Sonneratia* species, especially in *S. caseolaris* (Fig. B.81) and to a lesser extent in *S. alba*. Leaves on seedlings (and sometimes on larger plants below tidal levels) are narrowly lanceolate (12 to 13 by 1 to 5 cm). Chai (1982) refers to the condition as ''heterophylly'' but perhaps ''stenophylly'' (adaption of leaf shape to current flow in rheophytes – plants of moving waters) is the best term.

Pollination

The floral mechanism and floral biology in relation to pollinators are distinctive and have been much discussed. Flower buds enlarge rapidly and open in the early evening, the style at first projecting. The stamens are recurved but expand abruptly as the calyx segments diverge. Collectively they render the open flower conspicuous. At the same time a sour, buttery odor is emitted. The most frequently noted visitors are bats (e.g., van der Pijl 1936), which drink the copious nectar that fills the calyx cup. In west Malaysia several species of nectarivorous bats are responsible for the pollination of *Sonneratia*. *Macroglossus minimus* Geoffroy (Fig. 7.2A) is said to be dependent on *Sonneratia* as its major food source. It has never been recorded away from mangroves. Where bats are uncommon, hawk moths are an alternative visitor (Primack et al. 1981). Both kinds of pollinators touch both stamens and stigmas in their search for nectar, the bats in particular being clumsy in their approach to the flowers. It is not known whether flowers are self-compatible, but the seed set is always high in developed fruits. Stamens and petals fall from the flowers within 12 hours of opening.

The relationship between bats and flowers in west Malaysia has been extensively investigated (Start and Marshall 1976), and it has been shown to involve fruiting of the commercially important durian (*Durio zibethinus* Mull.). Durian is strongly seasonal in its fruiting; the main season is July following an extensive flowering in April. Durian flowers are pollinated almost entirely by a single species of bat, *Eonycterus spelaea* Dobson (Fig. 7.2B), a widely distributed nocturnal bat that roosts primarily in large limestone caves, for example, Batu caves in Selangor. The bats are extremely fast flyers and may range up to 50 km from their roosts each night in search of food in the form of pollen and nectar from a diversity of plants.

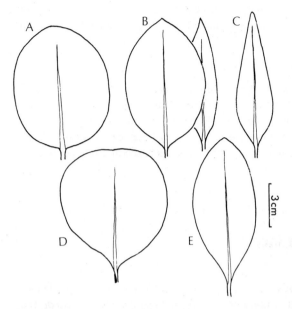

Figure B.81. *Sonneratia* spp. outline of leaves. (A) *S. ovata*; (B) *S. caseolaris*, adult and "juvenile" or stenophyllous form behind; (C) *S. apetala*; (D,E) *S. alba* from 2 collections. Identification of *Sonneratia* species from leaf morphology is sometimes possible, but there is considerable overlap.

This range includes the mangroves, and species of *Sonneratia*, especially *S. alba*, are important food sources for the bats at certain times of the year. The mangroves therefore may sustain bat populations when durians are not in flower. This interdependence among bat, mangrove, and durian is a demonstration of the complexity of the processes that underly conservation measures. The destruction of both caves and mangroves can adversely affect the production of durian, a fruit that plays an important role in the rural economy of Malaysia because it is a favorite of the city dweller (Lee 1980).

Start and Marshall (1976) comment on the contrasted floral phenologies of *Sonneratia* species; *S. caseolaris* is described as aseasonal, which implies that flowers are available continuously, even though in small numbers, whereas *S. alba* and *S. ovata* are seasonal in their flowering and flowers are not continuously available in one locality.

Hybrids in *Sonneratia*

This genus is a well-documented example of hybridization within mangroves (Muller and Hou-Liu 1966, see also Muller and van Steenis 1968). Hybrids were first recognized on the basis of reduced pollen fertility and subsequently by a mixture of parental characters. Populations in Brunei, northwest Borneo, are aggregates of *Sonneratia*, *S. alba*, and aggregates of *S. ovata*, *S. alba*, and *S. caseolaris* together with intermediate forms suggested to be the hybrids *S. alba* × *ovata* and *S. alba* × *caseolaris*. Flowering and fruiting were poorest in the latter. Meiosis is normal in the parental species ($n = 11$) but abnormal in the putative hybrids. The origin of these hybrids is not discussed. Hybrids in Queensland presumably between *S. alba* and *S. caseolaris* have been referred to as *S.* × *gulngai* by Duke (1984).

Vegetative spread

Sonneratia alba is described by Holbrook and Putz (1982) as having a capacity for clonal development by the rooting of reclining lower horizontal branches that root distally where they touch the mud. Repeated rooting produces extensive linear clones up to 37 m long, primarily oriented toward the sea. It is implied that this vegetative spread occurs only on rocky coastlines where seedlings do not become established.

Family: Sterculiaceae

A tropical and subtropical family of trees and shrubs with alternate simple leaves. It includes several commercially important species, notably *Thoebroma cacao* (cocoa). One genus is represented in the Indo-Malayan mangroves.

Heritiera Aiton. 1789
[*Hort. Kew.* 3:546]

> Including *Argyrodendron* F. v. Mueller and *Tarrietia* Bl.

A genus of 29 species (Kostermans 1959) mainly of rain forest trees with a distribution from eastern tropical Africa (2 species) and the remainder from India to the Pacific (Tonga and Niue), with *H. littoralis* introduced farther east in Polynesia and to Hawaii. The genus is characterized by its leaves with a scaly indumentum on the lower surface, the leaves either digitately compound or more usually simple (unifoliolate by reduction). The fruit is usually a wind-dispersed samara, but in water-dispersed species the wing is reduced to a vestigial keel.

A number of species have a coastal distribution. The genus is predominantly found in inland forests, but *H. littoralis* is a characteristic constituent of the back mangal, occupying almost the entire range of the genus; two others may be described as mangrove associates, both with a limited distribution. Several of the inland species are important timber species, for example, *H. sylvatica* Vid. in the Philippines.

> *Key to mangrove or salt-tolerant species of Heritiera*
>
> 1A. Petiole exceeding 2 cm (to 4 cm), fruit globose (Fig. B.83j) with a distal
>   slightly recurved beak; plants of back-mangal fringe and lowland forests in
>   northern Borneo .... *H. globosa* Kosterm.
> 1B. Petiole not exceeding 2 cm (1–2 cm), fruit ovoid, extended, and flattened

Figure B.82. *Heritiera littoralis* (Sterculiaceae). Buttresses and bark characteristics. Hinchinbrook Island, Queensland.

on dorsal side (Fig. B.83b), without a distal beak; plants of back mangal, distribution not as in 1A.. 2

2A. Fruit knobby with a ventral ridge together with a transverse, circular ridge; plants with a limited distribution in tidal swamps in the Ganges and Irrawaddy Deltas. Pneumatophores sometimes developed.....*H. fomes* Ait.

2B. Fruit smooth with a rudderlike crest, but without a transverse circular ridge; plants with a wide distribution in back mangal, from East Africa to Melanesia. Pneumatophores absent ....*H. littoralis* Buch.-Ham.

Heritiera littoralis Dryand. in Aiton 1789   (Figs. B.83, B.84)
[*Hort. Kew*. ed. 1, 3:546]

H. minor Lam. 1797 [*Encycl.* 3:299]

A complete synonymy and citation (almost 4 pages) are given in the monograph by Kostermans (1959). This species is the type of the genus.

Trees up to 25 m with a buttressed trunk to 60 cm in diameter, evergreen, monoecious. Bark fissured, dark or gray (Fig. B.82). Leaves spirally arranged,

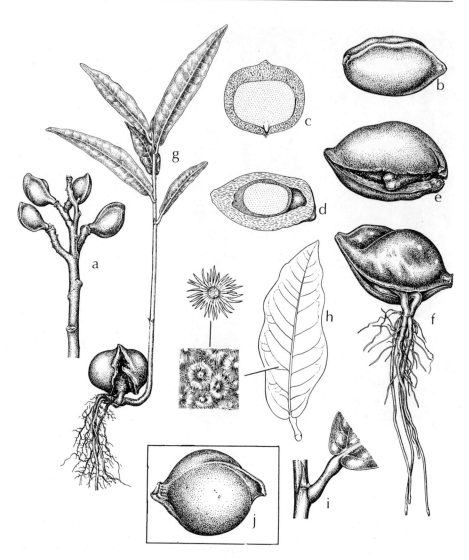

Figure B.83. *Heritiera* spp. (Sterculiaceae) leaf, fruit, and seedling. (a–i) *Heritiera littoralis*: (a) young fruits (× ½) on portion of inflorescence that is still erect; (b) detached mature fruit (× ½); (c) seed in T.S. (× ½); (d) fruit in L.S. (× ½); (e) fruit at early stage of germination (× ½), radicle protruding; (f) established fruit (× ½) before emergence of plumule; (g) young seedling (× ⅓); (h) leaf outline (× ¼), insets: details of abaxial epidermis (× 35) and single epidermal scale (× 90); (i) detail of leaf insertion (× 1) with bipulvinate petiole and one of the stipule pair. (j) *Heritiera globosa* fruit (× 2); cf. part b. (*H. littoralis* from Hinchinbrook Island, North Queensland, P. B. Tomlinson 12.3.76H; seedlings cultivated at Fairchild Tropical Garden; *H. globosa* from Kuching, Sarawak, P. Chai, *s.n.*)

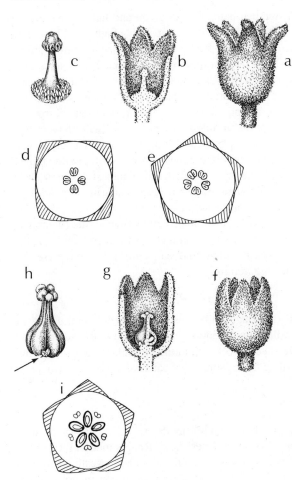

Figure B.84. *Heritiera littoralis* (Sterculiaceae). Floral morphology. (a–e) Male flower: (a) male flower (× 4); (b) male flower (× 4) in L.S.; (c) stamen column and disc (×9); (d) floral diagram of 4-merous flower; (e) floral diagram of 5-merous flower. (f–i) Female flower: (f) female flower (× 4); (g) female flower (× 4) in L.S.; (h) carpels (×9) with staminodes (arrow); (i) floral diagram. (Material from Labu Lagoon, Lae, Papua New Guinea. J. S. Womersley, s.n.)

simple (but regarded as a compound leaf with 1 leaflet, by reduction), stiffly coriaceous, oblong or ovate-elliptic, usually 10 to 20 by 5 to 10 cm (but up to 30 by 15 to 18 cm), with a usually obtusely pointed apex and somewhat cordate at the base. Margin entire. Veins prominent below. Leaves dark green above, grayish white beneath because of indumentum of overlapping, stellate scales (similar scales on young twigs). Petiole 1 to 2 cm long and 1 cm wide, shortly bipulvinate. Stipules in pairs at each node, short, subulate up to 1 cm long, caducous but conspicuous in resting buds and forming bud scales.

Flowers unisexual in complex tomentose panicles in axils mainly of distal foliage leaves on current growth, often subterminal, that is, arising immediately below resting terminal bud. Flowers 3 to 4 mm in diameter, 4 to 5 mm long, with a slender short pedicel, the male somewhat smaller; both sexes with a cup-shaped calyx, the cup shallower in male flowers, reddish hairy inside, green and hairy outside with 4 or 5 (-6) short pointed lobes. Receptacle within the cup with a flattened tuberculate disc, the tubercules represented by minute fleshy structures that resemble short-stalked colleters. Petals absent. Male flowers more numerous, each with an andro-gynophore arising from the center of the disc and consisting of stamens fused together around a central column representing the pistillode; stamens 4 or 5 (-6), anthers minute and forming a ring around the central column, dehiscent extrorsely by 2 longitudinal slits to release the small amount of pollen. Female flowers less numerous and usually terminating distal panicles of branches; disc scarcely devel-oped, staminodes minute below the carpels. Carpels 4 or 5 (-6) sessile, united loosely, each laterally compressed and with 1 (-2) basal ovule, styles as a short extension of the carpel, stigmas minute and recurved. Fruits maturing in pendulous clusters, usually only 1 developing from a single carpel in each flower; always 1-seeded, 6 to 8 cm long, 5 to 6 cm wide, ellipsoidal, flattened on 1 side (ventrally) and with a raised (dorsal) keel 5 mm (but sometimes up to 2 cm) high on the opposite side, epicarp woody, mesocarp fibrous, endocarp hard. Embryo represented by fused cotyledons, radicle directed ventrally.

Seedling

Germination hypogeal (Boerlage 1898), usually with the ventral side of the fruit downward, the fruit wall gaping ventrally, the radicle extruded first and soon branched, plumule extruded later, erect with initial plumular nodes supporting stipule pairs (leaf aborted). Seedling leaves somewhat narrower and longer (to much longer) than adult leaves.

Growth

The architecture corresponds to Rauh's model (Hallé et al. 1978) with somewhat differentiated branches. As is characteristic of Rauh's model, the growth of the axes is rhythmic and the shoots are distinctly articulate. Resting terminal buds develop a series of short bud scales whose scars remain conspicuous. Trees remain evergreen.

Reproductive biology

Nothing is known about the floral mechanism; flower visitors have not been reported. It is possible that the tuberculate disc at the base of the male floral cup is a nectary that is absent from the female flowers. Pollination could be promoted by insect visitors who do not distinguish flowers with and without rewards.

Fruits float with the ''keel'' upward (i.e., the ridge is not a functional keel) and germinate readily in muddy substrates at the upper limits of tidal influences.

Floral morphology

Kostermans comments that there could be some difficulty in establishing the number of anthers because of their small size, but concludes that there are 4 or 5 bithecate anthers per flower.

The morphology of the flower is not clear and some accounts are confused. The central column, which supports the minute anthers, is described as an androgynophore. Some descriptions imply a similar structure in the female flower, but this does not exist in the material I have examined. The function of the colleter-like organs on the disc of the male flower, which are conspicuous in fresh material, has not been elucidated.

Root system
This is not elaborated, except for a few short buttresses at the base of the trunk (Fig. B.82), which may be extended horizontally in a sinuous way, described by Parkinson (1934) as like those of *Xylocarpus granatum* but generally larger close to the trunk.

Ecology
Heritiera littoralis is a frequent member of the back mangal and may occupy the forest fringe or rocky shores. It seems intolerant of high salinities and does not occur in very exposed or poorly drained sites.

Geographic range
This species has almost the full range of the genus, extending from east Africa and Madagascar to the Pacific. In Australia it is common in Queensland and on the east coast to Cape York but has been recorded in neither Western Australia nor the Northern Territory. Its northern limit is Hong Kong and in its natural range extends eastward into the Pacific as far as New Caledonia.

Field characters
This tree is most easily identified by the silvery undersurface of the large leaves with their short bipulvinate petioles. The woody keeled fruits are highly distinctive, both on and off the tree.

Systematics and phylogeny
Kostermans (1959) suggests that the most closely related species is *H. dubia* Wallich. ex Kurz., which is an inland tree of the Khasia Hills at low elevations, probably distinguished by its dull rather than smooth and glossy fruit.

Following Kostermans, it is possible to describe within the family Sterculiaceae a reduction series in fruit type, beginning with the *Sterculia* type and ending with *H. littoralis* (and similar species). In this series there is a progressive change from animal dispersal of individual seeds from open carpels, to closed winged carpels (samaras) with wind-assisted dispersal, to closed, unwinged, and water-dispersed fruits (essentially large achenes). The sequence proceeds as follows:

1. *Sterculia*: Attractive seeds attached along the full margin of the open carpels.
2. *Firmiana*: Seeds along the full margin of the carpel, which is membranaceous and remains closed.
3. *Scaphium* and *Pterocymbium*: Only 1 seed at the base of the membranaceous wing-shaped carpel.

4. *Hildegardia*: One seed at the base of the membranaceous, wing-shaped carpel.
5. *Heritiera*: Some species with a 1-seeded, winged carpel (samara); others with the wing reduced; in other species that are clearly water-dispersed, notable *H. littoralis*, the wing is represented by a ventral keel (but may be developed as a vestigial wing). The seeds of some of the wind-dispersed species may be fairly large so that dispersal cannot be over very great distances. On the other hand, water-dispersed species always have large seeds (e.g., *H. fomes* and *H. littoralis*). The status of inland and upland species with large, wingless seeds (e.g., *H. dubia*) is problematical.

Heritiera fomes Buch.-Ham. 1800
[*Symes Kingdom Ava*, ed. 2, 3:319–20]

This species is characteristic of the forest of the Sundarbans of the Ganges Brahmaputra Delta (Khuma District) and in the Irrawaddy Delta of Burma. The Sundarbans is said to get its name from this tree (vernacular ''sundri''), which is characteristic of the region.

It is distinguished from *H. littoralis* in its fruit and is described by Parkinson (1934) as having the ability to produce ''erect pointed pneumatophores like those of *Xylocarpus gangeticus* [= *X. mekongensis*], but more numerous and longer.'' Kostermans refers to these pneumatophores as ''blind rootsuckers'' and ''perpendicular shoots from its roots,'' but it seems clear that they are morphologically root structures. Watson (1928) makes the same mistake in describing pneumatophores. Parkinson compares these 2 species of *Heritiera* with the 2 mangrove species of *Xylocarpus*; in both genera there are contrasted root structures in different species.

From the available descriptions of the habitat of this species, it would probably be classified as a mangrove if it were more widely distributed. It is said to grow ''down to the edge of the tidal creeks'' in Burma (Kostermans 1959).

Heritiera fomes is considered to be the best of the 2 commercial *Heritiera* species in India; it has properties superior to those of teak. The difficult access, which makes it uneconomical to exploit, may be an important factor in its conservation.

Heritiera globosa Kostermans 1959   (Fig. B.83j)
[*Reinwardtia* 4(4):484]

This species is known from Sarawak to Sabah and adjacent Indonesian Borneo but may have a wider distribution. It is distinguished from *H. littoralis* by its globose fruit, which has a shallow ventral crest extended at the distal end into a beak or vestigial wing, with the wing always slightly recurved in a characteristic manner. The seed apparently retains undigested endosperm at maturity, and the embryo has 2 distinct cotyledons but an obscure plumule and radicle. Details of germination are not known. It is described as occurring behind the tidal zone of the mangrove belt, but has been collected 70 km from the sea; even here, although the water is fresh, there is tidal fluctuation. The buttresses are said to be well developed, snakelike, and extending 2 to 4 m from the base of the trunk.

The existence of these 2 local species, which resemble the widely distributed *H. littoralis* in their morphology and to some extent in ecology, indicates that the former species is less isolated from terrestrial relatives than is usual among mangrove taxa.

Family: Tiliaceae

A family of about 50 genera and 400 species, chiefly tropical but with some familiar temperate genera (e.g., *Tilia*, linden, and basswood). Flowers are usually conspicuous with numerous stamens; the fruit is usually a capsule.

Brownlowia Roxburgh 1819
[*Pl. Corom.* 3:61, t. 265]

A genus of about 30 species widely distributed in Southeast Asia from Malaysia to the Solomon Islands. Two species are recorded in swamp forests and river banks and penetrate into back-mangal communities inundated by the highest tides. They are distinguished as follows:

1A. Tree to 10 m; leaves broad, cordate or ovate to 10 cm wide, petiole long (to 6 cm), slender with a distal fleshy or corky pulvinus; flowers in terminal, lax panicles ....*B. argentata* Kurz
1B. Shrub to 2 m; leaves narrow, lanceolate to 5 cm wide at most, petiole short (1 cm) without a distinct pulvinus; flowers in congested, usually axillary clusters ....*B. tersa* (L.) Kosterm.

Brownlowia argentata Kurz 1870
[*J. Asiat. Soc. Bengal* 39(2):67; see also Scheff, 1874 in *Nat. Tijd. Ned. Ind.* 34:94]

This species has the distribution of the whole genus from the Malay Peninsula through Indonesia (but not Java and the Lesser Sunda Islands) to New Guinea and the Solomon Islands.

A shrub or small tree growing to 10 m, often branched below, with scaly or stringy bark. Leaves cordate to ovate with a conspicuous pulvinus immediately below the insertion of the blade, the lower leaf surface and twigs silvery and covered with a close indumentum of minute peltate scales. Petiole slender, 4 to 6 cm long, the blade up to 15 by 10 cm. Flowers in erect, terminal, but lax panicles. Flowers about 5 mm long with a 5-lobed, scaly calyx; petals 5, orange, stamens numerous. Fruit a woody capsule or nut, obliquely globose-ellipsoidal, bilobed to heart shaped but asymmetrically inserted on a short (to 1 cm) thick stalk, fruit surface scaly like the leaves. Normally only 1 or 2 carpels per flower developing into fruits.

Brownlowia tersa (L.) Kosterm. 1959
[*Reinwardtia* 4(4):73]

A shrub, scarcely 2 m tall, with narrow, lanceolate to elliptic-lanceolate leaves, 8 to 12 by 2 to 3 cm, the apex gradually narrowed, often bluntly rounded. Petiole 1 to 2 cm without a pulvinus. Flowers small, in compact axillary clusters.

This species has a distribution in the Malay Archipelago and the Philippines. It could most often be confused with *Camptostemon* (both have a scaly leaf), but it is distinguished by the bilobed asymmetrical fruit and paniculate flower clusters.

References

Adams, D. C., and Tomlinson, P. B. 1979. *Acrostichum* in Florida. *Am. Fern J.* 69:42–6.

Airy-Shaw, H. K. 1975. The Euphorbiaceae of Borneo. *Kew Bull.* Add. Series 4:1–245.

Allen, P. H. 1956. *The rain forests of Golfo Dulce.* Gainesville: University of Florida Press.

Areschoug, F. W. C. 1902. Untersuchungen über den Blattbau der Mangrove-Pflanzen. *Bibl. Bot.* 56:1–90.

Arzt. T. 1936. Die Kutikula einiger afrikanischer Mangrove-Pflanzen. *Ber. d. bot. Gesell.* 54:247–60.

Atkinson, M. R., Findlay, G. P., Hope, A. B., Pitman, M. G., Saddler, H. D. W., and West, K. R. 1967. Salt regulation in the mangrove *Rhizophora mucronata* Lam. and *Aegialitis annulata* R. Br. *Aust. J. Biol. Sci.* 20:589–99.

Attims, Y., and Cremer, G. 1967. Les radicelles capillaires des palétuviers dans une mangrove de Côte d'Ivoire. *Adansonia*, Series 2, 7:547–51.

Aubréville, A. 1964. Problémes de la mangrove d'hier et d'aujourd'hui. *Adansonia, n.s.* 4:19–23.

Backer, C. A., and Bakhuizen van den Brink, R. C. 1963–68. *Flora of Java.* 3 vols. Groningen: N. V. P. Noordhoff.

Backer, C. A., and van Steenis, C. G. G. J. 1954. Sonneratiaceae. *Flora Malesiana* 1, 4:280–9.

Baillon, H. 1875. *Natural history of plants* (English ed.). London: L. Reeve.

Bakhuizen van den Brink, R. C. 1921. Revisio generis Avicenniae. *Bull. Jard. Bot. Buitenz.* Series 3, 3(2):199–226.

  1924. Revisio Bombacacearum. *Bull. Jard. Bot. Buitenz.* Series 3, 6(2):161–240.

Ball, M. C., and Farquhar, G. D. 1984a. Photosynthetic and stomatal responses of two mangrove species, *Aegiceras corniculatum* and *Avicennia marina*, to long term salinity and humidity conditions. *Plant Physiol.* 74:1–6.

  1984b. Photosynthetic and stomatal responses of the grey mangrove, *Avicennia marina*, to transient salinity conditions. *Plant Physiol.* 74:7–11.

Barlow, B. 1966. A revision of the Loranthaceae of Australia and New Zealand. *Aust. J. Bot.* 14:421–99.

Barth, H. 1982. The biogeography of mangroves. Chap. 3 in *Tasks for vegetation science*, eds. D. N. Sen and K. S. Rajpurohit, vol. 2. The Hague: Junk.

Baylis, G. T. S. 1940. Leaf anatomy of the New Zealand mangrove. *Trans. R. Soc. N. Z.* 70:164–70.

  1950. Root systems of the New Zealand mangrove. *Trans. R. Soc. N.Z.* 78:509–14.

Beauvisage, L. 1918. Etude anatomique de la famille des Ternstroemiacées. Thesis. Université de Toulouse.

Benecke, W., and Arnold, A. 1931. Kulturversuche mit Keimlingen von Mangrovepflanzen. *Planta* 14:471–81.

Bentham, G., and Hooker, J. D. 1862. *Genera plantarum*, vol. 1. London: Reeve.

Bhosale, L. J., and Shinde, L. S. 1983. Significance of cryptovivipary in *Aegiceras corniculatum* (L.) Blanco. Chap. 14 in *Biology and ecology of mangroves*, ed. H. J. Teas. *Tasks for vegetation science* 8. The Hague: Junk.

Biebl, R., and Kinzel, H. 1965. Blattbau und Salzhaushalt von *Laguncularia racemosa* (L.) Gaertn.f. und anderer Mangrovebaüme auf Puerto Rico. *Ost. Bot. Zeit.* 112:56–93.

Biswas, K. 1934. A comparative study of the Indian species of *Avicennia. Notes R. Bot. Gard. Edinburgh* 89:159–66.

Blasco, F. 1977. Outlines of ecology, botany and forestry of the mangals of the Indian subcontinent. Chap. 12 in *Ecosystems of the world*, vol. 1. *Wet coastal ecosystems*, ed. V. J. Chapman. Amsterdam: Elsevier Scientific.

Boerlage, J. G. 1898. Sur le manière de flotter et la germination des fruits du *Heritiera littoralis* Dryand. *Ann. Jard. bot. Buitenz.* Suppl. 2:137–42.

Booberg, G. 1933. Die malayische Strandflora. Eine Revision der Schimperschen Artenliste. *Bot. Jahrb.* 66:1–38.

Borassum Waalkes, J. van. 1966. Malesian Malvaceae revised. *Blumea* 14:1–213.

Bormann, F. H., and Berlyn, G. 1981. Age and growth rate of tropical trees: new directions for research. Bull. 94, New Haven: Yale University School of Forestry and Environmental Studies.

Bowman, H. H. M. 1916. Physiological studies in *Rhizophora. Proc. Nat. Acad. Sci. U.S.A.* 2:685–8.

  1917. Ecology and physiology of the red mangrove. *Proc. Amer. Philos. Soc.* 56:589–672.

Brenner, W. 1902. Ueber die Luftwurzeln von *Avicennia tomentosa. Ber d. bot. Gesell.* 20:175–89.

Breteler, F. J. 1969. The Atlantic species of *Rhizophora. Acta Bot. Neerl.* 18:434–44.

  1977. America's Pacific species of *Rhizophora. Acta Bot. Neerl.* 26:225–30.

Briggs, B. G., and Johnson, L. A. S. 1979, Evolution of the Myrtaceae – evidence from inflorescence structure. *Proc. Linn. Soc. N. S. W.* 102(4):157–256.

Brown, F. H. B. 1935. Flora of southeastern Polynesia. III. Dicotyledons. *Bull. Bishop Mus.* 130:1–386.

Brown, J. M. A., Outred, H., and Hill, F. C. 1969. Respiratory metabolism in mangrove seedlings. *Plant Physiol.* 44(2):287–94.

Brown, W. H., and Fischer, A. F. 1920. Philippine mangrove swamps. In Minor products of Philippine forests, ed. W. H. Brown, vol. 1. *Phil. Bur. Forest. Bull.* 22:1–125.

Brown, W. H., and Merrill, E. D. 1920. Philippine palms and palm products. In Minor products of Philippine forests, ed. W. H. Brown, vol. 1. *Phil. Bur. Forest. Bull.* 22:127–248.

Browne, P. 1756. *The civil and natural history of Jamaica.* London: published privately.

Budowski, G. 1965. Distribution of tropical American rain forest species in the light of successional processes. *Turrialba* (Costa Rica) 15:40–2.

Bunt, J. S., and Williams, W. T. 1981. Vegetational relationships in the mangroves of tropical Australia. *Mar. Ecol. Prog. Ser.* 4:349–59.

Bunt, J. S., Williams, W. T., and Duke, N. C. 1982. Mangrove distributions in northeast Australia. *J. Biogeogr.* 9:111–20.

Burkill, I. H. 1935. *A dictionary of the economic products of the Malay Peninsula.* 2 vols. Oxford: Oxford University Press.

Byrnes, N. B. 1977. A revision of Combretaceae in Australia. *Contrib. Queensl. Herb.* 20:1–88.

Camilleri, J. C., and Ribi, G. 1983. Leaf thickness of mangroves (*Rhizophora mangle*) growing in different salinities. *Biotropica* 15:139–141.

Candolle, A. C. de 1841. Second mémoire sur la famille des Myrsinaceae. *Ann. sci. nat.*, Series 2, 16:1–46.

1844. *Prodromus systematis naturalis*, part 8. Paris: Fortin Masson et Soc.

Carey, G. 1934. Further investigations on the embryology of viviparous seeds. *Proc. Linn. Soc. N.S.W.* 59:392–410.

Carey, G., and Fraser, L. 1932. The embryology and seedling development of *Aegiceras majus* Gaert. *Proc. Linn. Soc. N.S.W.* 57:341–60.

Chai, P. K. 1982. Ecological studies of mangrove forest in Sarawak. Ph.D. thesis. University of Malaysia, Kuala Lumpur.

Chandler, M. E. J. 1957. Note on the occurrence of mangrove in the London Clay. *Proc. Geol. Assoc.* 62:271–2.

Chapman, V. J. 1940. The functions of the pneumatophores of *Avicennia nitida* Jacq. *Proc. Linn. Soc. Lond.* 152(3):228–33.

1947. Secondary thickening and lenticels in *Avicennia nitida* Jacq. *Proc. Linn. Soc. Lond.* 158(1945–6):2–6.

1962a. Respiration studies of mangrove seedlings, I. *Bull. Mar. Sci. Gulf. Caribb.* 12(1):137–67.

1962b. Respiration studies of mangrove seedlings, II. *Bull. Mar. Sci. Gulf. Caribb.* 12(2):245–63.

1976. *Mangrove vegetation.* Valduz: Cramer.

1977a. *Ecosystems of the world*, vol. 1. *Wet coastal ecosystems.* Amsterdam: Elsevier Scientific.

1977b. Africa B, The remainder of Africa. Chap. 11 in *Ecosystems of the world*, vol. 1. *Wet coastal ecosystems*, ed. V. J. Chapman. Amsterdam: Elsevier Scientific.

1977c. Wet coastal formations of Indo-Malesia and Papua New Guinea. Chap. 13 in *Ecosystems of the world*, vol. 1. *Wet coastal ecosystems*, ed. V. J. Chapman, Amsterdam: Elsevier Scientific.

Christensen, B. 1978. Biomass and primary production of *Rhizophora apiculata* Bl. in a mangrove in southern Thailand. *Aquat. Bot.* 4:43–52.

1983. Mangroves – what are they worth? *Unasylva* 35:2–15.

Christensen, B., and Wium-Anderson, S. 1977. Seasonal growth of mangrove trees in southern Thailand, I. The phenology of *Rhizophora apiculata* Bl. *Aquat. Bot.* 3: 281–6.

Cintron, G., Lugo, A. E., Pool, D. J., and Morris, G. 1978. Mangroves of arid environments in Puerto Rico and adjacent islands. *Biotropica*, 10:110–21.

Clough, B. F., Boto, K. G., and Attiwill, P. M. 1983. Mangroves and sewage: a re-evaluation. Chap. 17 in *Biology and ecology of mangroves*, ed. H. J. Teas. *Tasks for vegetation science 8*. The Hague: Junk.

Collins, J. P., Berkelhamer, R. C., and Mesler, M. 1977. Notes on the natural history of the mangrove *Pelliciera rhizophorae* Tr. & Pl. (Theaceae). *Breynesia* 10/11:17–29.

Compere, P. 1963. The correct name of the Afro-American black mangrove. *Taxon.* 12:150–2.

Connor, D. I. 1969. Growth of grey mangrove (*Avicennia marina*) in nutrient culture. *Biotropica* 1:36–40.

Cook, M. T. 1907. The embryology of *Rhizophora mangle*. *Bull. Torrey Bot. Club* 34:271–7.

Corner, E. J. H. 1939. Notes on the systematy and distribution of Malayan phanerogams, III. *Gard. Bull. Straits Settl.* 10:239–328.

1976. *The seeds of dicotyledons*. 2 vols. Cambridge: Cambridge University Press.

1978. The freshwater swamp-forest of South Johore and Singapore. *Gard. Bull. Singapore*, Suppl. 1, 1–266.

Correll, D. S., and Correll, H. B. 1982. *Flora of the Bahama Archipelago*. Valduz: Cramer.

Cowan, R. S., and Polhill, R. M. 1981. Tribe 4, Detarieae. Pp. 117–34 in *Advances in legume systematics*, eds. R. M. Polhill and P. H. Raven. Royal Botanic Gardens, Kew: Ministry of Agriculture and Fisheries.

Craighead, F. C. 1971. *The trees of South Florida 1. The natural environments and their succession*. Coral Gables: University of Miami Press.

Cridland, A. A. 1964. *Amyelon* in American coal-balls. *Palaeontology* 7:186–209.

Cruden, R. W. 1977. Pollen-ovule ratios: a conservative indicator of breeding systems in flowering plants. *Evolution* 31:32–46.

Dacey, J. W. H. 1980. Internal winds in waterlilies: an adaptation for life in anaerobic sediments. *Science* 203:1253–5.

1981. Pressurized ventilations in the yellow waterlily. *Ecology* 62:1137–47.

Dahlgren, R. M. T. 1980. A revised system of classification of the angiosperms. *Bot. J. Linn. Soc.* 80:91–124.

Dandy, J. E., and Exell, A. W. 1938. On the nomenclature of three species of *Caesalpinia*. *J. Bot. Lond.* 76:175–83.

Davey, J. E. 1975. Notes on the mechanism of pollen release in *Bruguiera gymnorrhiza*. *J. S. Afr. Bot.* 41:269–72.

Davis, J. H. 1940. The ecology and geologic role of mangroves in Florida. *Pap. Tortugas Lab.* 32 *(Publ. Carn. Inst.* No. 517):303–41.

DeVogel, E. F. 1980. *Seedlings of dicotyledons*. Wageningen: Centre for Agricultural Publication and Documentation.

Ding Hou. 1957. A conspectus of the genus *Bruguiera* (Rhizophoraceae). *Nova Guinea* n.s. 8(1):163–71.

1958. Rhizophoraceae. *Flora Malesiana*, Series 1, 5:429–93.

1960. A review of the genus *Rhizophora*. *Blumea* 10(2):625–34.

1963. Celastraceae. *Flora Malesiana*, Series 1, 6(2):227–91.

1977. Anacardiaceae. *Flora Malesiana*, Series 1, 8:395–548.

Docters van Leeuwen, W. 1911. Ueber die Ursache der weiderholter Versweigung der Stültzwurzeln von *Rhizophora*. *Ber. d. bot. Gesell.* 29:476–8.

Duke, N. C. 1984. A mangrove hybrid, *Sonneratia* × *gulngai* (Sonneratiaceae) from north-eastern Australia. *Austrobaileya*. 2:103-5.

Duke, N. C., Birch, W. R., and Williams, W. T. 1981. Growth rings and rainfall correlations in a mangrove tree of the genus *Diospyros* (Ebenaceae). *Aust. J. Bot.* 29:135–42.

Duke, N. C., and Bunt, J. S. 1979. The genus *Rhizophora* (Rhizophoraceae) in north-eastern Australia. *Aust. J. Bot.* 27:657–78.

Duke, N. C., Bunt, J. S., and Williams, W. T. 1984. Observations on the floral and vegetative phenologies of north-eastern Australian mangroves *Aust. J. Bot.* 32:87–99.

Egler, F. E. 1948. The dispersal and establishment of red mangroves, *Rhizophora* in Florida. *Caribb. For.* 9(4):299–319.

1952. Southeast saline everglades vegetation, Florida, and its management. *Vegetatio* 3:213–65.

Essig, F. B. 1973. Pollination in some New Guinea palms. *Principes* 17:75–83.

Ewel, J. 1980. Tropical succession: manifold routes to maturity. *Biotropica* 12(suppl.):2–7.

Exell, A. W. 1954. Combretaceae. *Flora Malesiana*, Series 1, 4:533–89.

Exell, A. W., and Stace, C. 1966. Revision of Combretaceae. *Bol. Soc. Broteriana* 40(2):5–25.

Faber, F. C. von. 1913. Ueber transpiration und osmotischen druck bei den Mangroven. *Ber. d. bot. Gesell.* 31:277–81.

1923. Zur Physiologie der Mangroven. *Ber. d. bot. Gesell.* 41:227–34.

Fahn, A. 1979. *Secretory tissues in plants.* London: Academic Press.

Fahn, A., and Shimony, C. 1977. Development of glandular and nonglandular leaf hairs of *Avicennia marina* (Forsskal) Vierh. *Bot. J. Linn. Soc.* 74:37–46.

Fisher, J. B. 1978. A quantitative study of *Terminalia* branching. Chap. 13 in *Tropical trees as living systems*, eds. P. B. Tomlinson and M. H. Zimmermann. Cambridge: Cambridge University Press.

Fisher, J. B., and Honda, H. 1977. Computer simulation of branching pattern and geometry in *Terminalia* (Combretaceae), a tropical tree. *Bot. Gaz.* 138:337–84.

1979a. Branch geometry and effective leaf area: a study of *Terminalia*-branching pattern, 1. Theoretical trees. *Am. J. Bot.* 66:633–44.

1979b. Branch geometry and effective leaf area: a study of *Terminalia*-branching, 2. Survey of real trees. *Am. J. Bot.* 66:645–55.

Fisher, J. B., and Stevenson, J. W. 1981. Occurrence of reaction wood in branches of dicotyledons and its role in tree architecture. *Bot. Gaz.* 142:82–95.

Flowers, T. J., Troke, P. F., and Yeo, A. R. 1977. The mechanism of salt tolerance in halophytes. *Ann. Rev. Plant Physiol.* 28:89–121.

Fosberg, F. R. 1939. *Diospyros ferrea* (Ebenaceae) in Hawaii. *Occas. Pap. Bernice Pauahi Bishop Mus.* 15:119–31.

1975. Phytogeography of Micronesian mangroves. In *Proccedings of the International Symposium on Biology and Management of Mangroves*, eds. G. E. Walsh, S. C. Snedaker, and H. J. Teas. Gainesville: Institute of Food and Agricultural Sciences, University of Florida.

Fosberg, F. R., and Sachet, M.-H. 1972. *Thespesia populnea* (L.) Solander ex Correa and *Thespesia populneoides* (Roxburgh) Kosteletsky (Malvaceae). *Smithson. Contrib. Bot.* 7:1–13.

Foxworthy, F. W. 1910. Distribution and utilization of mangrove swamps in Malay. *Ann. Jard. Bot. Buitenz.* Series 2, suppl. 3, part 1:319–44.

Garcia de Lopez, I. 1978. Revision del género *Acrostichum* en la Republica Dominicana. *Moscosoa* 1:64–70.

Gentry, A. H.1982. Phytogeographic patterns as evidence for a Choco refuge. Pp. 112–36 in *Biological diversification in the tropics*, ed. Gh. T. Prance. New York: Columbia University Press.

Gill, A. M. 1971. Endogenous control of growth-ring development in *Avicennia*. *For. Sci.* 17:462–5.

Gill, A. M., and Tomlinson, P. B. 1969. Studies on the growth of red mangrove (*Rhizophora mangle* L.). I, Habit and general morphology. *Biotropica* 1(1):1–9.

1971a. Studies on the growth of red mangrove (*Rhizophora mangle* L.), 2. Growth and differentiation of aerial roots. *Biotropica* 3(1):63–77.

1971b. Studies on the growth of red mangrove (*Rhizophora mangle* L.), 3. Phenology of the shoot. *Biotropica* 3(2):109–24.

1975. Aerial roots: an array of forms and functions. Chap. 12 in *The development and function of roots*, eds. J. G. Torrey and D. T. Clarkson. London: Academic Press.

1977. Studies on the growth of red mangrove (*Rhizophora mangle* L.), 4. The adult root system. *Biotropica* 9(3):145–55.

Gill, L. S., and Kyauko, P. S. 1977. Heterostyly in *Pemphis acidula* Forst. (Lythraceae) in Tanzania. *Adansonia*. 17(2):139–46.

Goebel, K. 1886. Ueber die Luftwurzeln von *Sonneratia*. *Ber. d. bot. Gesell*. 4:249–55.

Gomez-Pompa, A., and Vazquez-Yanes, C. 1974. Studies on the secondary succession of tropical lowlands: the life cycle of secondary species. *Proc. First Int. Cong. Ecology*, pp. 336–42.

Graham, A. 1977. New records of *Pelliceria* (Theaceae/Pelliceriaceae) in the tertiary of the Caribbean. *Biotropica* 9:48–52.

Greenway, H., and Munns, R. 1980. Mechanisms of salt tolerance in nonhalophytes. *Ann. Rev. Plant Physiol*. 31:149–90.

Guppy, H. B. 1906. *Observations of a naturalist in the Pacific between 1896 & 1899, Plant dispersal*, vol. 2, pp. 627. London: Macmillan.

  1917. *Plant seeds and currents in the West Indies and Azores*, pp. 531. London: Williams & Norgate.

Haberlandt, G. 1910. *Eine botanische Tropenreise*. Leipzig: Wilhelm Engelmann.

  1928. *Physiological Plant Anatomy*. London: Macmillan.

Hallé, F., Oldeman, R. A. A., and Tomlinson, P. B. 1978. *Tropical trees and forests – an architectural analysis*. Berlin: Springer-Verlag.

Hartog, C. den. 1970. The sea-grasses of the World. *Verhandl. Kon. Ned. Akad. Wetensch. Nat*. 59(1): 1–275.

Hattink, T. A. 1974. A revision of Malesian *Caesalpinia*, including *Mezoneuron* (Leguminosae-Caesalpiniaceae). *Reinwardtia* 9(1):1–69.

Hemsley, W. B. 1879–88. *Botany in Biologia Centrali:-Americana*. 5 vols. London: Porter and Dulau.

Holbrook, N. M., and Putz, F. E. 1982. Vegetative seaward expansion of *Sonneratia alba* trees in a Malaysian mangrove forest. *Malay. For*. 45:278–81.

Hosakawa, T., Tagawa, H., and Chapman, V. J. 1977. Mangals of Micronesia, Taiwan, Japan, the Philippines and Oceania. Chap. 14 in *Ecosystems of the world, vol. 1. Wet coastal ecosystems*. Amsterdam: Elsevier Scientific.

Howard, R. A. 1981. Three experiences with manchineel (*Hippomane* spp., Euphorbiaceae). *Biotropica* 13:224–7.

Howe, M. A. 1911. A little known mangrove of Panama. *J. N. Y. Bot. Gard*. 12:61–72.

Jansonnius, H. H. 1950. The vessels in the wood of Javan mangrove trees. *Blumea* 6:465–9.

Jeník, J. 1970. Root system of tropical trees, 5, The peg roots and the pneumathodes of *Laguncularia racemosa* Gaertn. *Preslia* 42:105–13.

  1978. Roots and root systems in tropical trees: morphologic and ecologic aspects. Chap. 14 in *Tropical trees as living systems*, eds. P. B. Tomlinson and M. H. Zimmermann. Cambridge: Cambridge University Press.

Jiménez, J. A. 1984. A hypothesis to explain the reduced range of the mangrove *Pelliciera rhizophorae* Tr. & Pl. *Biotropica* 16:304–8.

Johnston, I. M. 1949. The botany of San Jose Island (Gulf of Panama). *Sargentia* 8:1–306.

Johnstone, I. M. 1981. Consumption of leaves by herbivores in mixed mangrove stands. *Biotropica* 13:252–9.

  1983. Mangrove succession and climax. Chap. 12 in *Biology and ecology of mangroves*, ed. H. J. Teas. *Tasks for vegetation science* 8. The Hague: Junk.

Jones, W. T. 1971. The field identification and distribution of mangroves in eastern Australia. *Queensl. Nat*. 20:35–51.

Jonker, F. P. 1959. The genus *Rhizophora* in Suriname. *Acta. Bot. Neerl.* 8:58–60.

Joshi, G. V., Pimplaskar, M., and Bhosale, L. J. 1972. Physiological studies in germination of mangroves. *Bot. Mar.* 15:91–5.

Juncosa, A. M. 1982a. Embryo and seedling development in the Rhizophoraceae. Ph.D. thesis. Duke University, Durham, North Carolina.

1982b. Developmental morphology of the embryo and seedling of *Rhizophora mangle* L. (Rhizophoraceae). *Am. J. Bot.* 69:1599–611.

1984a. Embryogenesis and seedling development in *Cassipourea elliptica* (Sw.) Poir. (Rhizophoraceae). *Am. J. Bot.* 71:170–9.

1984b. Embryogenesis and seedling developmental morphology of the seedling in *Bruguiera exaristata* Ding Hou (Rhizophoraceae). *Am. J. Bot.* 71:180–91.

Karsten, G. 1890. Ueber die Mangrove-Vegetation im malayischen Archipel. *Ber. d. bot. Gesell.* 8:49–55.

1891. Ueber die Mangrove-Vegetation im malayischen Archipel. Eine morphologisch-biologische Studie. *Bibl. Bot.* 22:71.

Kearney, T. H. 1954. Notes on Malvaceae V. *Leafl. West. Bot.* 7:118–19.

Keay, R. W. J. 1953. *Rhizophora* in West Africa. *Kew Bull.* 1953:121–7.

Kenneally, K. F., Wilson, P. G., and Semeniuk, J. 1978. A new character for distinguishing vegetative material of the mangrove genera *Bruguiera* and *Rhizophora* (Rhizophoraceae). *Nuytsia* 4:178–80.

Kipp-Goller, A. 1940. Ueber Bau und Entwicklung der viviparen Mangrovekeimlinge. *Z. Bot.* 35:1–40.

Knapp-van Meeuwen, M. S. 1970. A revision of four genera of the tribe Leguminosae-Caesalpinioideae-Cynometreae in Indomalesia and the Pacific. *Blumea* 18(1):1–52.

Kobuski, C. E. 1951. Studies in the Theaceae XXII, The genus *Pelliciera*. *J. Arnold Arbor.* 32:256–62.

Kostermans, A. J. G. J. 1959. Monograph of the genus *Heritiera* Aiton (Stercul.). *Reinwardtia* 4:465–83.

Lacerda, L. D. de 1981. Mangrove wood pulp, an alternative food source for the tree crab *Aratus pisonii*. *Biotropica* 13:317.

Lacerda, L. D. de, and May, D. V. 1982. Evolution of a new community type during the degradation of a mangrove ecosystem. *Biotropica* 14:238–9.

Larue, C. D., and Muzik, T. J. 1951. Does the mangrove really plant its seedlings? *Science* 114:661–2.

1954. Growth regeneration and precocious rooting in *Rhizophora mangle*. *Pap. Mich. Acad. Sci. Arts. Lett. (Part 1, Botany and Forestry)*. 39:9–29.

Lawrence, D. B. 1949. Self-erecting habit of seedling red mangroves (*Rhizophora mangle* L.). *Am. J. Bot.* 36(5):426–7.

Lee, D. 1980. *The sinking ark: environmental problems in Malaysia and southeast Asia*. Kuala Lumpur: Heinemann.

Lersten, N. R., and Curtis, J. D. 1974. Colleter anatomy in red mangrove, *Rhizophora mangle* (Rhizophoraceae). *Can. J. Bot.* 52:2277–80.

Lewis, D., and Rao, A. N. 1971. Evolution of dimorphism and population polymorphism in *Pemphis acidula* Forst. *Proc. R. Soc. Lon. B* 178:79–94.

Lewis, R. R. 1983. Impact of oil spills on mangrove forests. Chap. 19 in *Biology and ecology of mangroves*, ed. H. J. Teas. *Tasks for vegetation science* 8. The Hague: Junk.

Liebau, O. 1914. Beiträge zur Anatomie und Morphologie der Mangrove-Pflanzen, insbesondere ihres Wurzelsystems. *Beit. Biol. Pflanz.* 12:181–213.

Linnaeus. 1753. *Species plantarum*. Stockholm: Holmiae.

Lloyd, R. M. 1980. Reproductive biology and gametophyte morphology of New World populations of *Acrostichum aureum*. *Am. Fern. J.* 70:99–110.

Longman, K. A., and Jeník, J. 1974. *Tropical forest and its environment*. London: Longman Group.

Lot-Helgueras, A., Vázquez-Yanes, C., and Menéndez, F. 1975. Physiognomy and floristic changes near the northern limit of mangroves in the Gulf Coast of Mexico. In *Proceedings of the International Symposium on Biology and Management of Mangroves*, eds. G. E. Walsh, S. C. Snedaker, and H. J. Teas. pp. 52–61. Gainesville: Institute of Food and Agricultural Sciences, University of Florida.

Lotschert, W., and Liemann, F. 1967. Die Salzspeicherung im Keimling von *Rhizophora mangle* L. wahrend der Entwicklung auf der Mutterpflanzen. *Planta* 77:142–56.

Lugo, A. E. 1980. Mangrove ecosystems: successional or steady state. *Biotropica* 12(suppl.):65–72.

Lugo, A. E., and Snedaker, S. C. 1974. The ecology of mangroves. *Ann. Rev. Ecol. Syst.* 5:39–64.

MacArthur, R. H., and Wilson, E. O. 1967. *The theory of island biogeography*. Princeton: Princeton University Press.

MacNae, W. 1968. A general account of the fauna and flora of mangrove swamps and forests in the Indo-West Pacific region. *Adv. Mar. Biol.* 6:73–270.

MacNae, W., and Fosberg, F. R. 1981. Sonneratiaceae. In *A revised handbook to the flora of Ceylon,* vol. 3, ed. M. D. Dassanayake. New Delhi: American Publishing.

Markgraf, F. 1927. Die Apocynaceen von Neu-Guinea, 117. In C. Lauterbach, *Beiträge zur Flora von Papuasien,* 14. Leipzig: Max Weg. (Sonderabdruck aw Engler, *Bot. Jahrbücher* 61:1927.)

Markley, J. L., McMillan, C., and Thompson, G. A. 1982. Latitudinal differentiation in response to chilling temperatures among populations of three mangroves, *Avicennia germinans, Laguncularia racemosa,* and *Rhizophora mangle,* from the western tropical Atlantic and Pacific Panama. *Can. J. Bot.* 60:2704–15.

Marshall, A. G. 1983. Bats, flowers and fruit: evolutionary relationships in the Old World. *Biol. J. Linn. Soc.* 20:155–35.

Masters, H. M. 1872. *Camptostemon schultzii. Hooker's Ic. Pl.* 12:18 (table 111a).

McCoy, E. D., and Herk, K. L. 1976. Biogeography of corals, seagrasses and mangroves: an alternative to the center of origin concept. *Syst. Zool.* 25:201–10.

McMillan, C. 1971. Environmental factors affecting seedling establishment of the black mangrove on the central Texas coast. *Ecology* 52:927–30.

1975. Adaptive differentiation to chilling in mangrove populations. In *Proceedings of the International Symposium on Biology and Management of Mangroves,* eds. G. E. Walsh, S. C. Snedaker, and H. J. Teas, pp. 62–8. Gainesville: Institute of Food and Agricultural Sciences, University of Florida.

Mepham, R. H. 1983. Mangrove floras of the southern continents. Part 1: The geographical origin of Indo-Pacific mangrove genera and the development and present status of the Australian mangroves. *S. Afr. J. Bot.* 2:1–8.

Merrill, E. D. 1917. *An interpretation of Rumphius's Herbarium Amboinense.* Pub. 9, Dept. Agric. Nat. Resources Bureau of Science, Manila.

1918. Species Blancoanae. A critical revision of the Philippine species of plants described by Blanco and by Llanos. *Bur. Sci. Publ. Manila* 12:1–423.

Mez, C. 1902. Myrsinaceae. In Engler's *Das Pflanzenreich,* 4, 236. Leipzig: Wilhelm.

Miers, J. 1880. On the Barringtoniaceae. *Trans. Linn. Soc. Bot.,* Series 2, 1:47–118.

Moldenke, H. N. 1960. Materials towards a monograph of the genus *Avicennia* L. I and II. *Phytologia* 7:123–68, 179–252, 259–63.

    1967. Additional notes on the genus *Avicennia*, I and II. *Phytologia* 14:301–20, 326–36.

Moore, H. E. 1973. The major groups of palms and their distribution. *Gentes Herbarum* 11(2):27–141.

Mukherjee, J., and Chanda, S. 1973. Biosynthesis of *Avicennia* L. in relation to taxonomy. *Geophytology* 3:85–8.

Mullan, D. P. 1931a. Observations on the water-storing devices in the leaves of some Indian Halophytes. *J. Ind. Bot. Soc.* 10:126–32.

    1931b. On the occurrence of glandular hairs (salt glands) on the leaves of some Indian Halophytes. *J. Ind. Bot. Soc.* 10:184–9.

Muller, J. 1981. Fossil pollen records of extant angiosperms. *Bot. Rev.* 47:1–142.

Muller, J., and Hou Liu, S. Y. 1966. Hybrids and chromosomes in the genus *Sonneratia*. *Blumea* 14(2):337–43.

Muller, J., and van Steenis, C. G. G. J. 1968. The genus *Sonneratia* in Australia. *N. Queensl. Nat.* 34:6–8.

Nakanischi, S. 1964. An epiphytic community on the mangrove tree, *Kandelia candel*. *Hikobia* 4:124.

Ng, F. S. P. 1978. Strategies of establishment in Malayan forest trees. Chap. 5 in *Tropical trees as living systems*, eds. P. B. Tomlinson and M. H. Zimmermann. Cambridge: Cambridge University Press.

Noamesi, G. K. 1958. A revision of the Xylocarpeae (Meliaceae). Thesis. University of Wisconsin, Madison.

Ogura, Y. 1940. On the types of abnormal roots in mangrove and swamp plants. *Bot. Mag. Tokyo* 54:389–404.

Olexa, M. T., and Freeman, T. E. 1978. A gall disease of red mangrove caused by *Cylindrocarpon didymum*. *Plant Dis. Rep.* 62:283–5.

Päivökee, A. E. A. 1984. Tapping patterns in the Nipa palm (*Nypa fruticans* Wurmb.). *Principes* 28:132–7.

Pannier, F. 1959. El efecto de distintas concentraciones salinas sobre el desarrollo de *Rhizophora mangle* L. *Acta Cient. Venezolana* 10(3):68–78.

    1962. Estudio fisiologico sobre la viviparia de *Rhizophora mangle* L. *Acta Cient. Venezolana* 13(6):184–97.

    1965. Problemas de translocacion en *Rhizophora mangle*. 16th Congr. da s. B. B. Doc., 15, Itabuna, Bahia.

Pannier, F., and Pannier, R. F. 1975. Physiology of vivipary in *Rhizophora mangle* L. *Proc. Int. Symp. Biol. Mgt. Mangroves* 2:632–9.

Pannier, F., and Rodriquez, M. del P. 1967. The β-inhibitor complex and its relation to vivipary in *Rhizophora mangle* L. *Int. Rev. Ges. Hydrobiol.* 52:783–92.

Panshin, A. J. 1932. An anatomical study of the woods of the Philippine mangrove swamps. *Philipp. J. Sci.* 48(2):143–205.

Parkinson, C. E. 1934. The Indian species of *Xylocarpus*. *Ind. For.* 60:136–8.

Pax, F., and Hoffmann, K. 1912. Euphorbiaceae. In Engler's *Das Pflanzenreich* 52 (4.147.5 Hippomaneae). Leipzig: Wilhelm Engelmann.

    1931. Euphorbiaceae. In Engler and Prantl, *Die naturlichen Pflanzenfamilien*, ed. 2, Bd. 19c, 11–233. Leipzig. Wilhelm Engelmann.

Payens, J. P. D. W. 1967. A monograph of the genus *Barringtonia* (Lecythidaceae). *Blumea*. 15:157–263.

Pennington, T. D., and Styles, B. T. 1975. A generic monograph of the Meliaceae. *Blumea*. 22:419–540.

Percival, M., and Womersley, J. S. 1975. Floristics and ecology of the mangrove vegetation of Papua New Guinea. *Bot. Bull.*, No. 8., Dept. of Forests, Division of Botany, Lae, Papua New Guinea.

Pitot, A. 1958. Rhizophores et racines chez *Rhizophora* sp. *Bull. Inst. Fr. Afr. Noire* 20:1103–38.

Polhill, R. M. 1971. Some observations on generic limits in Dalbergieae-Lonchocarpineae Benth. (Leguminosae). *Kew Bull.* 25:259–73.

Prance, G. T. 1975. Revisao taxônomica das espécies amazônicas de Rhizophoraceae. *Acta Amazonica* 5(1):5–22.

Primack, R. B., Duke, N. C., and Tomlinson, P. B. 1981. Floral morphology in relation to pollination ecology in five Queensland coastal plants. *Austrobaileya* 4:346–55.

Primack, R. B., and Tomlinson, P. B. 1978. Sugar secretions from the buds of *Rhizophora*. *Biotropica* 10(1):74–5.

    1980. Variation in tropical forest breeding systems. *Biotropica* 12:229–31.

Rabinowitz, D. 1978a. Dispersal properties of mangrove propagules. *Biotropica* 10:47–57.

    1978b. Early growth of mangrove seedlings in Panama, and an hypothesis concerning the relationship of dispersal and zonation. *J. Biogeogr.* 5:113–33.

Radlkofer, L. 1932. Sapindaceae. In A. Engler, *Das Pflanzenreich*, Heft 98b, 4, 165. Leipzig: Wilhelm Engelmann.

Rains, D. W., and Epstein, E. 1967. Preferential absorption of potassium by leaf tissue of the mangrove *Avicennia marina*: an aspect of halophytic competence in coping with salt. *Aust. J. Biol. Sci.* 20:847–57.

Rao, A. N. 1971. Morphology and morphogenesis of foliar sclereids in *Aegiceras corniculatum. Isr. J. Bot.* 20:124–32.

Raymond, A., and Phillips, T. L. 1983. Evidence for an Upper Carboniferous mangrove community. Chap. 2 in *The biology and ecology of mangroves*, ed. H. J. Teas. *Tasks for vegetation science* 8. The Hague: Junk.

Rehm, A., and Humm, H. J. 1973. *Sphaeroma terebrans*: a threat to the mangroves of southwestern Florida. *Science* 182:173–4.

Reimold, R. J. 1977. Mangals and salt marshes of eastern United States. Chap. 7 in *Ecosystems of the world*, vol. 1. *Wet coastal ecosystems*, ed. V. J. Chapman. Amsterdam: Elsevier Scientific.

Retallack, G., and Dilcher, D. L. 1981. A coastal hypothesis for the dispersal and rise to dominance of flowering plants. Pp. 27–77 in *Paleobotany, paleoecology and evolution*, ed. K. J. Niklas. New York: Praeger.

Rheede (tot Drakenstein), H. 1678–1703. *Hortus indicus malabaricus*. Amsterdam: Amstelodami.

Ribi, G. 1982. Does the wood boring isopod *Sphaeroma terebrans* benefit red mangroves (*Rhizophora mangle*)? *Bull. Mar. Sci. Miami* 31:925–8.

Ridley, H. N. 1930. *The dispersal of plants throughout the world*. Kent: Ashford.

    1938. Notes on *Xylocarpus. Kew Bull.* 1938:288–92.

Rollet, B. 1981. *Bibliography on mangrove research, 1600–1975*. Paris: UNESCO.

Roth, I. 1965. Histogenese der Lentizellen am Hypokotyl von *Rhizophora mangle* L. *Ost. Bot. Z.* 112:640–53.

Rumphius, G. E. 1741–55. *Herbarium Amboinense*. Amsterdam: Amstelaedami.

Saenger, P., Specht, M. M., Specht, R. L., and Chapman, V. J. 1977. Mangal and coastal salt-marsh communities in Australasia. Chap. 15 in *Ecosystems of the world*, vol. 1, *Wet coastal ecosystems*. Amsterdam: Elsevier Scientific.

Salvoza, F. M. 1936. *Rhizophora. Nat. Appl. Sci. Bull. Un. Philipp.* 5:179–237.

Sandwith, N. Y. 1940. Contributions to the flora of tropical America: 44. Further notes on tropical American Bignoniaceae. *Kew Bull.* 1940:302–4.

Savory, H. J. 1953. A note on the ecology of *Rhizophora* in Nigeria. *Kew Bull.* 1953:127–8.

Schenk, H. 1889. Ueber die Luftwurzeln von *Avicennia tomentosa* und *Laguncularia racemosa. Flora* 72:83–8.

Schimper, A. F. W. 1891. *Botanische Mittheilungen aus den Tropen.* Heft 3: *Die Indomalayische Strandflora.* Jena: Gustav Fischer.

Schmid, R. 1980. Comparative anatomy and morphology of *Psiloxylon* and *Heteropyxis*, and the subfamilial and tribal classification of Myrtaceae. *Taxon* 29(5/6):559–95.

Schmidt, J. 1903. *Bidrag till kundskab om skuddene hos den gamle verdens mangrovetraeer.* Copenhagen.

Schnetter, M.-L. 1978. Der Einfluss von Ausserfaktoren auf die Struktur des Blattes von *Avicennia germinans* (L.) L. unter natürlichen Bedingungen. *Beitr. Biol. Pflanz.* 54:13–28.

Scholander, P. F. 1968. How mangroves desalinate water. *Physiol. Plant* 21:251–61.

Scholander, P. F., Hammel, H. T., Bradstreet, E. D., and Hemmingsen, E. A. 1965. Sap pressure in vascular plants. *Science* 148:339–40.

Scholander, P. F., Hammel, H. T., Hemmingsen, E. A., and Bradstreet, E. D. 1964. Hydrostatic pressure and osmotic potential in leaves of mangroves and some other plants. *Proc. Nat. Acad. Sci. U.S.A.* 52:119–25.

Scholander, P. F., Hammel, H. T., Hemmingsen, E. A., and Garay, W. 1962. Salt balance in mangroves. *Plant Physiol.* 37:722–9.

Scholander, P. F., Van Dam, L., and Scholander, S. I. 1955. Gas exchange in the roots of mangroves. *Am. J. Bot.* 42:92–8.

Schumann, K. 1891. Rubiaceae. In Engler and Prantl, *Die naturlichen Pflanzenfamilien*, ed. 1, t. IV. ab. 4, pp. 1–156. Leipzig: Wilhelm Engelmann.

Semeniuk, V. 1980. Mangrove zonation along an eroding coastline in King Sound, northwestern Australia. *J. Ecol.* 68:789–812.

    1983. Mangrove distribution in northwestern Australia in relationship to regional and local freshwater seepage. *Vegetatio* 53:11–31.

Semeniuk, V., Kenneally, K. F., and Wilson, P. G. 1978. *Mangroves of western Australia*, Handbook No. 12. Perth: Western Australian Naturalists Club.

Semple, J. C. 1970. The distribution of pubescent leaved individuals of *Conocarpus erectus* (Combretaceae). *Rhodora* 72:544–6.

Shimony, C., Fahn, A., and Reinhold, L. 1973. Ultrastructure and ion gradients in the salt glands of *Avicennia marina* (Forssk.) Vierh. *New Phytol.* 72:27–36.

Sidhu, S. S. 1968. Further studies on the cytology of mangrove species of India. *Caryologia* 21:353–7.

Simberloff, D., Brown, B. J., and Lowrie, S. 1978. Isopod and insect root borers may benefit Florida mangroves. *Science* 210:630–2.

Sinclair, J. 1968. Florae Malesianae precursores, 17. The genus *Myristica* in Malesia and outside Malesia. *Gard. Bull. Singapore* 23:1–540.

Sleumer, H. 1972. A taxonomic revision of the genus *Scolopia* Schreb. (Flacourtiaceae). *Blumea* 20(1):25–64.

Slooten, P. F. van. 1924. Contributions a l'étude de la flore des Indes néerlandaises II,

The Combretaceae of the Dutch East Indies. *Bull. Jard. Bot. Buitenz.* Series 3, 6(1):11–64.

Smith, A. C. 1981. *Flora vitiensis Nova*, vol. 2. Lawai, Kauai, Hawaii: Pacific Tropical Botanical Garden.

Snedaker, S. C. 1982. Mangrove species zonation: why? Chap. 1 in *Contributions to the ecology of halophytes*, eds. D. N. Sen and K. S. Rajpurohit. *Tasks for vegetation science*, vol. 2. The Hague: Junk.

Snedaker, S. C., Jimenez, J. A., and Brown, M. S. 1981. Anomalous aerial roots in *Avicennia germinans* (L.) L. in Florida and Costa Rica. *Bull. Mar. Sci. Miami* 31:467–70.

Soto, R., and Jiménez, J. A. 1982. Análisis fisonómico estructural del manglar de Puerto Soley, La Cruz, Guanacaste, Costa Rica. *Rev. Biol. Trop.* 30:161–8.

Sperry, J. S. 1983. Observations on the structure and function of hydathodes in *Blechnum lehmannii*. *Am. Fern J.* 73:65–72.

Stace, C. A. 1965a. Cuticular studies as an aid to plant taxonomy. *Bull. Br. Mus. Nat. Hist.* 4:3–78.

1965b. The significance of the leaf epidermis in the taxonomy of the Combretaceae, 1. A general review of tribal generic and specific characters. *J. Linn. Soc. Bot.* 59:229–52.

Start, A. N., and Marshall, A. G. 1976. Nectarivorous bats as pollinators of trees in West Malaysia. Pp. 141–50 in *Tropical trees, variation breeding and conservation*, eds. J. Burley and B. T. Styles. New York: Academic Press.

Stearn, W. T. 1958. A key to West Indian mangroves. *Kew Bull.* 1985:33–7.

Steenis, C. G. G. J. van. 1928. The Bignoniaceae of the Netherlands Indies. *Bull. Jard. Bot. Buitenz.* Series 3, 10:173–290.

1936. *Osbornia octodonta*, een weinig bekende mangrove-boom. *Trop. Nat.* 25:194–6.

1937. De soorten van het geslacht *Acanthus* in Nederlandsch-Indië. *Trop. Nat.* 26:202–7.

1948. Plumbaginaceae. *Flora Malesiana* 1, 4:107–12.

1949. Vicarism in the Malaysian flora. *Flora Malesiana* 1, 4:59.

1958. Ecology of mangroves. *Flora Malesiana* 1, 5:431–41.

1962. The distribution of mangrove plant genera and its significance for palaeogeography. *Proc. Kon. Net. Amsterdam*, Series C, 65:164–9.

1977. Bignoniaceae. *Flora Malesiana* 1(8):114–86.

Steenis-Kruseman, M. J. van. 1950. Malaysian plant collectors and collections. *Flora Malesiana* 1(1):3–63a.

Stehli, F. G., and Wells, J. W. 1971. Diversity and age patterns in hermatypic corals. *Syst. Zool.* 20:115–26.

Stern, W. L., and Voigt, G. K. 1959. Effect of salt concentration on growth of red mangrove in culture. *Bot. Gaz.* 121(1):36–9.

Stevens, P. F. 1980. A revision of the Old World species of *Calophyllum* (Guttiferae). *J. Arnold Arbor.* 61:117–699.

Studholme, W. P., and Philipson, W. R. 1966. A comparison of the cambium in two woods with included phloem. *Heimerliodendron brunonianum* and *Avicennia resinifera*. *N. Z. J. Bot.* 4(4):355–65.

Sussex, I. 1975. Growth and metabolism of the embryo and attached seedling of the viviparous mangrove, *Rhizophora mangle*. *Am. J. Bot.* 62:948–53.

Swingle, W. T. 1943. The botany of citrus and the wild relatives of its orange subfamily. In *The citrus industry*, vol. 1, eds. H. J. Webber and L. D. Batchelor. Berkeley: Los Angeles, Division of Agricultural Sciences.

Szyszylowicz, I. 1893. Theaceae. In Engler and Prantl, *Die natürlichen Pflanzenfamilien*, ed. 1, t. 3, pp. 175–92. Leipzig: Wilhelm Engelmann.

Tan, H., and Rao, A. N. 1981. Vivipary in *Opiorrhiza tomentosa* Jacq (Rubiaceae). *Biotropica* 13:232–3.

Tattersfield, F., Martin, J. T., and Howes, F. N. 1940. Some fish-poison plants and their insecticidal properties. *Kew Bull.* 1940:169–80.

Teas, H. J. 1982. An epidemic dieback gall disease of *Rhizophora* mangroves in the Gambia, West Africa. *Plant Dis. Rep.* 66:522–3.

Thom, B. G. 1967. Mangrove ecology and deltaic geomorphology: Tobasco, México. *J. Ecol.* 55:301–43.

1975. Mangrove ecology from a geomorphological viewpoint. Pp. 469-81 in *Proc. Int. Symp. Biol. Manag. Mangroves*, eds. G. Walsh, S. Snedaker, and H. Teas. Gainesville: University of Florida Press.

Tieghem, Ph. van 1898. Avicenniacées et Symphoremacées. Place de ces deux nouvelles familles dans la classification. *J. Bot. Paris.* 12:345–65.

Tobe, H., and Raven, P. H. 1983. An embryological analysis of Myrtales: its definition and characteristics. *Ann. Missouri Bot. Gard.* 70:71–94.

Tomlinson, P. B. 1961. Palmae. In *Anatomy of the Monocotyledons*, vol. 2, ed. C. R. Metcalfe. Oxford: Clarendon Press.

1971. The shoot apex and its dichotomous branching in the *Nypa* palm. *Ann. Bot. Lond.* 35:865–79.

1978. *Rhizophora* in Australasia – some clarification of taxonomy and distribution. *J. Arnold Arbor.* 59:156–69.

1980. *The biology of trees native to tropical Florida*. Petersham, Mass.: published privately.

1982a. Field collection and study of Old World (Indo-Pacific) mangroves. *Nat. Geog. Soc. Res. Rep.* 14:669–77.

1982b. *Helobiae (Alismatidae)*, vol. 7. *Anatomy of the monocotyledons*, ed. C. R. Metcalfe. Oxford: Clarendon Press.

Tomlinson, P. B., Bunt, J. S., Primack, R. B., and Duke, N. C. 1978. *Lumnitzera rosea* (Combretaceae) – its status and floral morphology. *J. Arnold Arbor.* 59:342–51.

Tomlinson, P. B., Primack, R. B., and Bunt, J. S. 1979. Preliminary observations in floral biology in mangrove Rhizophoraceae. *Biotropica* 11:256–77.

Tomlinson, P. B., and Wheat, D. W. 1979. Bijugate phyllotaxis in Rhizophoreae (Rhizophoraceae). *Bot. J. Linn. Soc.* 78:317–21.

Tomlinson, P. B., and Womersley, J. S. 1976. A species of *Rhizophora* new to New Guinea and Queensland, with notes relevant to the genus. *Contrib. Herb. Aust.* 19:1–10.

Tralau, H. 1964. The genus *Nypa* van Wurmb. *K.svensk Vetensk. Akad. Handl.*, Series 5, 10(1):5–29.

Trochain, J., and Dulau, L. 1942. Quelques particularités anatomiques d'*Avicennia nitida* (Verbernaceae) de la mangrove ouest africaine. *Bull. Soc. Hist. Nat. Toulouse* 77:271–81.

Troll, W. 1933a *Camptostemon schultzii* Mast. und *Camptostemon philippensis* (Vid.) Becc. als neue Vertreter der austral-asiatischen Mangrovevegetation. *Flora* n.f.128:348–60.

1933b. Ueber *Acrostichum aureum* L., *Acrostichum speciosum* Willd. und neotene Formen des lezteren. *Flora n.f.* 128:301–28.

1943. *Vergleichende Morphologie den hoheren Pflanzen*, bd. 1 t. 3, Vegetationsorgane. Berlin: Gebruder, Borntraeger.

Troll, W., and Dragendorff, O. 1931. Ueber die Luftwurzeln von *Sonneratia* L. und ihre biologische Bedeutung. *Planta* 13:311–473.

Tryon, R. M., and Tryon, A. F. 1982. *Ferns and allied plants, with special reference to tropical America.* New York: Springer-Verlag.

Uhl, N. W. 1972. Inflorescence and flower structure in *Nypa fruticans* (Palmae). *Am. J. Bot.* 59:729–43.

Ulken, A. 1983. Distribution of Phycomycetes in mangrove swamps with brackish waters and waters of high salinity. Chap. 12 in *Biology and ecology of mangroves*, ed. H. . Teas. *Tasks for vegetation science* 8. The Hague: Junk.

United Nations Educational, Scientific, and Cultural Organization. 1978. Secondary successions. Chap. 9 in *Tropical forest ecosystems*. A state-of-knowledge report prepared by UNESCO/UNEP/FAO. Paris: UNESCO.

Valeton, T. 1895. Les *Cerebera* du Jardin Botanique de Buitenzorg. *Ann. Jard. Bot. Buitenz.* 12:238–48.

Van der Pijl, L. 1936. Fledermäuse und Blumen. *Flora.* 31:1–40.

van Vliet, G. J. C. M. 1976. Wood anatomy of Rhizophoraceae. *Leiden Bot. Ser.* 3:20–75.

Verdcourt, B. 1979. A manual of New Guinea legumes. *Bot. Bull. Officer of Forests, Division of Botany*, Lae, Papua New Guinea.

Waisel, Y. 1972. *Biology of halophytes.* New York: Academic Press.

Walsh, G. E. 1974. Mangroves: a review. Pp. 51–174 in *Ecology of halophytes*. New York: Academic Press.

1977. Exploitation of mangal. Chap. 16 in *Ecosystems of the world, 1. Wet coastal ecosystems*, ed. V. J. Chapman. Amsterdam: Elsevier Scientific.

Walsh, G. E., Ainsworth, K. A., and Rigby, R. 1979. Resistance of red mangrove (*Rhizophora mangle* L.) to lead, cadmium and mercury. *Biotropica* 11:22–7.

Walsh, G. E., Snedaker, S. C., and Teas. H. J. (eds.) 1975. *Proceedings of the International Symposium on Biology and Management of Mangroves.* 2 vols. Gainesville: Institute of Food and Agricultural Sciences, University of Florida.

Walter, H., and Steiner, M. 1936. Oekologie der ost-afrikanschen Mangroven. *Z. Bot.* 30:65–93.

Ward, C. J., and Steinke, T. D. 1982. A note on the distribution and approximate areas of mangroves in South Africa. *S. Afr. J. Bot.* 1:51–3.

Watson, J. G. 1929. Mangrove forests of the Malay Peninsula. *Malay For. Rec.* no. 6.

Weber-El Ghobary, M. O. 1984. The systematic relationships of *Aegialitis* (Plumbaginaceae) as revealed by pollen morphology. *Plant Syst. Evol.* 144:53–8.

Webster, G. L. 1967. The genera of Euphorbiaceae in the southeastern United States. *Arnold Arbor.* 48:303–430.

1975. Conspectus of a new classification of the Euphorbiaceae. *Taxon* 24:593–601.

Wells, A. G. 1983. Distribution of mangrove species in Australia. Chap. 6 in *Biology and ecology of mangroves*, ed. H. J. Teas. *Tasks for vegetation science* 8. The Hague: Junk.

West, R. C. 1977. Tidal salt-marsh and mangal formations of middle and South America. Chap. 9 in *Ecosystems of the world, 1. Wet coastal ecosystems*. Amsterdam: Elsevier Scientific.

Westermaier, M. 1900. Zur Kenntnis der Pneumatophoren. *Botanische Untersuchungen im Amschluss an eine Tropenreise*, Heft 1. Freiburg.

White, C. T. 1926. A variety of *Ceriops tagal* C. B. Rob. (= C. *Candollean* W. & A.). *J. Bot. Lond.* 64:220–1.

Whitmore, T. C. 1983. Secondary succession from seed in tropical rain forests. *Common. For. Abstr.* 44:767–79.

Wilkinson, H. P. 1981. The anatomy of the hypocotyls of *Ceriops* Arnott (Rhizophoraceae), recent and fossil. *Bot. J. Linn. Soc.* 82:139–64.

Winograd, M. 1983. Observaciones sobre el hallozgo de *Pelliciera rhizophorae* (Theaceae) en el Caribe Colombiano. *Biotropica* 15:297–8.

Wium-Anderson, S. 1981. Seasonal growth of mangrove trees in southern Thailand, 3. Phenology of *Rhizophora mucronata* Lamk. and *Scyphiphora hydrophyllacea* Gaertn. *Aquat. Bot.* 10:371–6.

Wium-Anderson, S., and Christensen, B. 1978. Seasonal growth of mangrove trees in southern Thailand, 2. Phenology of *Bruguiera cylindrica, Ceriops tagal, Lumnitzera littorea* and *Avicennia marina. Aquat. Bot.* 5:383–90.

Woodroffe, C. D. 1983. Development of mangrove forests from a geological perspective. Chap. 1 in *Biology and ecology of mangroves*, ed. H. J. Teas. *Tasks for vegetation science* 8. The Hague: Junk.

Wright, D. F. 1977. A North Queensland mangrove pollen flora. B. Sc. Hons. Botany thesis. James Cook University, Townsville, Queensland.

Wyatt-Smith, J. 1953a. The Malayan species of *Brugueira. Malay. For.* 16(3):156–61.

1953b. The Malayan species of *Sonneratia. Malay.* 16:213–16.

1954. The Malayan species of *Avicennia. Malay. For.* 17:21–5.

1960. Field key to the trees of mangrove forests in Malaya. *Malay. For.* 23:126–32.

Yamashiro, M. 1961. Ecological study on *Kandelia candel* (L.) Druce, with special reference to the structure and falling of the seedlings. *Hikobia* 2(3):209–14.

Zahran, M. A. 1977. Africa A, Wet formations of the African Red Sea coast. Chap. 10 in *Ecosystems of the world*, 1. *Wet coastal ecosystems*, ed. V. J. Chapman. Amsterdam: Elsevier Scientific.

Zamski, E. 1979. The mode of secondary growth and the three-dimensional structure of the phloem in *Avicennia. Bot. Gaz.* 140:67–76.

1981. Does successive cambia differentiation in *Avicennia* depend on leaf and branch initiation? *Isr. J. Bot.* 30:57–64.

Zimmermann, M. H. 1983. *Xylem structure and the ascent of sap*. Berlin: Springer-Verlag.

Index

Bold face numbers refer to main description; *ialic* numbers refer to illustrations